Zeolite Characterization and Catalysis

Arthur W. Chester • Eric G. Derouane
Editors

Zeolite Characterization and Catalysis

A Tutorial

Springer

Editors
Arthur W. Chester
Dept. of Chemical and Biochemical Engineering
Rutgers University
Piscataway, NJ 08854
USA
achester@rci.rutgers.edu

Prof. E.G. Derouane
Formerly Departamento de Química
Bioquímica e Farmácia
Faculdade de Ciências e Tecnologia
Universidade do Algarve
8005-139 FARO
Portugal

ISBN 978-1-4020-9677-8 e-ISBN 978-1-4020-9678-5
DOI: 10.1007/978-1-4020-9678-5
Springer Dordrecht Heidelberg London New York

Library of Congress Control Number: 2009937221

Cover illustration: Atlas of Zeolite Structures, 5th Ed., Elsevier, 2001

Cover design: WMXDesign GmbH, Heidelberg, Germany

Printed on acid-free paper

Springer is part of Springer Science+Business Media (www.springer.com)

Preface

The idea for putting together a tutorial on zeolites came originally from my co-editor, Eric Derouane, about 5 years ago. I first met Eric in the mid-1980s when he spent 2 years working for Mobil R&D at our then Corporate lab at Princeton, NJ. He was on the senior technical staff with projects in the synthesis and characterization of new materials. At that time, I managed a group at our Paulsboro lab that was responsible for catalyst characterization in support of our catalyst and process development efforts, and also had a substantial group working on new material synthesis. Hence, our interests overlapped considerably and we met regularly. After Eric moved back to Namur (initially), we maintained contact, and in the 1990s, we met a number of times in Europe on projects of joint interest. It was after I retired from ExxonMobil in 2002 that we began to discuss the tutorial concept seriously. Eric had (semi-)retired and lived on the Algarve, the southern coast of Portugal. In January 2003, my wife and I spent 3 weeks outside of Lagos, and I worked parts of most days with Eric on the proposed content of the book.

We decided on a comprehensive approach that ultimately amounted to some 20+ chapters covering all of zeolite chemistry and catalysis and gave it the title *Zeolite Chemistry and Catalysis: An integrated Approach and Tutorial.* Over the next several years, we sought authors for these chapters among both industry and academia. Inclusion of industrial authors was important, since so much of early zeolite science was developed in the industry, before it became a major academic subject. But many industrial authors had difficulty with finding the time and feared that their company proprietary restrictions would hamper them. So many times we had to go back and find new authors.

When Eric suddenly passed away last year (2008), our author list was essentially complete, but we had only six chapters in hand. Other chapters did not appear to be close to completion, and I was afraid that existing material could "age out." The publisher then agreed to a more limited book based on the existing chapters. Review of the existing contents led me to change the title to the more limited *Zeolite Chemistry and Catalysis; A Tutorial.* But late in 2008, with all seven proposed chapters in hand, I realized that the characterization coverage would be much more complete if we had a chapter covering the very important NMR techniques now utilized broadly. After recommendations from the authors, Dr. Michael Hunger graciously agreed to write such a chapter under a very short deadline (2 months!).

Thus the first five chapters of the book provide tutorials in the major areas of zeolite characterization: X-ray powder diffraction, NMR, temperature programmed desorption and adsorption calorimetry, electron microscopy, and infrared spectroscopy. All these techniques provided major contributions to the development of zeolite science, particularly XRD, IR, and measurement of acid–base properties in the early days (1955–1980). In the 1980s, electron microscopy and NMR started to become more prominent and now are equally important. The only major zeolite characterization area not directly addressed here is electron diffraction, although it was planned in the original contents.

The subsequent chapters deal with theory and catalysis. It has become quite common now with improved theory and high-speed computers to predict zeolite properties and reactivities on strictly theoretical grounds, as outlined in Chap. 6. The two remaining chapters outline the principles and practice of C1 chemistry, a field made possible by zeolites, and the breadth of zeolite catalysis in the chemical industry. What is missing is a description of zeolite catalysis in the refining industry, which drove the initial development of zeolite catalysts, but much of that is actually described in Chap. 8.

The book should be useful in allowing new practitioners, whether students or practicing scientists in other fields, to quickly become familiar with the principles of zeolite science and to apply that understanding to their own fields. Newly developed catalytic materials – ordered mesoporous materials, MOFs (metal-organic frameworks), framework phosphates, and hierarchal nanomaterials – all have their roots in zeolite science, and so an understanding of the basics is important.

The book is dedicated to Eric Derouane and a memoriam is included following the Preface, written by his friends Fernando Ramôa Ribeiro and Jacques C. Védrine.

A Brief Introduction to Zeolites

Zeolites are porous crystalline framework materials containing pores of molecular size (5–12 Å or 0.5–1.2 nm). The term *zeolite* is derived from the Greek words for "boiling stone," from the ability of these materials to absorb water and release it upon heating. Conventional zeolites are based on silicate frameworks in which substitution of some of the Si with Al (or other metals) leads to a negative charge on the framework, with cations (usually Na or other alkaline or alkaline earth metals) within the pore structure. This leads to another important property, ion exchange, where the metal ions in the pore structure can be replaced by other cations (e.g. metal, ammonium, quaternary ammonium).

The zeolitic frameworks are networks composed of tetrahedral T atoms (T=Si, Al, etc.) linked by oxygen ions. Common building blocks of zeolite structures consist of 3, 4, 5, and 6 membered rings (*n*-MR). Each *n*-MR consists of *n* T atoms linked in a ring by O ions and thus actually has 2*n* atoms; thus a 6-MR has 12 total atoms. The structures are arranged such that they form larger rings that represent the molecular pores – commonly 8-, 10- and 12-MR, although structures with 9-, 14-, 18-, and 20-MR pores are known. The 8-, 10-, and 12-MR containing zeolites are commonly known as *small, intermediate,* and *large* pores. Small pore zeolites will generally allow *n*-paraffins to be adsorbed, while large pore zeolites allow all highly branched paraffins to be absorbed as well. Intermediate pore zeolites are just that, allowing some branched chain but not highly branched paraffins to be adsorbed. Thus zeolites are part of the larger class of materials called *molecular sieves*, which allow mixtures of molecules of differing structures to be separated.

Zeolites occur naturally and are generally formed in alkaline environments from volcanic sediments and materials. The first zeolite discovered and identified as such was stilbite; common abundant zeolites are analcime, clinoptilolite, erionite, heulandite, laumontite, and mordenite. Many of these materials have valuable properties as sorbents and even catalysts, but the natural forms often have faults and irregularities in their structures that limit their application. It is the development of laboratory methods of synthesizing zeolites that led to the many commercial applications of zeolites.

The first synthetic zeolite was made from Na, Si, and Al at Linde's laboratories in Tonawanda, NY. It was termed *zeolite A*, being the first, and was found to be composed of sodalite cages arranged to give 8-MR pores. Zeolite A was capable of adsorbing water and *n*-paraffins as well. Its first commercial application was as a drying agent and it is still commonly used. Sometime later, Linde synthesized zeolites X and Y; these zeolites had the structure of the natural zeolite faujasite and were also composed of sodalite cages arranged such that a 12-MR pore structure existed. The major difference between X and Y are their SiO_2/Al_2O_3 ratios in their framework. For X, this was 2–3; for Y, it was 3.5–5.5, which conferred greater hydrothermal stability, which was important in applying it to catalytic cracking.

It was then found that organic "templates" could be used to make new zeolite structures. This approach was pioneered by Mobil and led to ZK-5 (from Dabco), Beta (from tetraethylammonium ion), and ZSM-5 (from tetrapropylammonum ion), among others. Like natural zeolites, synthetic zeolites are generally named by their inventors. Since it was mostly companies that were involved in early zeolite synthesis, most names derived from them: ZSM for Zeolite Socony Mobil, LZ for Linde Zeolite, ECR for Exxon Corporate Research, and SSZ for Standard Selective Zeolite (from Chevron). This practice has continued into the present, nowadays with designations based on universities as well (for example, ITQ for Intstito de Technologia Quimica in Valencia). The International Zeolite Association has also developed a "Structure Code," which is applied to particular structure types but does not indicate chemical composition. Thus, for example, zeolite A, X and Y, and ZSM-5 are known as LTA, FAU, and MFI (for Linde Type A, Faujasite, and Mobil Five) respectively. Proposals for structure codes are made by those determining the structure and are approved by the IZA.

These three zeolites are probably the most important in terms of commercial development of zeolites. Their structures are illustrated in Fig. 0.1. The vertices in the structures represent the T atoms (Si or Al) and the lines between vertices represent the O atom bridges. This is a conventional way to represent zeolite structures, since showing all the oxygen atoms only produces a confused representation.

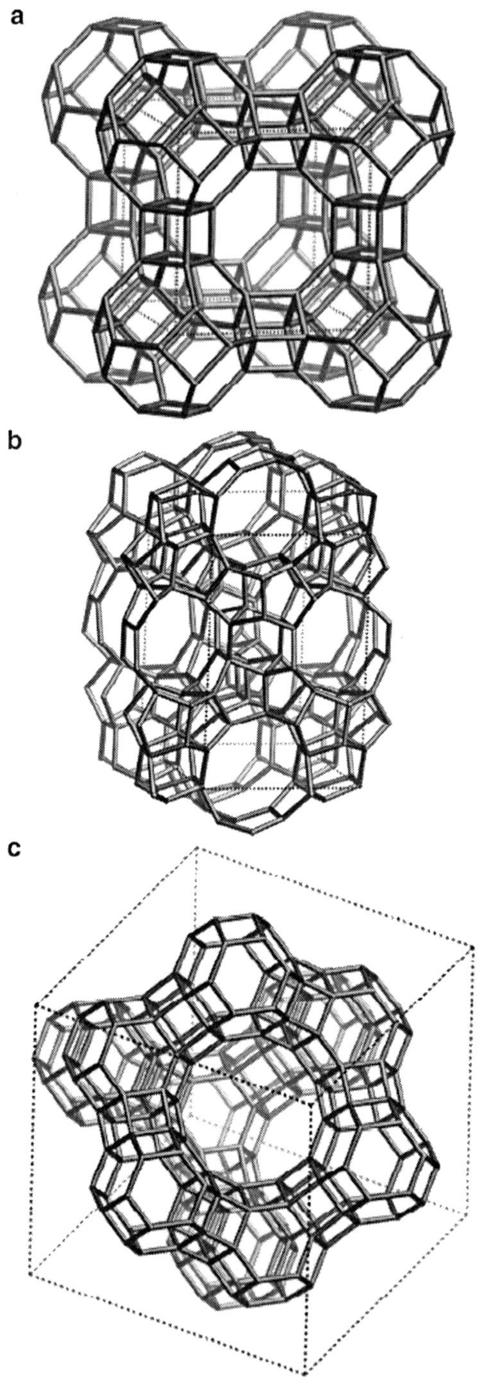

Fig. 0.1 The structures of (**a**) zeolite A (LTA), (**b**) ZSM-5 (MFI), and (**c**) faujasite (FAU). Vertices represent T atoms (Si or Al); *lines* between vertices are the O bridges. Structures taken from Baerlocher Ch, Meier WM, Olson DH (ed) (2001), Atlas of zeolite framework types, 5th edn, Elsevier, Amsterdam

Zeolite A is constructed from sodalite cages connected by 4-MR, leading to a cubic structure and three orthogonal 8-MR pores with a diameter of 4.1 Å – large enough for only small molecules, bimolecular gases, water and *n*-paraffins. Zeolite A is usually made with a SiO_2/Al_2O_3 ratio of 2, indicating equal numbers of Si and Al atoms, although different versions have been made with higher ratios. Zeolite A is generally used in adsorption and separation applications.

Faujasite (also zeolites X or Y) is also constructed from sodalite cages, but connected through 6-MR, leading to a crystallographically cubic structure in which the sodalite cages are tetrahedrally arrayed and resulting in three large orthogonal pores of 7.4 Å diameter. Most organic molecules, with some exceptions, fit into these pores. Zeolite Y has a SiO_2/Al_2O_3 ratio of 4–6 and is used in a very large scale catalytic cracking applications.

ZSM-5, on the other hand, is based on cages made of 4-, 5-, and 6-MR resulting in two elliptical pores of 5.1×5.5 and 5.3×5.6 Å normal to each other. Small and intermediate organic molecules can be adsorbed, but not larger molecules. ZSM-5 has a much higher SiO_2/Al_2O_3 ratio than the other zeolites mentioned, anywhere from around 20 to almost infinity. It is most useful in conversion of small olefins and alcohols (particularly methanol) to gasoline range hydrocarbons, as well as in shape selective cracking applications such as dewaxing.

There were 176 known structures as of 2007. An excellent source of information in general on zeolites is the Web page of the IZA (http://www.iza-online.org/). Detailed information on the structures of all known zeolite structure types is available, plus information on catalysis, synthesis, and other aspects of zeolite science.

The definition of zeolites has undergone some changes over time. Zeolites were thought to be inherently aluminosilicates, since all known examples had that composition. In the 1970s, however, Union Carbide synthesized porous zeolite-like aluminum phosphates with structures identical in some instances to known zeolites as well as new structures ($AlPO_4$s). Materials with silicon and other metals substituted for Al or P were also made that had acidity and catalytic activity (SAPOs and MAPOs). Because they were not aluminosilicates, carbide claimed that they were nonzeolitic molecular sieves (NZMSs) as a way get stronger patent claims. Similarly, carbide was able to synthesize a form of ZSM-5 that they claimed had no Al and was therefore a silicate and not a zeolite (silicalite). In fact, these latter materials had Al from the silica sources used and had SiO_2/Al_2O_3 ratios as low as 200. Nowadays, with the commercial interests out of the picture, all of these materials are recognized as part of zeolite science (in point of fact, carbide always published papers on $AlPO_4$s, SAPOs, and silicalites in the journal *Zeolites*).

Zeolites are useful in catalysis because of their acidity. Acidity arises from the Si-OH-Al grouping formed by ion exchange with acid or, more typically, by thermal decomposition of exchanged ammonium ions to form the acid group and gaseous ammonia. Zeolitic acidity is much stronger than that formed in amorphous aluminosilicates, which is usually based on the Al-OH group. Aspects of zeolite acidity are explored in detail in Chap. 3.

The most important use of zeolites, particularly by volume, is in catalytic cracking, in which the faujasite zeolites X and Y were applied by Mobil Oil in early 1960s. Prior to this, catalysts were amorphous aluminosilicates prepared by coprecipitation or cogelation, or were made from acidified natural clays. Catalytic cracking is a cyclic process in which the catalyst generates coke during the reaction and must be regenerated before reuse. Early units used swing reactors that were alternately on reaction and then on regeneration, but the more efficient cyclic units, both moving bed and fluid bed (FCC) were developed during the Second World War.

The principles of cyclic catalytic cracking are shown schematically in Fig. 0.2. Starting at the bottom of the figure and going counter-clockwise:

- Gas oil feed and regenerated (hot) catalysts are mixed and the vapor–solid mixture is transported into a reaction zone. Reaction occurs at a temperature set by the regenerated catalyst temperature and the feed initial temperature.
- Reaction products (shown to the right) are separated and sent to a downstream processing plant; spent (coked) catalyst is sent through a steam stripper into a regenerator.
- The coke on the catalyst, which contains C, H, N, and S deposited from the feed, is oxidized, heating the catalyst and releasing the gaseous compounds shown, including steam, which permanently deactivates the catalyst by dealuminization. The heated catalyst is then combined with feed and the cycle begins again.
- Fresh catalyst is added continuously or batch wise in the regenerator in order to maintain catalyst activity.

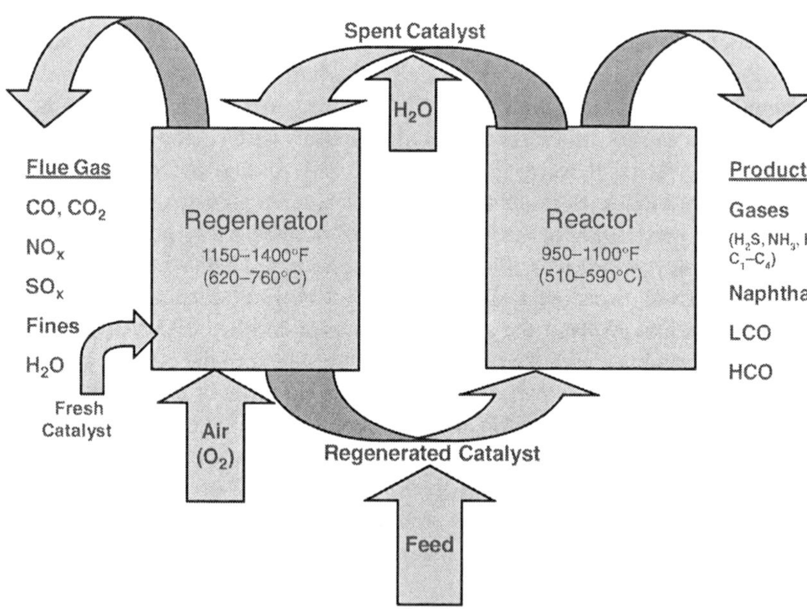

Fig. 0.2 Schematic of a cyclic catalytic cracking unit

Early proposed cracking catalysts were prepared from the low SiO_2/Al_2O_3 zeolite X, but were hydrothermally unstable. Mobil researchers discovered that exchanging zeolite X (and later zeolite Y) with mixed rare earth ions led to higher hydrothermal stability and activity (REX and REY). Early demonstrations indicated significantly higher conversions, higher gasoline selectivities, and lower coke yields than obtained with amorphous catalysts, even though early zeolite cracking catalysts contained only 5% zeolite! Although zeolite cracking catalysts did give lower octane gasoline, this could be corrected by using higher reaction temperatures, and, within 10 years of the first demonstration, zeolites were adopted throughout the industry, with REY the dominant zeolite component at up to 25–30% of the total composition, the remainder being a matrix formulated for low activity but high binder strength and low attrition. In the late 1970s, a new component USY (for ultrastable Y) began to be used to improve octane, although it had poorer hydrothermal stability. USY is a framework dealuminized version of Y made by decomposition of the $NaNH_4$ form of Y in the presence of steam at controlled temperatures. Modern catalysts generally contain rare earth-modified USY and also contain additive components for oxidizing CO to CO2 in the regenerator, reducing S and N oxides in the flue gas and for passivating metal contaminants found in the feed (particularly resids).

The success of zeolites in catalytic cracking (including hydrocracking, jointly developed by Union Carbide and Union Oil), led to significant programs at Mobil, Union Carbide, Exxon, BP, and ICI to synthesize new zeolites with improved properties. The most significant early success, as mentioned above, was the high silica zeolite ZSM-5, which was found to have revolutionary applicability in a wide number of applications: catalytic dewaxing, improving gasoline octane in FCC, conversion of methanol to gasoline (MTG) or olefins (MTO), olefin oligomerization, xylene isomerization, ethylbenzene synthesis, toluene disproportionation, and selective toluene disproportionation (directly to *p*-xylene), to name the most prominent. All of the named processes – and others – have been commercialized over different forms of ZSM-5. It is currently used in FCC to generate high volumes of propylene – an offshoot of its octane enhancement properties.

New applications continue to be discovered for existing and new zeolites. Newer materials being developed – like MOFs and mesoporous materials – may have similar successes in the future. The zeolite science taught in this volume should be of help and guide for such successes.

Cherry Hill,
NJ, USA

Arthur W. Chester
April, 2009

In Memory of Our Friend and Colleague Eric G. Derouane, 1944–2008

Eric Derouane in Paris during the 14th International Congress in Catalysis (ICC) in July 2004

Fernando Ramôa Ribeiro, Instituto Superior Tecnico, Universidade Técnica de Lisboa, Palácio Centeno, Alameda de Santo António dos Capuchos, n°1, 1169-047 Lisboa, Portugal (ramoa.ribeiro@ist.utl.pt).

Jacques C. Védrine, Laboratoire de Réactivité de Surface, Université Pierre et Marie Curie, 4 Place Jussieu, Paris, 75252, France (jacques.vedrine@upmc.fr).

And his colleagues and friends from Namur, Belgium, from Caen, Montpellier, Mulhouse, Paris, France, from Aveiro, Faro, Lisboa, Porto, Portugal, from Moskva, Russia and from Liverpool, United Kingdom.

Professor Eric Derouane died on 17th March 2008 from a heart attack in his home in Luz, Lagos, Portugal. With him, the Catalysis Community has lost one of its strongest and most brilliant scientists.

Born on 4th July 1944 at Péruwelz (Hainaut), Belgium, Eric Derouane obtained a Licence degree at the University of Liège, B (1965), a Master of Arts (MA) degree in Chemistry in Prof. J. Turkevich's laboratory at Princeton University, USA (1966) and a Doctorat ès Sciences (>PhD) at the University of Liège, B (1968), including a year (1966–1967) in France at the "Service de Physique du Solide et de Résonance Magnétique, CEN Saclay" in Prof. A. Abragam's laboratory. He stayed a year (1969–1970) in USA at Stanford University as a visiting Scholar in Prof. M. Boudart's laboratory. He became Research Assistant of the "Fonds National de la Recherche Scientifique" (FNRS) and Lecturer at the University of Liège, B (1969–1973). In 1973, he was appointed Professor at the "Facultés Universitaires Notre-Dame de la Paix" (FUNDP) in Namur, B, where he created the Laboratory of Catalysis, of which he remained Director until 1995. He was on sabbatical in 1979 as Research Fellow with J. Sinfelt at Exxon Res. & Develop. Corp., Linden, USA, and in 1982–1984 as Research Scientist, Head of Exploratory Catalysis Synthesis Group at Mobil Res. & Develop. Corp., Central Research Laboratory, at Princeton, USA.

In 1995, he became Full Professor at the University of Liverpool and was appointed Director of the Leverhulme Centre for Innovative Catalysis (LCIC).

In 2003, he obtained the prestigious Gulbenkian Professorship at the University of Algarve in Faro, P, where he was the Director of the Chemical Research Centre. Later, he also became invited Professor at the "Instituto Superior Tecnico" (IST) of the Technical University of Lisbon, where he had extensive cooperation with the group led by Prof. F. Ramôa Ribeiro.

His main fields of investigation dealt with catalysis over zeolites in general, supported metals, novel materials and mixed oxides in particular, and alkane upgrading and fine chemicals more specifically. One of Eric's most striking qualities was his acute interest for every new scientific discovery and for industrial applications of his findings.

Eric worked with unusual efficiency. He had a high intellectual mobility and was always attracted by new materials and new concepts. Among them, one can mention the new zeolite ZSM-5/MFI in the early 70s, leading to 30 year collaboration with J.C. Védrine; cuprate-type superconductors and confinement effects and molecular traffic control in microporous zeolitic materials. He also studied reaction mechanisms using isotopic labelling and in situ MAS-NMR in the 80s, combinatorial catalysis and high throughput technology in the late 90s.

During his 20 years of dedicated service to the University of Namur, Eric developed new concepts, which had an important impact on catalysis and zeolite communities. In 1986, he was elected Head of the Chemistry Department. He then embarked upon an impressive re-structuring program to improve its efficiency. The model which he initiated is still in service today. His laboratory was recognized as an outstanding school of scientific research and education in catalysis.

Quite early on, Eric realized the importance of interdisciplinarity, which led him to play a key role in the creation of the Institute for Studies in Interface Sciences (ISIS) at Namur in 1987, which incorporated laboratories of physics and chemistry for 20 years. Eric Derouane also paid heed to technology transfer to industry. After his experience gained through his sabbatical positions at Exxon and Mobil, he developed collaborations with industrial partners and served as consultant for many companies.

At Liverpool, the aim of the LCIC was to promote creative fundamental catalytic science and often to take-up industrial challenges. Eric defined innovation as "the creation of new or better products or processes, implying creativity, usefulness, and application." Towards this end, the LCIC had industrial affiliates as partners. Under his leadership, the LCIC became the largest catalysis centre in the UK and a centre of scientific exchanges and collaborations. Eric established links with many UK and international laboratories. He created in 1997 an European Associated Laboratory "Laboratory for high specificity catalysis" between LCIC/University of Liverpool and Institut de Recherches sur la Catalyse, Lyon / CNRS, of which J.C. Védrine became the Director in 2003.

In 1999, he co-founded with Prof. S. Roberts the *spin-off* Liverpool-based company "Stylacats," of which he became the Director. He provided wise suggestions and ideas, which led the company to pioneer new technologies, particularly catalysts for asymmetric hydrogenation, microwave-induced reactions and enzyme mimetics.

At the University of Faro, Eric developed a research project jointly with the Instituto Tecnico de Lisboa on Friedel-Crafts reactions. He also collaborated closely on various research projects with Prof. F. Ramôa Ribeiro's zeolite group of the Instituto Superior Tecnico of the Technical University of Lisbon.

Eric co-authored over 400 scientific papers, 11 books and 61 patents.

Eric Derouane has contributed greatly to the development and strengthening of the European Catalysis Community. He created in 1975 the European Association in Catalysis (EUROCAT), a consortium of more than 30 European laboratories under the auspices of the Council of Europe and promoted standardisation of characterisation of catalysts: For this purpose, catalysts such as Euro-Pt1 to -Pt4 (Pt/SiO_2 and $Pt\text{-}Re/SiO_2$), Euro-Ni1 & -Ni2 (Ni/SiO_2), Eurocat zeolite (TS1-type), Eurocat oxides (V_2O_5/TiO_2 and $V_2O_5\text{-}WO_3/TiO_2$) were synthesized by industrial companies (Johnson Matthey from United Kingdom, Unilever from the Netherlands, Rhône Poulenc from France, industrial partner from Austria) and distributed to the 30 laboratories in Europe for characterisation. This work led to several articles by G. Bond, J. Coenen, P. Wells and others in Applied Catalysis in the '80s or several special issues of Catalysis Today by J.C. Védrine and others in the '90s to help any scientist to calibrate his/her characterisation techniques by having standard and well-characterized samples.

This Eurocat group paved the way for the creation of the European Federation of Catalysis Societies (EFCATS) and of the François Gault lectureship for which Eric played a decisive role. He was elected President of EFCATS in 1995 for 2 years.

Eric Derouane was the Editor-in-chief of J. Mol. Catal.A: Chemical from 1982 till his death. He was the member of the Editorial Boards of many scientific journals and member of the scientific committees of many congresses and colloquia. He co-organized several congresses himself, particularly with Prof. F. Lemos and F. Ramôa Ribeiro in Portugal on several NATO Advanced Studies Institutes on topics including "the conversion of light alkanes," "combinatorial catalysis and high throughput catalyst design and testing," "principles and methods for accelerated catalyst design and testing" and "sustainable strategies for the upgrading of natural gas." The content of these summer schools was published in NATO editions by Riedel & Co.

Eric's contributions to catalysis have been recognized by many awards and academic honors, including the Wauters Prize (1964), the Mund Prize (1967) of the "Société Royale de Chimie," the Stas-Spring Prize (1971) and the Adolphe Wetrems Prize (1975) of the "Académie Royale de Belgique," the Rosetta Briegel-Barton Lecturership at the University of Oklahoma (1973), the Prize of the "Cercle of Alumni de la Fondation Universitaire de Belgique" (1980), the Ciapetta Lectureship of the North American Catalysis Society (1981), the Catalysis Lectureship of the Société Chimique de France (1993) and the prestigious Francqui Prize, B (1994), the highest honor for all Sciences in Belgium.

He was made "Officier de l'Ordre Léopold" in Belgium (1990), corresponding Member of the "Académie Royale des Sciences, des Lettres et des Beaux Arts de Belgique" (1991), member of the "New York Academy of Sciences" and Associate Member of the "European Academy of Arts, Sciences and Humanities." He was conferred Doctor Honoris Causa by the Technical University of Lisbon (1996).

Eric attracted many students and scholars to his laboratories in Namur, Liverpool and Faro. His energy, clear mind and broad knowledge impressed his students, researchers and colleagues. He was an outstanding and demanding professor, always ready to share his knowledge with his students. His courses were always clear, highly structured and easily understandable. Many of his former students and post-docs occupy today prominent positions in universities and industries. All of them will remember his brilliant and rigorous scientific approach and no doubt, will greatly miss him.

In honour of him, many of Eric's friends and colleagues decided to pay their tribute to him by participating to a two days symposium organized in Lisbon at the Instituto Superior Tecnico of the Technical University of Lisbon on 25–26 September 2008 and organized by Prof. Jacques C. Védrine and by Prof. Fernando Ramôa Ribeiro and his team. Almost one hundred participants attended this symposium from all over the world, in particular from Belgium, Canada, Denmark, France, Italy, Poland, Portugal, Russia, South Africa, Spain, United Kingdom, etc. Many messages of sympathy were received from all over the world as recognition of his worldwide influence. The scientific contributions for this symposium were in all his fields of interest with particular emphases to zeolites, confinement effect, molecular traffic control and catalytic reaction mechanism using MAS-NMR technique, both in heterogeneous and homogeneous catalysis and both from industrial and academic scientists, in excellent coherence with his own scientific career. A special issue of his journal: "*Journal of Molecular Catalysis A: Chemical*" will assemble all contributions and will appear by mid-2009.

Contents

Contributors

Aline Auroux
Institut de Recherches sur la Catalyse et l'Environnement de Lyon, UMR 5256 CNRS/Université Lyon1, 2 avenue Albert Einstein, 69626 Villeurbanne Cedex, France, aline.auroux@ircelyon.univ-lyon1.fr

Allen W. Burton
Chevron Energy Technology Company, 100 Chevron Way, Building 50, Office 1254, Richmond, CA 94802, USA, buaw@chevron.com

Ljiljana Damjanović
Institut de Recherches sur la Catalyse et l'Environnement de Lyon, UMR 5256 CNRS/Université Lyon1, 2 avenue Albert Einstein, 69626 Villeurbanne Cedex, France
Faculty of Physical Chemistry, University of Belgrade, Studentski trg 12-16, 11000 Belgrade, Serbia

Matteo Guidotti
CNR-Istituto di Scienze e Tecnologie Molecolari, IDECAT-CNR Unit, Dip. Chimica IMA "L. Malatesta", via G. Venezian, 2120133 Milano, Italy, m.guidotti@istm.cnr.it

Michel Guisnet
UMR CNRS 6503, Université de Poitiers (France), Departamento de Engenharia Quimica, IST-Technical University of Lisbon, Av. Rovisco Pais, 11049-001 Lisboa, Portugal, michel.guisnet@univ-poitiers.fr

Michael Hunger
Institute of Chemical Technology, University of Stuttgart, 70550 Stuttgart, Germany, michael.hunger@itc.uni-stuttgart.de

Christine E. Kliewer
ExxonMobil Research & Engineering, 1545 Route 22 East, Annandale, NJ, 08801, USA, chris.e.kliewer@exxonmobil.com

Xinsheng Liu
BASF Catalysis Research, R&D Center, 25 Middlesex-Essex Turnpike, Iselin, NJ 08830, USA, xinsheng.liu@basf.com

Evgeny A. Pidko
Schuit Institute of Catalysis, Eindhoven University of Technology, 513, NL-5600 MB Eindhoven, The Netherlands, e.a.pidko@tue.nl

Rutger A.van Santen
Schuit Institute of Catalysis, Eindhoven University of Technology, 513, NL-5600 MB Eindhoven, The Netherlands

Michael Stöcker
SINTEF Materials and Chemistry, 124 Blindern, 0314, Oslo, Norway, michael.stocker@sintef.no

Chapter 1
Powder Diffraction in Zeolite Science

An Introductory Guide

Allen W. Burton

Abstract This tutorial discusses the fundamental principles of X-ray diffraction and its applications in zeolite science. The early sections review the physics of diffraction, crystal symmetry, and reciprocal space. We discuss how the intensity of diffracted radiation is affected both by geometric effects involving detection (the Lorentz-polarization factor) and by the arrangement of atoms within the crystal (the structure factor). The differences between powder diffraction and single-crystal diffraction are then described, and differences between X-ray and neutron diffraction are also discussed. Later sections describe the effects of symmetry, lattice substitution, crystallite size, residual strain, preferred orientation, and X-ray absorption. Special emphasis is placed on the proper application of the Scherrer analysis in reporting crystalize size. The principles of structure solution from direct methods and Patterson methods are then introduced, and a description of Rietveld analysis is given. Finally the effects of stacking disorder on a powder diffraction pattern are presented.

1.1 Introduction

X-ray diffraction (XRD) is an essential tool in the identification and characterization of zeolites at various stages in their syntheses, modifications, and uses as catalysts. Because the typical student is expected to use several characterization methods in his studies of zeolites, attaining a thorough understanding of powder diffraction can initially be daunting for the investigator who wishes to use XRD as more than a "fingerprint" for phase identification. A detailed molecular understanding

---

A.W. Burton
Chevron Energy Technology Company, 100 Chevron Way, Building 50, Office 1254, Richmond, CA 94802, USA
e-mail: buaw@chevron.com

A.W. Chester and E.G. Derouane (eds.), *Zeolite Characterization and Catalysis*,
DOI 10.1007/978-1-4020-9678-5_1,

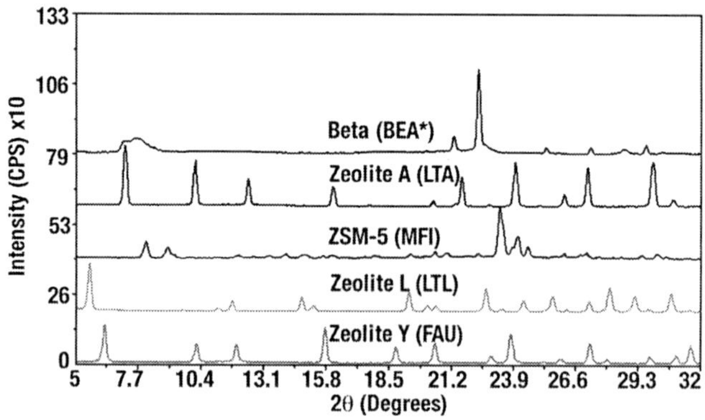

Fig. 1.1 Powder XRD patterns of selected zeolites with important commercial applications

of zeolite structures is essential in explaining their catalytic, adsorption, and ion-exchange properties. XRD allows us to examine the long-range atomic structure of crystalline materials. For zeolites this includes the framework topology and the positions of extra-framework cations and/or adsorbed molecules. Cations strongly influence adsorption and catalysis in zeolites through the interactions they have with guest species.

Figure 1.1 shows powder XRD patterns for five zeolites that have important commercial applications. Note that the patterns of each zeolite are very distinct from one another. Much information may be gleaned from a powder diffraction pattern: the topological and long-range structure of a material, the approximate crystal size of the material, strain or stress in the material, the approximate extent of heteroatom substitution, crystallinity, or the presence of stacking disorder. This chapter begins by discussing fundamental concepts of crystal symmetry, diffraction, reciprocal space, and scattering. Differences between single crystal diffraction and powder diffraction will then be discussed. We will then describe how crystallite size, strain, absorption, preferred orientation, and instrumental broadening affect a powder pattern. Next we will briefly describe methods used to solve and refine crystal structures. Finally, the qualitative effects of stacking disorder will be discussed.

Throughout this chapter, theory is provided to enhance the reader's understanding, but many practical examples are also given to illustrate important concepts. For the beginning student, I highly recommend the practical guides by Bish and Post [1], Jenkins and Snyder [2], and Chung and Smith [3]. The books by Cullity [4], Warren [5], Klug and Alexander [6], and Giacovazzo [7] provide excellent combinations of instructive theory and applications. Warren [5] and Guinier [8] provide thorough mathematical treatments of diffraction, and they give excellent discussion on the effects of disorder in crystalline materials. For explanations of methods involved in crystal structure determination, I recommend the book by Stout and Jensen [9] and the monograph by David et al. [10]

1.2 Lattice Spacings, Unit Cells, Crystal Symmetry, and Space Groups

It is instructive to clarify terms that are often used in the discussion of crystal structures (the book by Burns and Glazer provides useful discussion [11]). A *lattice* is an infinite array of points in space in which each point has identical surroundings to the others. In the case of a crystal structure, the surroundings of each lattice point are defined by the periodic arrangement of atoms in space. Each lattice point can be related to another by an integral number of translations in the *a*, *b*, and *c* directions. The vectors *a*, *b*, and *c* enclose a volume known as the *primitive unit cell*. Primitive unit cells contain only one lattice point, but it is sometimes more convenient to use unit cells with multiple lattice points. For example, a body-centered (i.e., a unit cell with a lattice point also at 1/2*a*, 1/2*b*, 1/2*c* if the origin is chosen as a lattice point) unit cell possesses two lattice points. The *unit cell parameters* define the magnitudes of the unit cell vectors and the angles between them.

The crystal structure is described by associating each lattice point with a group of atoms within the unit cell known as the *basis*. The basis includes the atoms and their positions. From the basis, the unit cell composition can be directly determined and is often listed, for example, in the Atlas of Zeolite Framework Types [12] or in the Collection of Simulated XRD Patterns for Zeolites [13]. The *space group* is the set of symmetry operations that take a three-dimensional periodic object into itself. The *asymmetric* unit is the smallest region of space that fills all space when these symmetry operations are applied. If the atoms within the asymmetric unit are specified, then the basis of atoms in the unit cell is produced by the symmetry operations of the space group. In a typical crystallographic report, the asymmetric unit, unit cell parameters, and space group are provided. These can be used as input to software such as Cerius [14] or ORTEP [15] to provide three-dimensional views of the structure with as many unit cells in the three dimensions that the investigator wishes to visualize. In the patent literature, inventors of new materials occasionally list the coordinates for the entire basis of atoms in the unit cell of the crystal structure. The space group symmetry and unit cell parameters of the material are sometimes not provided in the patent, but in some cases they can be inferred.

Each space group symbol provides enough information to determine all the symmetry operations performed by the space group. For example, the structure of ITQ-3 (ITE) has the orthorhombic space group symmetry *Cmcm*. The "*C*" indicates the unit cell is face-centered on the *a*–*b* plane at $c = 0$ (i.e., there is a lattice point at 0,0,0 and 1/2, 1/2, 0). The first "*m*" indicates there is a mirror plane perpendicular to the *a*-axis, the "*c*" indicates there is a glide plane perpendicular to the *b*-axis, and the second "*m*" indicates there is a mirror plane perpendicular to the *c*-axis. A glide plane is an operation in which a reflection across a plane is followed by a translation; in the current example, the "*c*" glide plane indicates a reflection across a plane perpendicular to b followed by a translation 1/2*c*. Other glide planes include the "*n*" and "*d*" glide planes. In an *n*-glide plane that is perpendicular to the *c*-axis, the reflection across the plane is followed by translations of 1/2*a* and 1/2*b*.

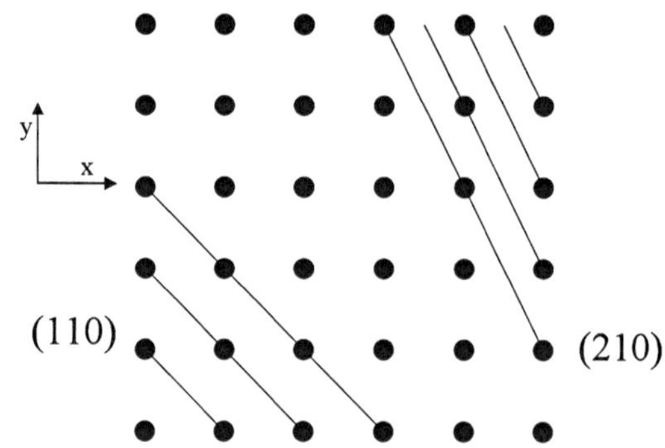

Fig.1.2 Illustration of the (*110*) and (*210*) Miller plane. The *z*-direction is perpendicular to the plane of the paper

For higher symmetry (hexagonal, tetragonal, cubic) lattices in which the *a*- and *b*-axes are symmetry-equivalent, the order of the space group symbols carries different meaning with regard to the positions of symmetry axes or planes. For example, let us consider the cubic space group *I432*. In this body-centered space group, there are four-fold rotation axes along the *x*-, *y*-, and *z*-axes. There are threefold rotation axes along the diagonals of the unit cell and twofold rotation axes along the diagonals of each face of the unit cell. The twofold rotation axes do not help to define the space group since they can be derived from the other symmetry operations.

The lattice points of a given crystal structure define an infinite number of families of parallel planes. Each family of parallel planes in the lattice is defined by the set of Miller indices $h\ k\ l$. The lattice points repeat in the x, y, and z (not necessarily perpendicular) directions with distances of a, b, and c respectively. If one lattice point is chosen as the origin, then the intercepts of all planes (along the x, y, and z axes) can be expressed as ma, nb, and pc where m, n, and p are integers or infinity. In terms of the Miller indices

$$h \propto \frac{1}{m},\ k \propto \frac{1}{n},\ l \propto \frac{1}{p}$$

so $h = np/t$, $k = mp/t$, and $l = mn/t$, where t is the greatest common divisor of np, mp, and mn. Figure 1.2 shows examples of the (210) and (110) planes for a lattice in which a and b are equal (but not necessarily identically equal).

1.3 Fundamentals of Diffraction and Reciprocal Space

Figure 1.3 shows two infinite parallel planes separated by a distance d. If monochromatic radiation (for diffraction this will typically be X-rays, neutrons, or electrons) of wavelength λ strikes those planes at an angle θ, constructive interference will occur when the path difference between the diffracted (or reflected) waves

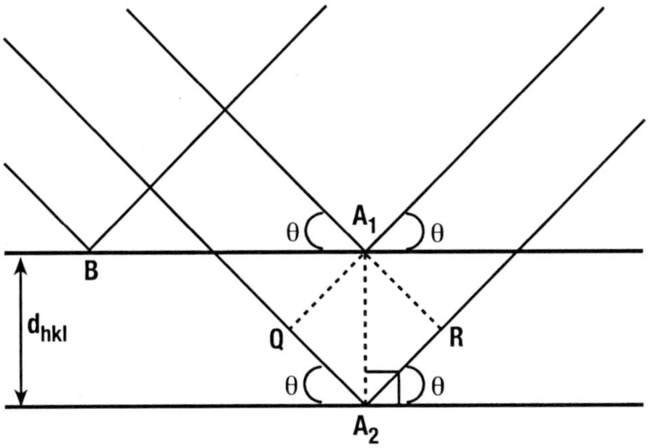

Fig.1.3 Illustration of diffraction from two parallel planes. Constructive interference occurs when the path difference (i.e., $QA_2 + A_2R = 2d_{hkl}\sin\theta$) of the reflected waves is equal to an integral number of wavelengths. Figure adapted from [2]

from each plane is equal to an integral number, n, of wavelengths. The rays in Fig. 1.3 are shown such that the distance A_1A_2 is equal to the distance between the parallel planes. It is straightforward to show that constructive interference occurs when

$$n\lambda = 2d\sin\theta \tag{1.1}$$

This condition applies regardless of where the parallel rays strike the planes. For example, parallel rays that strike the first plane at points B and A_1 will both be in phase with the rays that reflect from the second plane when the Bragg equation is satisfied. Can you prove this? Note that the lower limit for a detectable "d" spacing is $\lambda/2$ since $\sin\theta \leq 1$. This explains why X-rays are suitable for diffraction studies of crystalline materials. For visible light ($\lambda \sim 5{,}500$ Å), the lowest detectable d-spacing is about 2,750 Å. This is clearly larger than the unit cell dimensions of most crystalline materials (zeolite cell dimensions are typically found in the range of 5–80 Å).

It can be shown that the distance, d_{hkl}, between a set of hkl planes in an orthorhombic lattice (a lattice where symmetry constraints force the x, y, and z directions to be mutually perpendicular, but the dimensions are not identically equal) can be determined from the relation

$$\frac{1}{{d_{hkl}}^2} = \frac{h^2}{a^2} + \frac{k^2}{b^2} + \frac{l^2}{c^2}$$

Similar equations may be derived for other lattice systems. Hence, from the measured positions of diffraction peaks, information on the unit cell parameters can be determined.

1.3.1 Ewald Sphere of Reflection and Reciprocal Space

In some discussion of diffraction, it is more convenient to use the "reciprocal space lattice" rather than the real space lattice. The reciprocal lattice is more applicable to the interpretation of diffraction data because, in essence, the diffraction pattern is a manifestation of the reciprocal lattice. We can define a space of vectors of magnitude $1/d_{hkl}$ (i.e., the "reciprocal" of the d-spacing) that are perpendicular to their respective *hkl* plane. Each *hkl* reciprocal space vector is expressed as $\mathbf{r}^* = \mathrm{h}\mathbf{a}^* + \mathrm{k}\mathbf{b}^* + \mathrm{l}\mathbf{c}^*$. The reciprocal lattice spacing $\mathbf{a}^*$ is defined such that $\mathbf{a}^* \cdot a = 1$ and $\mathbf{a}^* \cdot b = \mathbf{a}^* \cdot c = 0$ (i.e., the reciprocal lattice vector $\mathbf{a}^*$ is perpendicular to the real space vectors **b** and **c**). Analogus expressions apply to $\mathbf{b}^*$ and $\mathbf{c}^*$. Note that each of the reciprocal space lattice vectors is parallel to the real space lattice vector in lattices with orthogonal cell parameters.

The reciprocal lattice can be used to understand Bragg's law in a pictorial fashion using the concepts of the Ewald (or reflection) sphere and the limiting sphere. In Fig. 1.4 we have drawn two spheres that are tangent to one another at point Q: the Ewald sphere has radius $1/\lambda$ and the limiting sphere has radius $2/\lambda$. Point *P* is the center of the smaller sphere and corresponds to the position of our crystal. The center of the larger circle is at point O, which is where we will assign the origin of the reciprocal lattice. When diffraction occurs, the angle between the primary beam along QP and the diffracted beam along PR is 2θ. The intersection of

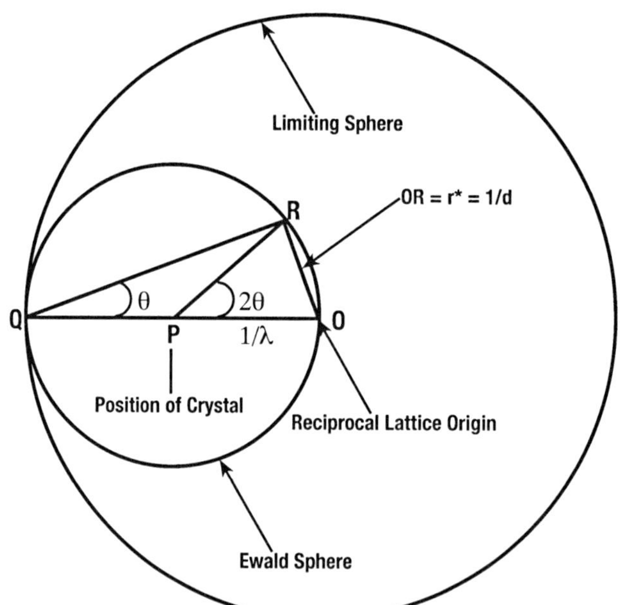

Fig. 1.4 Illustration of Ewald sphere and the limiting sphere of reflection. The images on the next page show points of the reciprocal lattice as it rotates

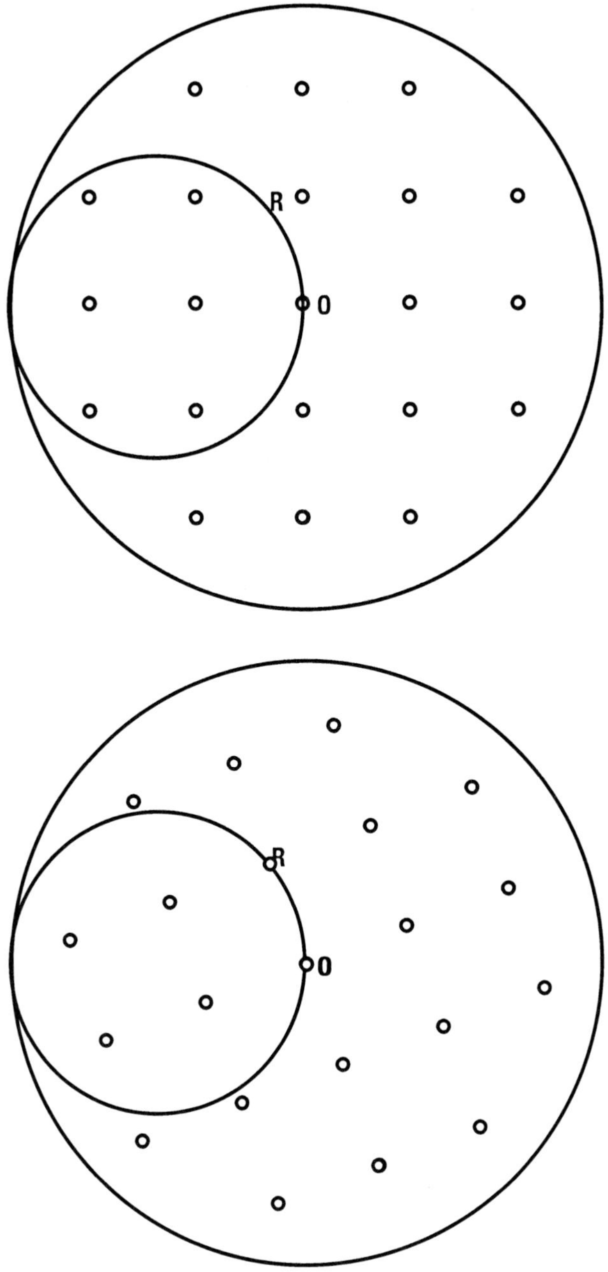

Fig. 1.4 (Continued)

the diffracted beam with the Ewald sphere occurs at point R. The angle $\angle PQR$ equals θ (can you prove that?). Note that $OR = OQ \sin\theta = 2\sin\theta/\lambda$. If we state that $OR = \mathbf{r}^*$ (which is true since O is the origin of the reciprocal lattice), then the previous equation is equivalent to Bragg's law since $\mathbf{r}^* = 1/d$. Therefore, if the reciprocal lattice is rotating about point O (as the crystal is being rotated), then diffraction occurs when a reciprocal lattice point intersects the Ewald sphere. If the reciprocal lattice vector has magnitude $r^* > 2/\lambda$, then $\mathbf{r}^*$ lies outside the limiting sphere and it is impossible for its reciprocal lattice point to intersect the Ewald sphere. This is equivalent to the statement near the beginning of this section that d_{hkl} must be greater than $\lambda/2$ in order for diffraction to occur. By completely rotating the reciprocal lattice in all possible directions, all lattice spacings with $\mathbf{r}^* < 2/\lambda$ will pass through the Ewald sphere. In subsequent sections, the reciprocal lattice will be used to explain other concepts of diffraction.

1.4 Single Crystal Diffraction

It is from the *hkl* families of planes that diffraction of X-rays occurs within a given crystal. Figure 1.5 shows a single crystal diffraction pattern of zeolite RUB-3 (RTE) along the *hk0* zone. [16] This image contains the diffraction spot for each measurable *hk0* reflection. From this diffraction pattern, we are able to infer the *d*-spacings

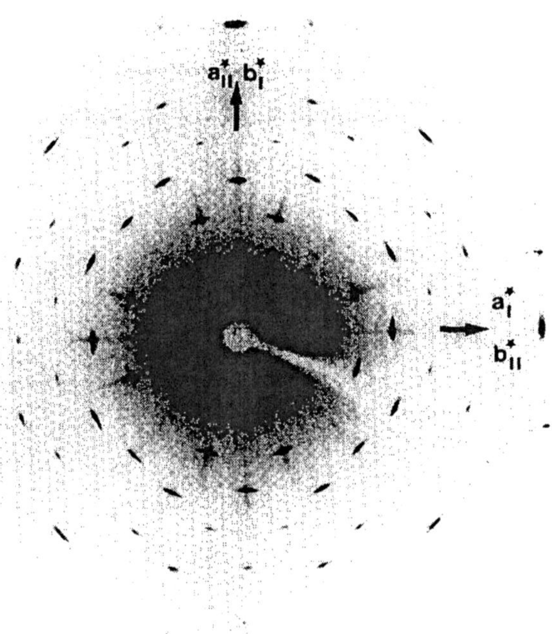

Fig.1.5 Precession photograph of the *hk0* layer of RUB-3. Reprinted from [16] with permission from Elsevier

within the *a–b* plane and therefore the likely cell dimensions along the a- and b-directions. If the film is perpendicular to the primary beam, the diffraction angle θ for each spot is equal to $\tan^{-1}(r/D)/2$, where D is the distance from the sample to film and r is the distance of the diffraction spot from the spot for the direct beam.) By rotating a crystal, we are able to record diffraction patterns along other (*hkl*) zones and thereby create a mapping of the diffracted intensity in three-dimensional space.

Many single crystal diffraction studies today are performed with four-circle diffractometers. By rotating a crystal around three axes as it is being irradiated, all measurable *hkl* planes may be placed in a position to diffract and a recording can be made for the measured intensities of each position of the detector in three-dimensional space.

1.5 Intensity From Diffraction

In order to obtain information about atomic structure from intensity data, we must first understand all the factors that contribute to the measured intensity of an *hkl* reflection. Relative peak intensities are a function not only of the atomic structure of the material, but also of angle-dependent geometric factors that result from the detection of the diffracted radiation and from the polarization of the radiation source. This section explains these factors. The integrated intensity for a given Bragg reflection is given by

$$I(hkl) = KL_p(\theta)|F(hkl)|^2 \tag{1.2}$$

K is a constant that depends on the intensity and wavelength of the incident beam, the distance from the detector to the sample, the properties of the electron (for X-ray scattering), the unit cell volume, and the total volume of the irradiated sample. $L_p(\theta)$ is the *Lorentz-polarization factor* at the diffraction angle for the *hkl* reflection, and $F(hkl)$ is the *structure factor* for the *hkl* reflection.

1.5.1 The Lorentz-Polarization Factor

The Lorentz-polarization factor contains corrections which account for (1) the polarization (or lack thereof) of the radiation source and (2) geometric effects involved in the detection of the diffracted radiation (often referred to as the Lorentz component). The polarization component can be expressed generally as $K_1 + K_2\cos^2(2\theta)$. This term arises because the primary beam possesses electric field components that lie perpendicular and parallel, respectively, to the plane containing the primary beam and the scattered radiation. An electric field induces oscillatory motion in the electrons of an atom. It is this motion that gives rise to the scattered radiation. If the radiation is unpolarized, then $K_1 = K_2 = 1/2$.

For scattering of neutrons, there is no polarization of the radiation, and so the polarization term is replaced by a factor of 1. When single crystal monochromators are used, additional terms are required for the polarization correction because the monochromator partially polarizes the X-ray beam.

For a single crystal irradiated by unpolarized radiation, the entire Lorentz-polarization factor is expressed as

$$L_p = \frac{1 + \cos^2(2\,\theta)}{2\ \sin(2\theta)} \tag{1.3}$$

The factor $1/\sin(2\theta)$[1] arises because the time required for a reciprocal lattice point to pass through the Ewald sphere (i.e., to be in diffracting position) is not constant; that is, the time varies with the position of the lattice point in reciprocal space and the direction in which it approaches the Ewald sphere. As a result, the reciprocal lattice points remain within detection range for different periods of time. In my experience, I admittedly have initially found this to be a difficult concept to grasp. It is useful to visualize the Ewald sphere as a shell that has some infinitesimal thickness ΔR as shown in Fig. 1.6. In a typical single crystal experiment, the crystal is rotated around an axis perpendicular to the plane of the incident and diffracted beams at a constant angular velocity ω. The reciprocal lattice (centered at point O) therefore also rotates with this same angular velocity. Our goal is to find the amount

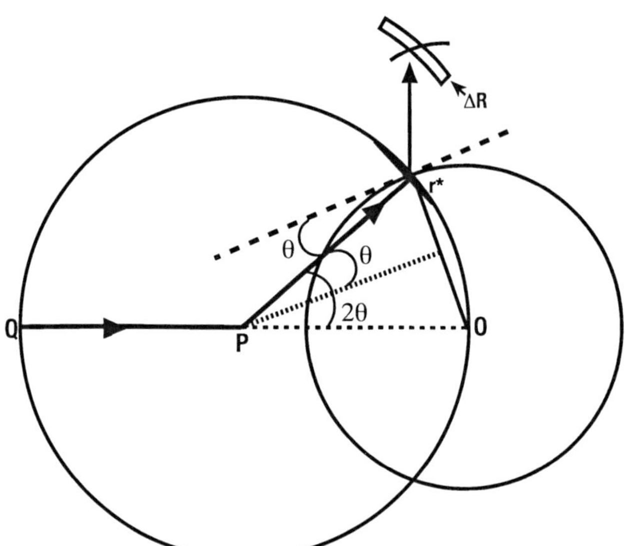

Fig.1.6 Illustration of reciprocal lattice point passing through the Ewald sphere of reflection

[1]This assumes that the axis of rotation of the crystal with respect to the incident beam is 90°.

of time that a reciprocal lattice point remains within the shell of the Ewald sphere. The linear velocity of the reciprocal lattice point is $\mathbf{r}^*\omega$. The time required for the lattice point to pass through the shell is equal to the shell thickness (ΔR) divided by the component of the linear velocity that lies along the radius of the Ewald sphere. This component of the velocity is $\mathbf{r}^*\omega\cos\theta = (2\omega\sin\theta\cos\theta)/\lambda = \omega\sin 2\theta/\lambda$. The time is therefore proportional to $1/(\sin 2\theta)$.

For a randomly oriented powder, the Lorentz-polarization factor is given by

$$L_p = \frac{1 + \cos^2(2\,\theta)}{\sin(\theta)\,\sin(2\theta)} \tag{1.4}$$

In this case an additional factor of $\cos\theta/\sin(2\theta)$ has been multiplied by the single crystal factor. The additional factor has two different sources: (1) it represents the fraction of crystallites in the powder that are oriented in such a way to diffract at the Bragg angle ($\cos\theta$ term) and (2) it accounts for the fact that the intensity of each reflection at a given Bragg angle is distributed over a circular ring with a circumference that changes with the diffraction angle ($1/\sin 2\theta$). From Fig. 1.7 we can determine the origin of the $\cos\theta$ term. For a perfectly random distribution of crystallites, the normal vectors of each *hkl* plane will uniformly cover the surface of a sphere of radius R around the sample. When diffraction occurs from these *hkl* planes, the angle between the primary beam and their normal is $(90-\theta)$. The key then is to find the fraction of normal vectors that are at an angle $(90-\theta)$ to the primary beam. Since a given *hkl* vector uniformly covers the surface of this sphere,

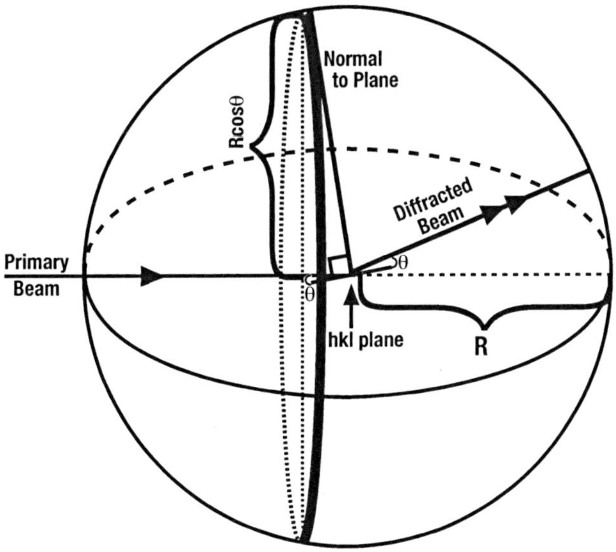

Fig.1.7 Illustration used to estimate fraction of crystallites that are in a position to diffract at a given angle θ. This fraction gives rise to the $\cos\theta$ factor in the Lorentz polarization correction

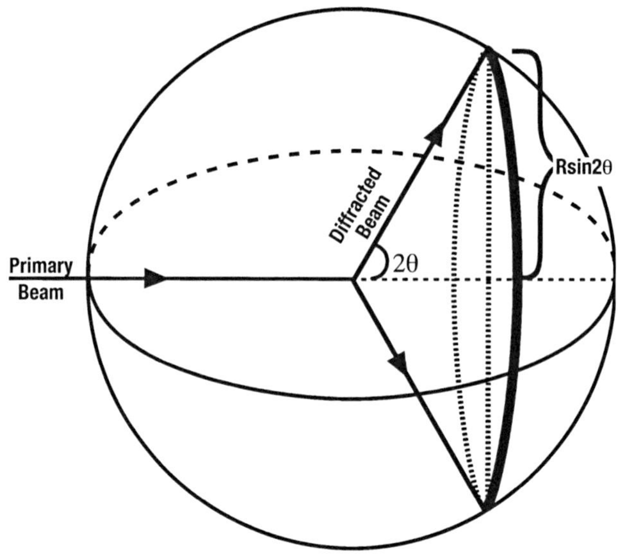

Fig.1.8 Illustration of the powder ring distribution that gives rise to the $1/\sin 2\theta$ term in the Lorentz polarization factor

this can be found by determining the area of a circular band of infinitesimal width that surrounds the sphere at an angle $(90-\theta)$. The radius of this band is $R\cos\theta$, and the area of the band is therefore proportional to $\cos\theta$.

The $1/\sin 2\theta$ term arises from the fact that the intensity of a reflection is distributed over the circle of a diffraction cone (Fig. 1.8) with radius proportional to $\sin 2\theta$, and so the intensity per unit length is proportional to $1/\sin 2\theta$. These circular bands sometimes are referred to as "Debye" rings. An analogy would be that of a circular rubber band with an even distribution of particles on its surface whose radius changes with the diffraction angle. Because the detector measures intensity only at a single point along the circle of the cone and does not measure the integrated sum of intensities around the entire circular band, the measured intensity is, in effect, diluted by a factor proportional to the circumference of the circle.

After these factors are multiplied by the single crystal factor, we obtain the entire Lorentz contribution for powder diffraction:

$$\frac{\cos\theta}{\sin 2\theta \sin 2\theta} = \frac{\cos\theta}{2\cos\theta \sin\theta \sin 2\theta} = \frac{1}{2\sin\theta \sin 2\theta}$$

Figure 1.9 shows a graph of the Lorentz-polarization factor for powder diffraction. Here it can be seen that the Lorentz-polarization correction is very large at low angles and that it reaches a minimum around 100°. This is one reason that low-angle peaks tend to be of greater intensity than high angle peaks in powder diffraction patterns. Also, when one is comparing powder patterns of the same material collected with different wavelengths, it is important to remember that the relative

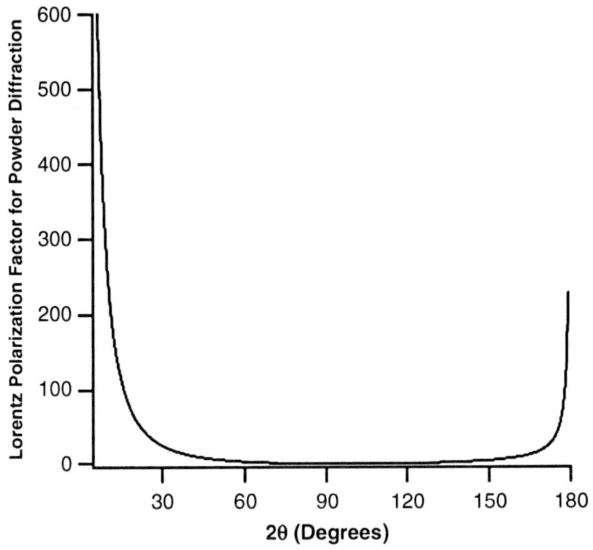

Fig. 1.9 The Lorentz polarization factor as a function of 2θ for powder diffraction

intensities of the same *hkl* peak will vary between the data sets because they possess different L_p values. This is particularly relevant for the low-angle peaks, where a change of a few degrees has significant effects on the L_p correction.

1.5.2 The Structure Factor

The structure factor for a given *hkl* reflection is given by

$$F(hkl) = \sum_j p_j f_j \exp\left(\frac{-B_j \sin^2\theta}{\lambda^2}\right) \exp\left[2\pi i\left(h x_j + k y_j + l z_j\right)\right] \tag{1.5}$$

where the sum is over all atoms in the unit cell, f_j is the scattering factor of atom j, p_j is the fractional occupancy of atom j, and x_j, y_j, and z_j are the fractional coordinates of atom j.

The term $\exp(-B_j \sin^2 \theta/\lambda^2)$ is a correction to the scattering factor that accounts for the thermal motion of the atom. The B_j term is a measure of the root-mean-square amplitude of the vibration of the atom. Care must be used in interpreting the f_j; in many references this term actually designates

$$f_j \exp\left(\frac{-B_j \sin^2\theta}{\lambda^2}\right)$$

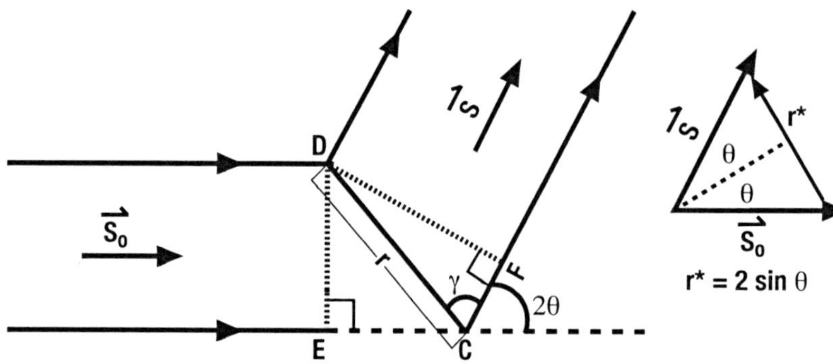

Fig.1.10 Scattering of parallel waves from two point sources (D and C)

Note that the thermal motion of the atoms will most strongly affect the intensities of reflections at higher angles. As the temperature is lowered, the thermal factor will decrease. Therefore, the relative intensities of high angle peaks will tend to increase as the temperature is lowered (assuming no changes in symmetry).

This sum initially may seem complex; but after discussion of the scattering process and wave addition, it will make more sense. In Fig. 1.10, there are two scatterers located at points C and D, which are separated by a vector **r**. Our goal is to find the amplitude of the scattered waves that exists at an arbitrary angle 2θ between the directions of the primary beam and the scattered radiation. The scattered radiation emanates from each point as spherical waves that then interfere with each other. We denote the unit vector in the direction of the primary beam by $\mathbf{s_o}$ and the unit vector in the chosen direction of the scattered waves by **s**. Segments ED and DF are perpendicular to $\mathbf{s_o}$ and **s**, respectively. The difference in path length between the radiation scattered at point D and the radiation scattered at point C is equal to $CF+EC$. Note that $CF = CD\cos\gamma = \mathrm{r}\cdot\mathbf{s}$ (since s is a unit vector). Likewise, $EC = -\mathrm{r}\cdot\mathbf{s_o}$. The path difference is therefore $\mathrm{r}\cdot(\mathbf{s}-\mathbf{s_o})$. The phase difference (in radians) between the radiation scattered by points C and D therefore is

$$\frac{2\pi}{\lambda}\overline{r}\cdot(\overline{s}-\overline{s_o})$$

If we define $\mathbf{r^*}$ as $(\mathbf{s}-\mathbf{s_o})/\lambda$, we can show from Fig. 1.10 that $\mathbf{r^*} = 2\sin\theta/\lambda$, which equals $1/d_{hkl}$ at the hkl Bragg angle (this is the same form we developed for Bragg's law). If C_o and D_o are the amplitudes of the waves scattered by points C and D, respectively, then the sum of the amplitudes is $C_o+D_o\ \exp(2\pi i\mathbf{r^*}\cdot\mathbf{r})$. We can generalize this expression for N scatterers (in a unit cell, for example) by the following sum:

$$F(r*) = \sum_{j=1}^{N} f_j \exp\left(2\pi i r^* \cdot r_j\right) \tag{1.6}$$

We have replaced the amplitude C_o term for each scatterer (or atom) by a "scattering factor" f_j. The term $(2\pi i r^* \cdot r_j)$ denotes how "out of phase" the scattered radiation will be relative to radiation scattered from a point at the origin of our unit cell. In the following sections, we will discuss how the scattering factor varies with both the atom type and the scattering angle.

We previously noted that for a given *hkl* reflection the reciprocal lattice vector **r*** is equal to (h**a***+k**b***+l**c***). The atomic position $\mathbf{r_j}$ of atom j in a unit cell can be expressed in fractional coordinates as $x_j\mathbf{a}+y_j\mathbf{b}+z_j\mathbf{c}$. The dot product $\mathbf{r^*} \cdot \mathbf{r}$ therefore equals $(hx_j+ky_j+lz_j)$. Substitution of this result into the above expression for F(**r***) yields the equation for the structure factor at the beginning of this section (with the exception of the correction term for thermal vibration of each atom).

In summary, the equation for the structure factor accounts for the fact that the atoms in the unit cell do not scatter in phase. Note that if every atom could be placed at the same position, then there would be no phase difference. The structure factor is a complex number that equals the sum of a "real" component "A" and an "imaginary" component "B"; i.e., $F(hkl) = \mathrm{A}+\mathrm{B}i$. The square of the modulus of $F(hkl)$ (which we can extract from the intensity data using (1.2)) is $(\mathrm{A}+\mathrm{B}i)(\mathrm{A}-\mathrm{B}i) = \mathrm{A}^2+\mathrm{B}^2$. We can then express the structure factor in exponential form as

$$\sqrt{A^2 + B^2}\,\exp(i\phi)$$

The "phase" ϕ of the structure factor is $\tan^{-1}$(B/A). A key point here is that it is not possible to measure the phase ϕ from the diffraction data. In the section on direct methods, we will discuss how these phases can be estimated and used to determine the locations of atoms within a unit cell.

1.6 X-Ray Diffraction

In X-ray diffraction, X-rays are scattered by the electrons that surround each atom. Because the electrons are distributed throughout the volume of the atom and because atomic dimensions are of the same magnitude as the wavelengths of X-rays, X-rays scattered from different regions of an atom do not scatter in phase. That is, there is partial interference of X-rays scattered from electron density in different regions of the same atom. As a result, there is a net decrease in the scattered amplitude of the atom that varies with the diffraction angle. At low angles the scattering factor for an atom is nearly proportional to the number of its electrons since there is no phase difference for $\theta = 0$ and since the amount of scattering is proportional to the number of scatterers. Heavy atoms therefore are strong X-ray scatterers. However, there is a decrease in the X-ray scattering factor with an increase in sin θ/λ since the interference effects are greater at higher angles. This is another reason that the intensities of higher angle reflections are, in general, of lower magnitude than those at low angles (e.g., see Fig. 1.1). Figure 1.11 shows how the atomic scattering factor varies with $\sin\theta/\lambda$ for the elements O, Al, Si, Ge, and Ag.

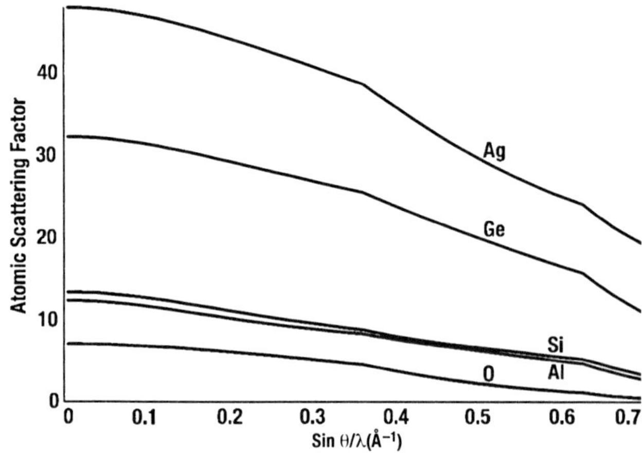

Fig. 1.11 Angular dependence of X-ray scattering factor for selected elements. The curves were created from data in the International Tables for Crystallography

As shown in Fig. 1.11, light elements do not scatter as effectively as heavy elements. The weaker contributions of light atoms are further magnified by the fact that the intensity of a reflection is proportional to the *square* of the magnitude of its structure factor. In a zeolite, the most common elements are silicon, aluminum, and oxygen. Each of these atoms has between 8 and 14 electrons, the actual number depending upon the "degree" of ionicity. A lithium cation, on the other hand, formally possesses only two electrons. In a lithium-exchanged zeolite the relative contributions of the lithium cation to the diffraction intensities are therefore expected to be very weak. Hence, it is difficult to locate lithium and refine its structural parameters using powder X-ray diffraction data. It is for this reason that structure refinements of lithium- or deuterium-exchanged zeolites are usually performed on neutron (rather than X-ray) diffraction data. Also, for materials that have a significant proportion of heavy atoms, neutron diffraction studies are often carried out in conjunction with X-ray diffraction in order to facilitate the location and refinement of the lighter elements.

1.6.1 Example of Structure Factor Calculation

We will illustrate the calculations of structure factors and peak positions by considering sodalite (SOD) with a unit cell composition of $Na_6(H_2O)_8Al_6Si_6O_{24}$, cubic space group symmetry *P-43n*, and a unit cell parameter of 8.848 Å. In this example, we calculate the peak position, Lorentz-polarization factor, and the structure factor for the 200 reflection. The atomic coordinates for the asymmetric unit are provided by the website for the International Zeolite Association

Table1.1 Coordinates, occupancy factors, and thermal factors for atoms in asymmetric unit of sodalite

Atom Name	Form Factor	*x*	*y*	z	PP	*B*(iso)
NA	Na	0.1504	0.1504	0.1504	0.75	3.71
SI1	Si	0.25	0.0	0.5	1.0	0.79
AL1	A1	0.25	0.5	0.0	1.0	1.18
O1	O	0.1366	0.4338	0.1490	1.0	1.5
H2O	O2-(H2O)	0.3753	0.3753	0.3753	1.0	3.0

within the Collection of Simulated Patterns. The coordinates are listed in Table 1.1.

If we examine the space group table for *P-43n* (218) in the International Tables for Crystallography [17], we find that application of the symmetry operations to the general position *x,y,z* (that is, a position that is not located on a special symmetry operator) yields 24 symmetry-related sites with coordinates

(1) x,y,z (2) $-x,-y,z$ (3) $-x,y,-z$ (4) $x,-y,-z$ (5) z,x,y (6) $z,-x,-y$ (7) $-z,-x,y$ (8) $-z,x,-y$ (9) y,z,x (10) $-y,z,-x$ (11) $y,-z,-x$ (12) $-y,-z,x$ (13) $y+1/2, x+1/2$, $z+1/2$ (14) $-y+1/2,-x+1/2$,$z+1/2$ (15) $y+1/2,-x+1/2$,$-z+1/2$ (16) $-y+1/2$, $x+1/2,-z+1/2$ (17) $x+1/2$, $z+1/2,y+1/2$ (18) $-x+1/2,z+1/2,-y+1/2$ (19) $-x+1/2,-z+1/2,y+1/2$ (20) $x+1/2,-z+1/2,-y+1/2$ (21) $z+1/2,y+1/2,x+1/2$ (22) $z+1/2,-y+1/2,-x+1/2$ (23) $-z+1/2,y+1/2,-x+1/2$ (24) $-z+1/2,-y+1/2,x+1/2$. It is useful to verify that the reported unit cell composition is consistent with the atomic coordinates and their occupancy factors. Framework atom O1 is located on a general position and thus there are 24 framework O atoms in the unit cell. Si1 and Al1 are both positioned on a −4 axis, and so there are 24/4 = 6 of each atom in the unit cell. The extra-framework species Na and H_2O (modeled as O^{2-}) are each located on a 3-fold axis, and so there are 24/3 = 8 positions for each atom in the unit cell. However, the Na partially occupies only 75% of these sites so there are actually 6 Na atoms within the unit cell.

The *d*-spacing for the 200 reflection in a cubic unit cell is $a/2 = 4.424$ Å. The 2θ angle (from Bragg's law) for CuKα radiation therefore equals $2\ \sin^{-1}(\lambda/2d) = 2\ \sin^{-1}(1.5418/8.848) = 20.07°$. The Lorentz-polarization factor is

$$\frac{1+\cos^2(2\,\theta)}{\sin(\theta)\ \sin(2\theta)} = \frac{1+\cos^2(20.07^{\circ})}{\sin(10.035^{\circ})\ \sin(20.07^{\circ})} = 31.5$$

For the 200 reflection, the structure factor $F(hkl) = \sum_j p_j f_j \exp\left(\frac{-B_j \sin^2\theta}{\lambda^2}\right) \exp\left[2\pi i\left(h x_j + k y_j + l z_j\right)\right]$ reduces to

$$F(200) = \sum_j p_j f_j \exp\left(\frac{-B_j \sin^2\theta}{\lambda^2}\right) \exp\left[4\pi\, i x_j\right]$$

$$= \sum_j p_j f_j \exp\left(\frac{-B_j \sin^2\theta}{\lambda^2}\right) \left[\cos 4\pi\, x_j + i \sin 4\pi\, x_j\right] \qquad (1.8)$$

To simplify the calculations, we will consider the individual contribution of each atom type to the sum. This is performed by adding the contributions of each symmetry-equivalent atom within the unit cell. If we first examine the coordinates for the symmetry related sites of a general position, we find that for every atom with an "x" as the first coordinate, there are an equal number of atoms that have an "$-x$" coordinate at the first position. Since $\sin(-4\pi A) = -\sin(4\pi A)$, this means that all the sin terms will identically cancel in (1.8). Also, note that there are 8 terms that have x, $-x$, $x+1/2$, or $-x+1/2$ as the first coordinate. Since $\cos(4\pi A) = \cos(-4\pi A) = \cos(4\pi(A+1/2))$, the sum of these eight terms is $8\cos(4\pi x)$. The same reasoning applies to the group of sites that have (y, $-y$, $y+1/2$, or $-y+1/2$) as their first coordinate and to the group of sites that have (z, $-z$, $z+1/2$, or $-z+1/2$) as their first coordinate. Hence, for the framework O atom, the contribution to the 200 structure factor may be recast as

$$8f_O \exp\left(\frac{-B_O \sin^2\theta}{\lambda^2}\right)[\cos 4\pi x_O + \cos 4\pi y_O + \cos 4\pi z_O]$$

where x_O,y_O,z_O are the coordinates for atom O in Table 1.1. For the 200 reflection, $\sin\theta/\lambda = \sin(10.035°)/1.5418 \text{ Å} = 0.11 \text{ Å}^{-1}$. From Fig. 1.11 (or from the X-ray scattering tables), we find that $f_O = 7.1$. The total contribution of the framework O atoms is therefore

$$8^*7.1\exp(-1.5^*0.11^2)[\cos(4\pi^*0.1366) + \cos(4\pi^*0.4338) + \cos(4\pi^*0.1490)] = 12.9$$

For the Si atoms, the six symmetry-equivalent positions are (1/4, 0, 1/2), (3/4, 0, 1/2), (1/2, 1/4, 0), (1/2, 3/4, 0), (0, 1/2, 1/4), and (0, 1/2 3/4). The contribution from the Si atoms is therefore

$$f_{Si}\exp\left(\frac{-B_{Si}\sin^2\theta}{\lambda^2}\right)[\cos\pi + \cos 3\pi + \cos 2\pi + \cos 2\pi + \cos 0 + \cos 0]$$
$$= 2f_{Si}\exp\left(\frac{-B_{Si}\sin^2\theta}{\lambda^2}\right) = 2^*11.85\exp(-0.79^*0.11^2) = 23.5$$

In a similar fashion, we can show that the contribution from the Al atoms is 21.6.

The contribution from the symmetry-equivalent Na atoms is

$$0.75^*8f_{Na}\exp\left(\frac{-B_{Na}\sin^2\theta}{\lambda^2}\right)[\cos 4\pi x_{Na}]$$
$$= 0.75^*\, 8^*\, 9.61\exp(-3.71^*\, 0.11^2)\cos(4\pi^*\, 0.1504) = -17.3$$

… and from the extra-framework water O atoms the contribution is

$$8f_{Ow}\exp\left(\frac{-B_{Ow}\sin^2\theta}{\lambda^2}\right)[\cos 4\pi\, x_{Ow}]$$
$$= 8^{*}\,8.0\exp\left(-3.08^{*}\,0.11^2\right)\cos(4\pi^{*}\,0.3753) = 0.3$$

The structure factor for the unit cell is $12.9+23.5+21.6-17.3+0.3=41$. It can be appreciated that the calculation of structure factors is a function well suited for computers.

Sometimes, structure factors are reported as a normalized "unitary structure factor" defined as $U_{hkl} = F_{hkl}/F_{000}$. The value of the unitary structure factor gives an indication of how strong the reflection is. Note that F_{000} is simply the total number of electrons in the unit cell. In the current example,

$$U_{200} = 41/500 = 0.08$$

1.7 Neutron Diffraction

Neutrons are scattered from the nuclei of atoms rather than from their electrons. Thus, in contrast to the diffraction of X-rays, neutrons are effectively scattered from points (compared to the wavelength of the radiation) rather than from electron clouds. Therefore, no angular dependence exists for the neutron scattering factor. As a consequence, the relative intensities of high angle reflections do not, in general, tend to be as weak as the relative intensities observed at high angles in X-ray diffraction. This is illustrated quite well in Fig. 1.12, which compares a powder neutron diffraction pattern and a synchrotron X-ray diffraction pattern of rutile [18]. These data were collected at the same wavelength, and (in the words of Cheetham and Wilkinson) "the angular dependence of the X-ray scattering amplitudes … is clearly seen in the bottom pattern."

Neutron scattering factors do not vary systematically with atomic number (Fig. 1.13), but instead depend on the properties of the nucleus. Isotopes of the same element therefore possess different scattering lengths. Also, in contrast to X-ray scattering, some neutron scattering factors are negative (which implies a 180° phase shift after scattering). This is the case for the isotopes ^{2}H and ^{7}Li. Since most (isotopes of) elements have positive scattering factors, a negative scattering can be advantageous when attempting to locate these atoms in difference Fourier maps. A handful of isotopes even possess an "imaginary" component (which implies a phase shift between 0 and 180° after scattering). Neutron scattering factors vary over a smaller range of relative values than X-ray scattering factors do. Neutron scattering factors typically lie between 15 and – 10 fm. Therefore, the scattering from "weak" neutron scatterers is not masked as heavily as the X-ray scattering effects from light

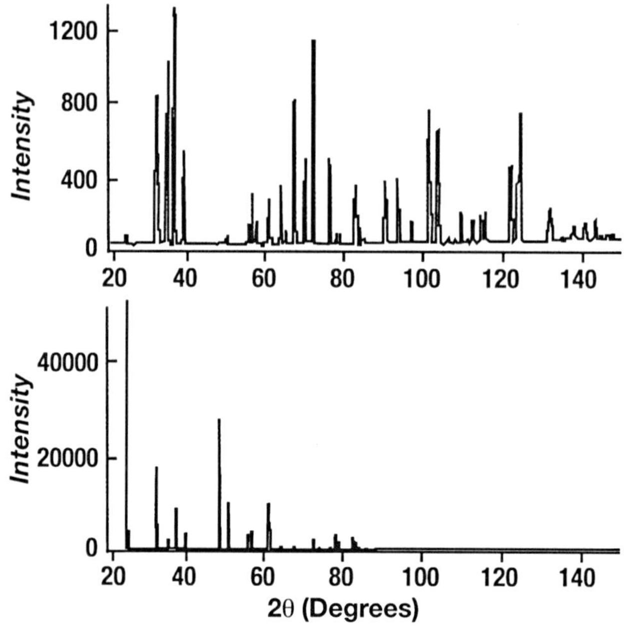

Fig. 1.12 A comparison of powder diffraction patterns for rutile (TiO_2) from neutron (top) and synchrotron X-ray radiation (bottom). The X-ray data were collected at Daresbury Laboratory (UK) with a mean wavelength of 1.37225 Å: the neutron data were collected at Lucas Heights (Australia) with a mean wavelength of 1.377 Å. (Reproduced from [18] with permission of Angew. Chem. Int. and Anthony K. Cheetham)

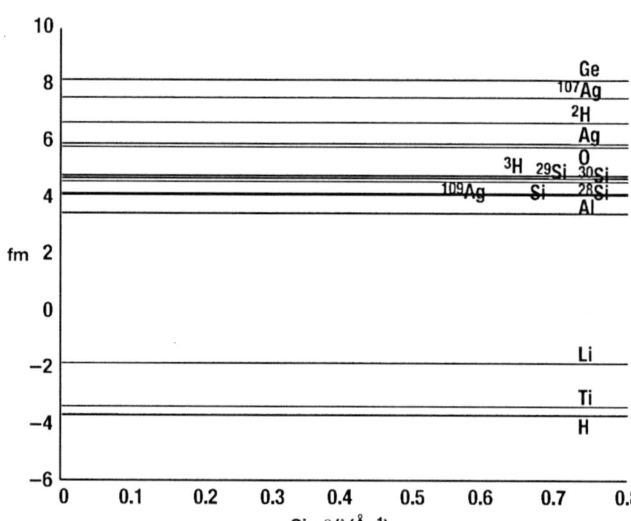

Fig. 1.13 Neutron scattering lengths for selected elements and isotopes

elements. However, there are still a few elements that are difficult to "see" by neutron diffraction. In fact, in a neutron diffraction experiment, the sample is usually placed within a cylindrical can made of vanadium because vanadium possesses a negligible scattering length, and most of the scattering that does occur is incoherent. With the exception of its background contribution, the vanadium can is therefore nearly invisible to the neutron beam. Figure 1.13 shows the neutron scattering factors for selected isotopes of some of the same elements shown in Fig. 1.11. The scattering length of each element is a weighted average of its isotopes according to their natural abundance.

For isotopes with nuclei that possess nonzero spin states, there is also an incoherent scattering length that must be considered. This is especially true in the case of ^{1}H, which possesses an enormous incoherent scattering length of 25.2 fm. A large degree of incoherent scattering is generally undesirable in a diffraction experiment because it diminishes the intensity of the diffracted beam and contributes to the background of the diffraction pattern. Therefore, it may be advantageous to examine samples enriched with an isotope that possesses a small incoherent scattering length (like ^{2}H) compared to that for the isotope most prevalent for a particular element.

1.8 Powder Diffraction

Many materials, especially zeolites, can be prepared only in the form of polycrystalline powders. The crystallite dimensions in zeolite powder samples typically vary between 0.1 and 10 μm, and therefore they are not amenable to typical single-crystal measurements. In powder samples, there are millions of crystallites that are randomly oriented with respect to the beam. In a single crystal diffraction experiment, a particular *hkl* reflection will yield a discrete diffraction spot (as in Fig. 1.5) at an angle 2θ to the primary beam. In the powder diffraction experiment, the same *hkl* (and its symmetry-related reflections) will instead yield a ring of diffracted intensity at an angle 2θ (Figs. 1.8 and 1.14) because of the random orientation of the crystallites. For a perfectly random distribution of crystallites, the intensity within the "powder ring" or "Debye ring" will be uniform. Spottiness or uneven intensity (Fig. 1.14) indicates the sample has preferred orientation or graininess (Sect. 1.12).

In a powder diffactometer study, the sample, X-ray source, and detector remain within the same plane during the experiment (Fig. 1.15). Unlike the Debye-Scherrer camera, which uses film to measure intensity data in two dimensions (Fig. 1.14), the Bragg-Brentano detector moves along a single dimension (as shown in Fig. 1.15) and records intensity for each 2θ angle. Most modern laboratory powder diffractometers use Bragg-Brentano geometry. For a Bragg-Brentano diffractometer in the $\theta/2\theta$ setup (Fig. 1.15b), the X-ray tube remains fixed while the specimen plate varies as θ and the detector varies as 2θ. In the θ/θ setup (Fig. 1.15a), the X-ray tube varies with θ, the specimen remains fixed (although it can rotate about an axis perpendicular to the plate or about the goniometer axis), and the detector varies as θ.

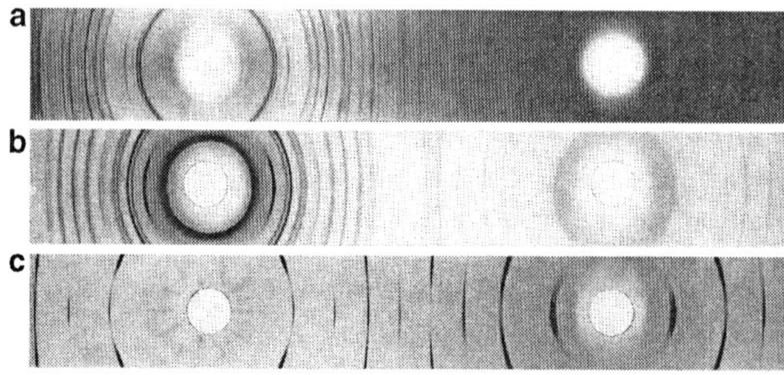

Fig.1.14 Preferred orientation effects in powder photographs. (Cu$K\alpha$ 40 kv, 18 ma.) A. Shape orientation. Needle-like crystals of PbO. B. Shape orientation. Plate-shaped crystals of talc. C. Deformation orientation. Drawn tungsten wire Reprinted with permission from [40]

The output is a plot of intensity vs. 2θ as shown in Fig. 1.1. A powder diffraction pattern is thus a one-dimensional projection onto 2θ of three-dimensional single-crystal data. These powder diffraction patterns serve as the "fingerprint" of the material. Although the Bragg-Brentano geometry allows easy collection of data, effects due to preferred orientation cannot be as readily identified as they can in Debye-Scherrer images.

For powder diffraction, (1.2) can be recast as

$$I(hkl) = m_{hkl} L_p(\theta) |F(hkl)|^2 \tag{1.7}$$

where m_{hkl} is a "multiplicity" factor that arises from the fact that there are several sets of *hkl* planes that have different orientations in the crystal but have identical *d*-spacings and $|F|^2$ values (as required by symmetry). For example, in a cubic lattice *100*, *010*, *001*,$\bar{1}00$, $0\bar{1}0$, and $00\bar{1}$ form equivalent sets of lattice planes and hence m_{100} is equal to 6. Likewise, m_{111} for a cubic lattice is equal to 8. In extracting the structure factors from the intensity data, we therefore have to divide the intensity of each measured reflection by the appropriate multiplicity. Sometimes the geometry of the lattice requires perfect overlap of reflections that possess different structure factors. For example, in any cubic lattice the *511* and *333* reflections possess identical lattice spacings.

Powder diffraction data present difficulties for structure solutions or refinements since diffraction peaks with close lattice spacings overlap. The degree of overlap in powder diffraction data is determined both by the diffraction peak widths and by the angular separation of the peak positions. The extraction of intensities from overlapping reflections is often problematic, particularly at high 2θ values. This is especially true for structures with unit cells that have orthogonal lattice parameters that are near integral multiples of one another. In structure determination of novel phases, peak overlap is especially problematic because not only is it difficult to unambiguously determine the intensity of each reflection, but it is also more difficult to estimates the phases of the reflections because the phasing process is intimately connected to the relative intensities of the peaks (see Section on direct methods).

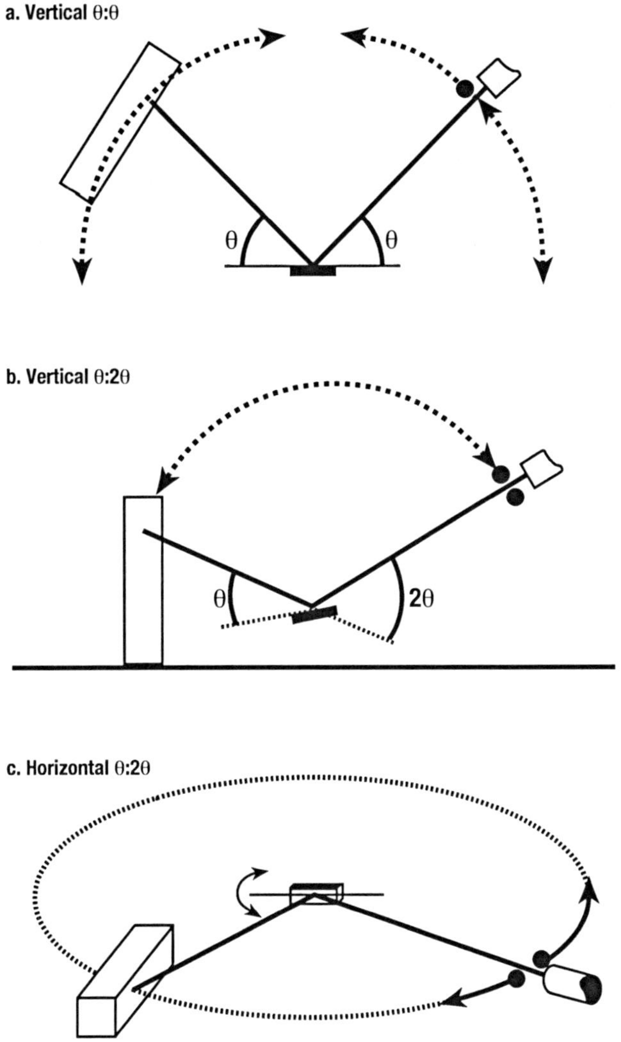

Fig. 1.15 Various configurations of the Bragg-Brentano parafocusing diffractometer. From [2], p. 22, Fig. 1.3.

1.8.1 Differences in XRD Patterns of As-Made vs. Calcined Materials

Figure 1.16 compares the powder XRD patterns of an as-made and calcined sample of SSZ-44. A common observation is that the relative intensities of low angle peaks significantly increase after removal of the organic structure directing agent or the extra-framework water. Why does this occur?

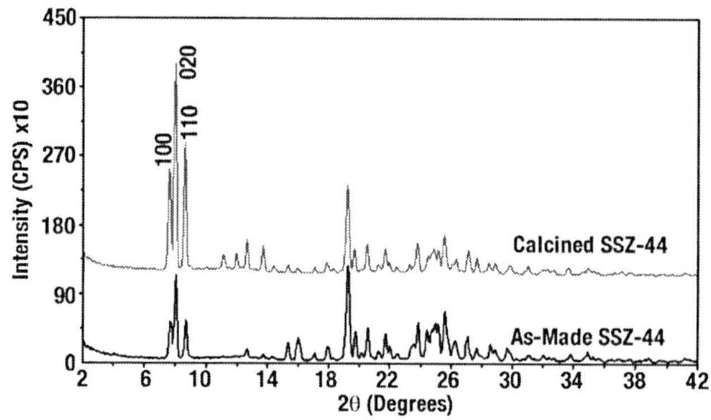

Fig.1.16 Comparison of the as-made (with organic SDA) sample of SSZ-44 and the calcined sample of SSZ-44

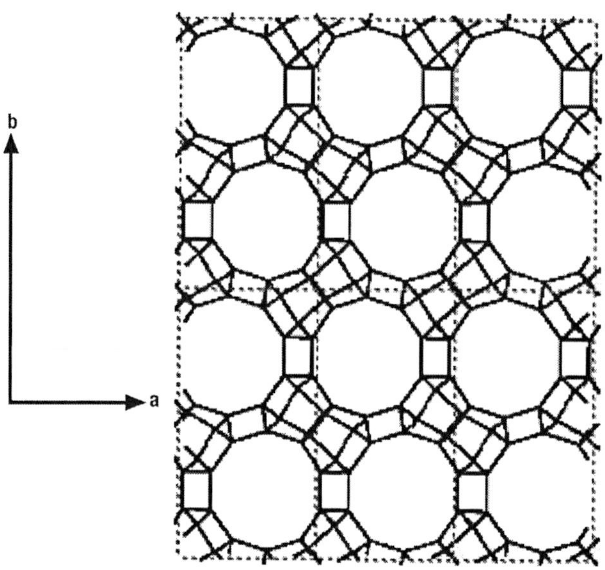

Fig.1.17 Framework model of SSZ-44 (O atoms omitted) with three-unit cells along the x-axis and 2-unit cells along the y-axis

In a typical zeolite, most of the framework atoms are in positions approximately midway between the centers of pores (see Fig. 1.17 showing a projection of the pores in SSZ-44). When an organic molecule or water fills the void volume of the zeolite, they contribute scattering density that is close to the centers of the pores. If we take the case of the 100 reflection, the structure factor is given by

$$F(hkl) = \sum_j p_j f_j \exp\left(\frac{-B_j \sin^2\theta}{\lambda^2}\right) \left[\cos(2\pi x_j) + i \sin(2\pi x_j)\right]$$

For the purposes of calculating the modulus of a structure factor, the assignment of a unit cell origin is arbitrary. If we place the origin at the middle of the pore, then most of the extra-framework atoms will make positive contributions to the cosine term of the structure factor since their x_j are close to 0. However, because the framework atoms are approximately halfway between the pore centers, their x_j are close to 1/2. They therefore make negative contributions to the cosine term. The important point is that the extra-framework scattering makes a contribution that is opposite in sign to that of the framework atoms. This will tend to decrease the modulus of the structure factor. If the extra-framework atoms are removed, then the modulus increases for these low order reflections. In the case of SSZ-44, we might expect that the same reasoning should not apply to the 020 peak, but note that symmetry of the structure is such that there are two pores in the unit cell along the b-direction. The structure factor for the *020* reflection is

$$F(hkl) = \sum_j p_j f_j \exp\left(\frac{-B_j \sin^2\theta}{\lambda^2}\right) \left[\cos(4\pi y_j) + \mathrm{i} \sin(4\pi y_j)\right]$$

For the SSZ-44 structure, the middles of the pores along the *b* direction are about 1/4 of a unit cell from the framework atoms, and so again the extra-framework atoms decrease the modulus of the structure factor.

We can envision the scattering contributions of the atoms as residing on a sine curve whose periodicity changes with the *hkl* indices. For the low order reflections, the middle of the pore is located at the maximum of the sine curve while the framework atoms are located near the minimum. The intensities of higher order reflections are less affected by calcinations or dehydration because the framework or extra-framework species are not systematically located near the maxima or minima, respectively, of the sine curves.

1.9 Systematic Absences in Diffraction Patterns

The symmetry operations determine which Bragg reflections are not permitted in the diffraction pattern. That is, the space group defines the reflection conditions or systematic *hkl* absences in the diffraction pattern. For example, in the diffraction pattern of a body-centered lattice, all reflections must satisfy the condition that $h+k+l=2n$, where n is an integer. Although the systematic absences are not unique to a particular space group, they do narrow down the possible candidates when the structure is unknown. The reflection conditions for all space groups are listed in the Volume A of the International Tables for Crystallography [17].

We will now revisit the space group symmetry *Cmcm* shared by the zeolites ITQ-3 (ITE) [19], SSZ-55 (ATS) [20], mordenite (MOR) [21], terranovaite (TER) [22] and MCM-65 [23]. C-centering implies that every general position x, y, z has a symmetry-related position $x+1/2, y+1/2, z$. The contribution to the structure factor for the symmetry related atoms at these positions is

$$f\left(\exp^{2\pi i(hx+ky+lz)} + \exp^{2\pi i(h(x+1/2)+k(y+1/2)+lz)}\right)$$
$$= f\left(\exp^{2\pi i(hx+ky+lz)}\right)\left(1 + \exp^{\pi i(h+k)}\right)$$

This sum is identically zero when $\exp[\pi i(h+k)] = -1$. This implies that if $(h+k) = 2n+1$, then the structure factor will be zero because of symmetry requirements. In the Collection of Simulated Patterns, note that for structures with C-centered lattices there are no reflections listed for which $(h+k)$ is an odd number.

In space group *Cmcm*, there is also a c-glide plane at $y = 0$. For the general position (x,y,z), this operation generates a symmetry equivalent position at $(x, -y, z+1/2)$. The contribution to the structure factor for symmetry-related atoms at these positions is

$$f\left(\exp^{2\pi i(hx+ky+lz)} + \exp^{2\pi i(hx+k(-y)+l(z+1/2))}\right)$$
$$= f\left(\exp^{2\pi i(hx+ky+lz)}\right)\left(1 + \exp^{\pi i(-4ky+l)}\right)$$

The term $\left(1 + \exp^{\pi i(-4ky+l)}\right)$ is zero when $k = 0$ and $1 = 2n+1$. Hence for reflections $h0l$, l must be equal to $2n$.

1.10 Phase Identification, Unit Cell Refinement, and Lattice Substitution

Powder diffraction is used as a routine characterization method to verify the identity and purity of the phase(s) obtained from a zeolite synthesis. After collection of powder diffraction data, (1) the patterns can be visually compared with reference pattern(s) for a given material or (2) the peaks in the experimental pattern may be fitted with profile functions to obtain the 2θ positions of the reflections. In identifying zeolite phases from a diffraction pattern, there is no substitute for empirical familiarity with the chemistry of the reaction gel. For example, in our syntheses of high-silica zeolites using zeolite Y as an aluminum source in gels with silica to alumina ratios of 30, we commonly observe mordenite, analcime, or gismondine-type phases as cocrystallization products. These are low-silica phases that do not require the presence of an organic for their crystallization. Likewise, for similar syntheses that use gallium instead of aluminum sources, mazzite and mordenite are common phases. When performing syntheses of low-silica zeolites in the presence of sodium, the investigator needs to be acutely aware of phases such as zeolites Y (or X), A, P, sodalite, analcime, mordenite, or ferrierite. Likewise, for low-silica preparations in the presence of potassium, zeolite L, merlinoite, offretite, and erionite are common products. Common high-silica phases crystallized in the presence of an organic include ZSM-12, ZSM-5, ZSM-48, and beta. For exploratory research with novel organic structure directing agents, it is useful to be familiar with the zeolite phases that are typically crystallized in the presence of molecules of similar size and shape.

If the material anticipated from a synthesis is a known phase, then the observed peak positions are compared with those of the expected unit cell. Comparison can be made with a reference diffraction pattern, but it is important to remember that both peak positions and intensities may vary with heteroatom substitution, organic content, the exchange of extraframework cations, or the presence of framework-bound fluoride. Because tetrahedral Al-O (1.74 Å), Ga-O (1.82), and Ge-O (1.74) bond distances are longer than Si-O (1.61) distances, the unit cell dimensions of zeolites with these heteroatoms are larger than those of their all-silica analogues. Likewise, borosilicate zeolites generally have cell dimensions smaller than those of the completely siliceous structures. These differences are manifested in the powder diffraction patterns by shifts in the peak positions either to lower angle (for increasing Al incorporation) or to higher angle (for increasing B incorporation). Figure 1.18 shows patterns of **MTT**-type zeolite samples prepared in gels with different concentrations of aluminum or boron. Note that the peak shifts are more apparent at higher angles because greater changes in *d*-spacing occur at lower angles for the same change in angle. This can be understood by taking the differentials of the Bragg equation and solving for the change in diffraction angle:

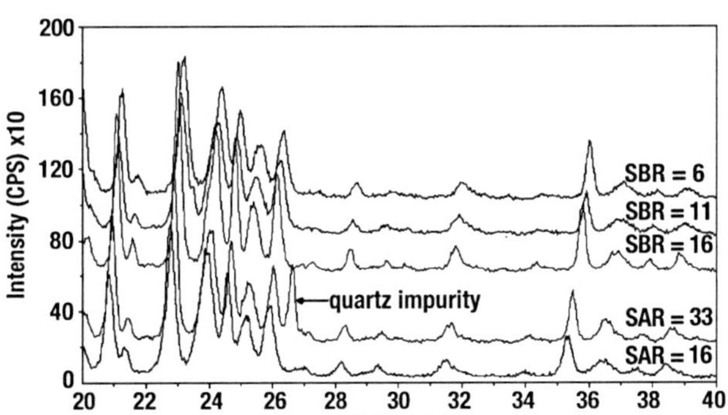

Fig. 1.18 Powder diffraction patterns for **MTT**-type zeolites prepared with varying amounts of boron or aluminum. SBR indicates the silica to boron oxide ratio, and SAR indicates the silica to alumina oxide ratios in the gel preparations

$$\partial\left(\frac{\lambda}{2d}\right) = \partial \sin\theta \Rightarrow \frac{-\lambda(\Delta d)}{2d^2} = (\cos\theta)\Delta\theta \Rightarrow \Delta\theta = \frac{-\lambda(\Delta d/d)}{2d\cos\theta}$$
$$= \frac{-(\Delta d/d)\sin\theta}{\cos\theta} \tag{1.9}$$

Therefore, the greater the diffraction angle, the greater (the magnitude of) the difference in diffraction angle will be for an equivalent fractional change ($\Delta d/d$) in d-spacing.

As shown in Fig. 1.19, linear correlations between framework heteroatom concentrations and unit cell dimensions have been demonstrated in FAU [24] and MFI-type [25] zeolites. These correlations are sometimes used to estimate the ratio of framework to extra-framework aluminum species in zeolites where the total aluminum concentration is known.

For rigorous verification of phase purity, diffraction peaks may be fitted with an appropriate peak shape function to determine their 2θ positions. An internal standard may be used to correct for offset error in the peak positions due to misalignment. An *hkl* is then assigned to each of the well-resolved peaks in the pattern. A least-squares refinement is next performed to optimize the unit cell parameters which best match the experimental data in the anticipated lattice setting. This can be performed with several readily available programs for unit cell optimization such as Lapod [26]. If the data are collected from an in-house laboratory diffractometer, peaks below 15° 2θ are usually excluded because of errors caused by peak asymmetry, and peaks above 35° 2θ are excluded because of $K\alpha_1/\alpha_2$ splitting. Peaks at higher angle usually suffer from a large degree of peak overlap, making correct assignment of peak positions more difficult. The presence of "unindexed" peaks may indicate there is an impure phase. If the instrument and sample are well aligned, the lattice parameters are usually refined to within three decimal places (measured in Å). Trouble in indexing a pattern could also suggest that the specified lattice setting is incorrect. For example, ion exchange, dehydration, calcination, or temperature change may induce structural changes that lead to subtle differences in symmetry.

Figure 1.20a shows neutron powder diffraction patterns of a dehydrated sample of LiX (Si/Al = 1.0) collected at (a) 296 K and (b) 20 K [27]. At room temperature LiX (Si/Al = 1) has cubic symmetry (as do most FAU structures), but upon cooling to 230–220 K the structure undergoes a transition to an orthorhombic phase. Figure 1.20b shows how the unit cell parameters vary with temperature. The lower symmetry is manifested by "broadening" of the peaks in the diffraction pattern of the cooled sample. The perceived broadening is actually due to (a) the appearance of previously symmetry-forbidden reflections in the cubic phase and (b) variation in the previously symmetry-equivalent cubic cell parameters. Because of the lower symmetry, there are *hkl* planes that have d-spacings that are close to (but nonetheless different from) one another. For example, in the cubic setting the *100*, *010*, and *001* reflections are all symmetry-equivalent, but they yield three distinct peaks in the orthorhombic case.

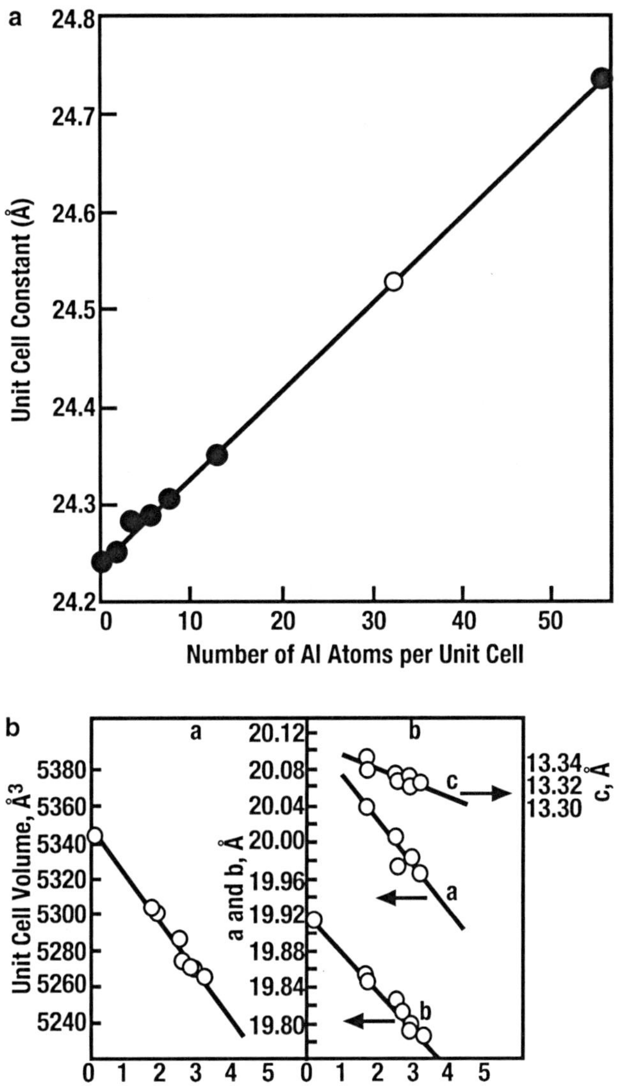

Fig. 1.19 (**a**) Plot of unit cell constant vs. number of Al framework atoms per unit cell for zeolite FAU [24]. Reprinted with permission from Elsevier. (**b**) Plot of unit cell volume (left]) and unit cell constants (right) vs. number of B framework atoms per unit cell for zeolite ZSM-5 [25] Reprinted with permission from Elsevier

Instead of using individual peak positions as input to an indexing program, an investigator may instead choose to perform a LeBail fit [28] of the powder diffraction data. The LeBail method is a least-squares routine that is applied to the whole pattern. The background, peak shape function parameters, unit cell parameters, and zero offset can be refined together. This is similar to a Rietveld refinement

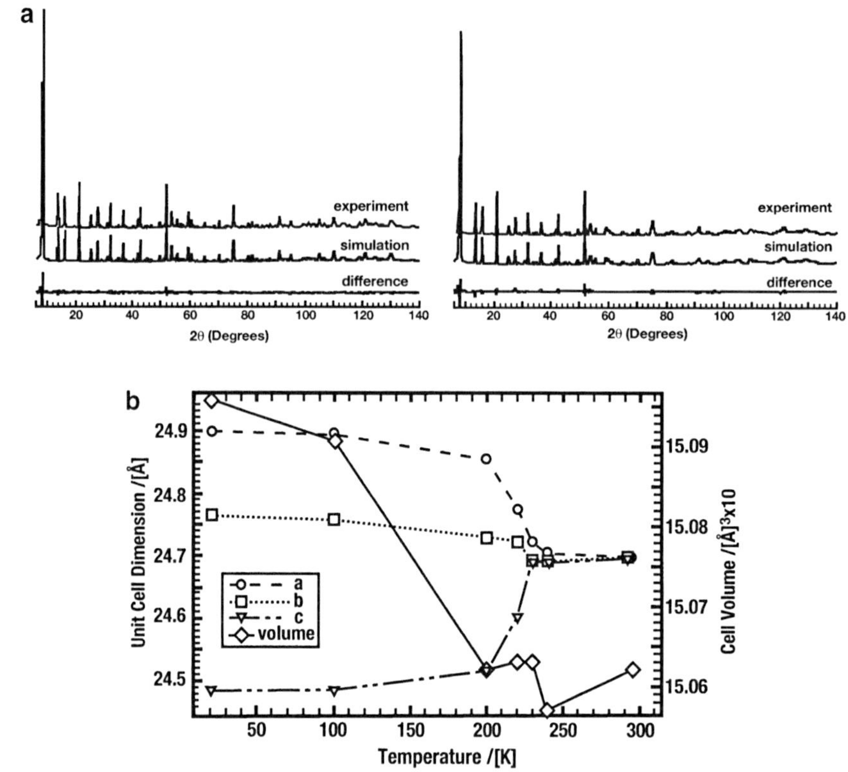

Fig. 1.20 (**a**) On left: Neutron diffraction data of LiX-1.0 (T=296 K). On right: neutron diffraction data of LiX-1.0 (T=20 K) and (**b**) Temperature dependence of the unit cell dimensions of LiX-1.0. Figure reprinted with permission from [27]

(Sect. 1.13.5.3), except that the intensities are refined independently of atomic positions to give the best fit to the observed pattern. The LeBail method is frequently used to extract intensities for structure solutions of powder data, but it can also be implemented to refine the unit cell parameters of a known or unknown material. Although the LeBail method requires more time, it provides several advantages over using individual peak positions to determine the unit cell parameters. Because the entire pattern is used, the investigator does not need to exclude certain reflections from the indexing process because of severe overlap. The LeBail method also uses an analytical peak shape function to fit the entire profile rather than using independent peak parameters for the individual *hkl* peaks. The use of an analytical profile function (with a limited number of parameters) provides a more rigorously accurate method for fitting peaks by avoiding the skewed parameters that are often found in the fits of overlapping peaks. Figure 1.21 shows a LeBail fit of the powder diffraction pattern of the novel zeolite SSZ-65 in space group P6/mmm.

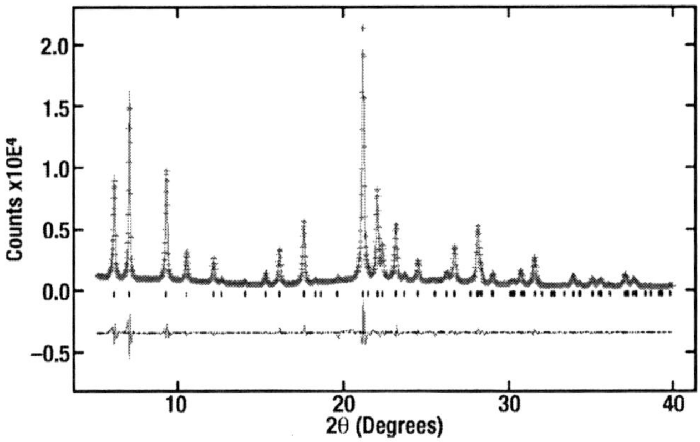

Fig. 1.21 LeBail fit for zeolite SSZ-65 in space group P 6/mmm with $a = 16.791$ and $c = 12.608$ Å. The cross-hatches are the experimental data, the overlapping trace is the fitted curve, and the bottom trace is the difference between the fitted and experimental points. The tick marks indicate the positions of Bragg peaks for the unit cell

1.11 Factors Affecting Peak Profiles in Powder Diffraction

To this point, we have not discussed how the intensity of a given peak is distributed around the exact Bragg angle. The peaks widths and shapes of a powder diffraction pattern are a convolution of effects due to crystallite size, crystallite strain, and instrumental broadening.

1.11.1 Crystallite Size

The Scherrer equation relates diffraction peak width to the average (by mass or volume) crystallite dimension:

$$\beta_s(2\theta) = \frac{K\lambda}{T\cos\theta} \tag{1.10}$$

where β_s is the crystallite size contribution to the integral peak width in radians, K is a crystal shape constant near unity, and T is the average thickness of the crystal in a direction normal to the diffracting plane *hkl*. In interpreting the Scherrer formula, it is important not to confuse the crystallite size with the aggregate particle size. For small crystallites, there is non-zero intensity for small angle ranges around the Bragg angle.

To derive the intensity distribution around a Bragg maximum, we will use the explanation presented by Guinier ([8], pg. 120). In our discussion of Bragg's law, we determined the conditions required for constructive interference from parallel

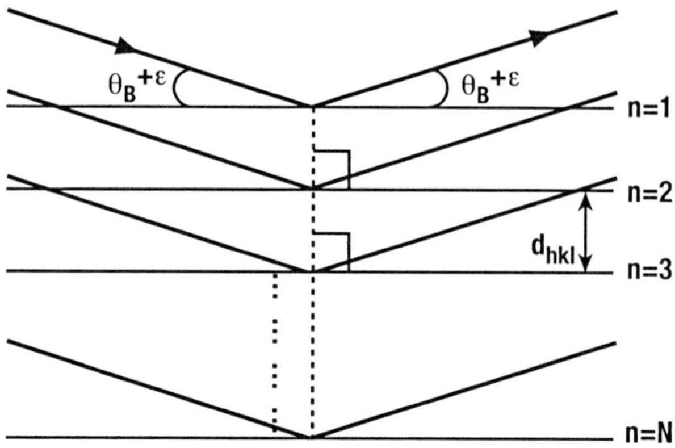

Fig. 1.22 Diffraction from N parallel planes for small deviations (ε) from the exact Bragg angle

planes. For small changes in angle around the Bragg maxima, there still exists finite intensity because there is not complete destructive interference of the reflected rays unless there are an infinite number of parallel planes. Consider the case presented in Fig. 1.22, where there are N parallel planes, each separated from the next by a distance d_{hkl}. The Bragg angle for these parallel planes is θ_B. We want to consider the case where radiation strikes the planes at an angle $\theta+\varepsilon$ and then reflects from the planes at the same angle. The path difference of reflected rays from neighboring planes is $2d\sin(\theta_B+\varepsilon) = 2d(\sin\theta_B\cos\varepsilon+\sin\varepsilon\cos\theta_B)$. Since we are interested in small changes in angle about the Bragg maxima, $\sin\varepsilon \cong \varepsilon$ and $\cos\varepsilon \cong 1$. Substitution of these terms into the path difference yields $2d\sin\theta_B+2d\varepsilon\cos\theta_B = \lambda+2d\varepsilon\cos\theta_B$. The phase difference in radians between these two planes is therefore $2\pi(2d\varepsilon\cos\theta_B)/\lambda$. For the $(m+1)$th plane, the phase difference with the first plane is m times the phase difference between consecutive planes: $4\pi md\varepsilon\cos\theta_B/\lambda$.If the amplitude of the radiation scattered from each plane is A, then the total scattered amplitude from N parallel planes is given by

$$\sum_{1}^{N} A\exp(\mathrm{i}\phi) = \sum_{m=0}^{N-1} A\exp\left(\frac{4\pi imd\varepsilon\cos\theta_B}{\lambda}\right)$$
$$= \sum_{m=0}^{N-1} A\left(\cos\frac{4\pi md\varepsilon\cos\theta_B}{\lambda} + \mathrm{i}\sin\frac{4\pi md\varepsilon\cos\theta_B}{\lambda}\right) \quad (1.11)$$

The sum in (1.11) can be simplified using the following relationships:

$$\sum_{m=1}^{N}\cos(m-1)x = \frac{\sin(Nx/2)}{\sin(x/2)}\cos\frac{(N-1)x}{2}$$

and

$$\sum_{m=1}^{N} \sin(m-1)x = \frac{\sin(Nx/2)}{\sin(x/2)} \sin\frac{(N-1)x}{2}$$

The intensity of the scattered radiation is proportional to the square of the total amplitude

$$A^2 \frac{\sin^2(Nx/2)}{\sin^2(x/2)} = A^2 \frac{\sin^2(2\pi Nd\varepsilon \cos\theta_B/\lambda)}{\sin^2(2\pi d\varepsilon \cos\theta_B/\lambda)} \tag{1.12}$$

We are interested in determining the peakwidth of the Bragg reflection. This will depend on the way in which we define the width of the peak. A common method is to measure the width of the peak at half of its maximum intensity (the full width at half maximum, or FWHM). For a function of the form

$$\frac{\sin^2 N\phi}{\sin^2\phi}$$

this occurs when $N\phi = 0.44\pi$. This means that the FWHM occurs for $\varepsilon = 0.22\lambda/(Nd \cos\theta_B)$. If the angle of incidence increases from θ_B by ε, then the angle between the incident and diffracted rays increases by 2ε. The peakwidth is measured between $2\theta_B+2\varepsilon$ and $2\theta_B-2\varepsilon$. Therefore the peakwidth is equal to $4\varepsilon = 0.88\lambda/(Nd\cos\theta_B)$. Since Nd is simply the dimension of the crystal measured in a direction normal to the *hkl* planes, we can replace Nd by T. This gives us the equation at the start of this section.

Another commonly used peakwidth is the integral peakwidth. This is the value that must be multiplied by the maximum intensity to obtain the integrated area of the peak. With this definition, the value of K is found to be 1 rather 0.88. Figure 1.23 shows plots of the function

$$\frac{1}{N^2}\frac{\sin^2 N\phi}{\sin^2\phi}$$

for values of $N = 10$, 50, and 100.

In this plot, the

$$\frac{\sin^2 N\phi}{\sin^2\phi}$$

function has been normalized by the factor $1/N^2$ to allow the reader to examine the effect of N on the peak width. The plots illustrate how diffraction peaks have nonzero intensity at non-Bragg angles and how the broadening increases as the

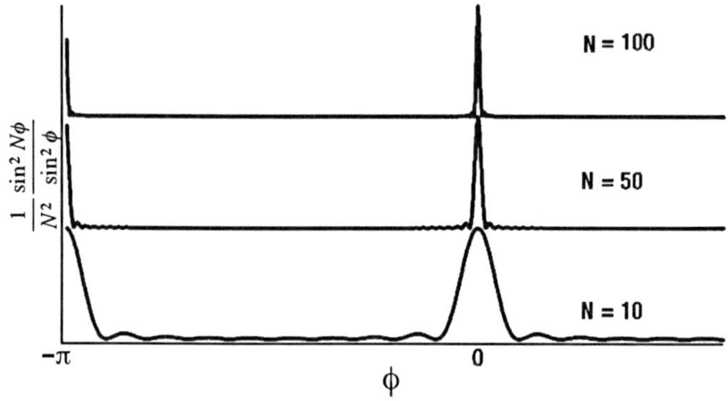

Fig. 1.23 Plot demonstrating dependence of peak broadening around the exact Bragg angle for crystallites of finite size (ϕ equals $2\pi d\varepsilon \cos \theta_B/\lambda$)

crystal becomes smaller. The ripples around the central maxima are rarely observed in experimental data because they have negligible intensity for sufficiently large crystallites. The positions for the ripples also vary with the crystal size and therefore are "averaged out" in powder samples where there is a distribution of crystallite sizes.

It may seem counterintuitive that the Scherrer equation explicitly depends on the crystal dimensions ($Nd = T$) rather than on the number of unit cell or lattice plane repeats N. However an analogy may be drawn with diffraction gratings. In this case the angular widths of the intensity maxima have the same dependence on the product of the number of gratings (N) and the distance (d) between them. [29]

From experimental data, how do we determine the contribution of the crystal size broadening to the peak width? We first need to determine the broadening due to the instrument. This is best achieved by collecting diffraction patterns of a standard material that (1) has crystallites that are 5–20 μm in size, (2) has no detectable strain (Sect. 1.10.2), (3) has diffraction peaks that are well-resolved from one another, and (4) has negligible sample transparency. Note that instrumental broadening varies with diffraction angle and the slit configuration of the instrument. After the diffraction data have been collected for the standard and the peaks have been fitted, a calibration curve can be created that relates the instrumental peak broadening to the diffraction angle. LaB_6 is the standard that is typically chosen to determine instrumental broadening. However, the first diffraction peak for LaB_6 occurs after 21° 2θ. Therefore, if the investigator wishes to use low order reflections for Scherrer analysis of zeolite materials, it could be useful to find another standard that has well-separated diffraction peaks at lower angles and has peak widths similar to those for LaB_6 at the higher angles.

For peaks that are fit with Lorentzian functions, the non-instrumental broadening is simply determined by subtracting the instrumental peakwidth from the measured peak width:

$$\beta(2\theta) = \beta_i(2\theta) + \beta_s(2\theta) \tag{1.13a}$$

where the subscripts i and s indicate contributions from the instrument and sample, respectively. For peaks that are modeled with Gaussian functions, the relationship is

$$\beta^2 (2\theta) = \beta_i^{\,2} (2\theta) + \beta_s^{\,2}(2\theta) \qquad (1.13b)$$

X-ray diffraction peaks are highly Lorentzian in shape, while neutron diffraction peaks are highly Gaussian. However, some peak shape functions (the Voigt or pseudo-Voigt) use a combination of a Lorentzian component and a Gaussian component. Therefore, it is important that the investigator choose a suitable peak shape function and then properly account for the Lorentzian and Gaussian contributions. Note that as the crystallite size increases, its contribution to the observed peak width becomes smaller. At some point, it becomes difficult to accurately estimate the crystallite size within experimental error when β_s is so small compared to the measured peakwidth. For 1 μm crystals, the peak width anticipated from the Scherrer equation at low angle is only 0.009° 2θ. It is for this reason that ([2], p.89) that many authors place an upper limit of 1–10 μm on the crystallite size that can be accurately estimated from Scherrer analysis. With enhanced instrumental resolution available at a synchrotron source, this upper limit can be increased. However, synchrotron sources are rarely used for such measurements since it is more convenient to obtain electron micrographs of the crystallites.

Certain zeolites often possess highly isotropic or spherical crystallite dimensions, especially materials with cubic symmetry like zeolites A, X, or Y. However, it is not uncommon for crystallites to have highly anisotropic dimensions. Many one-dimensional pore zeolites tend to possess needle- or rod-shaped crystals. In catalytic or adsorptive applications, it may be desirable to have crystallites that are short in a particular direction in order to create the most efficient access to the micropores. Characterizing the crystallite dimensions of a zeolite is therefore relevant in the interpretation of their catalytic behavior. Needle-shaped crystallites that are very thin in the x- and y-directions will have *hk0* peaks that are much broader than the *00l* peaks while the opposite is true for plate-like crystallites. Therefore, when performing a Scherrer analysis, one must be mindful of the reflection or group of reflections that are chosen to characterize the crystallite size. The dimension inferred from the Scherrer analysis is the average dimension of the crystal along a normal to the *hkl* plane. For parallelpipedons that have edges parallel to the crystallographic axes, the average crystallite dimensions can be directly determined from the *h00*, *0k0*, and *00l* reflections. The peak widths of general *hkl* reflections, however, are a more complex function of the *hkl* indices and the crystallite dimensions; but in cases where $h \cong k \cong l$, the peak breadth is most affected by the shortest dimension.

Figure 1.24 shows powder diffraction patterns of three different high-silica samples with the MTT framework topology, and Fig. 1.25 shows the corresponding SEM images of those samples. The MTT framework has a one-dimensional channel system with 10-ring pores. The unit cell dimensions of its orthorhombic cell are

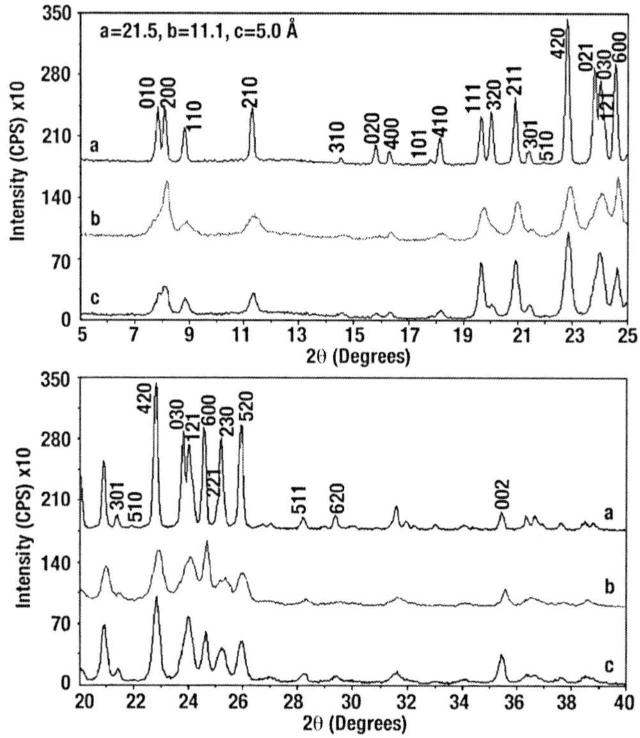

Fig. 1.24 Powder diffraction patterns of **MTT** samples with different crystallite dimensions

approximately $a = 21.5$, $b = 11.1$, and $c = 5.0$ Å. Not surprisingly, the crystal morphologies for this zeolite are often needle-, rod- or lath-like. The top pattern in Fig. 1.24 is very sharp compared to the other two patterns. The SEM image for this specimen shows that the crystals form aggregates that are about 7–8 μm in length, 3 μm in width, and 1.5–2 μm thick. Closer examination of the images reveals that the aggregates are composed of overlapping grains that are 3–5 μm in length, and about 0.5 μm in the other two dimensions. Because each of these dimensions is close to the upper limit of crystallite size detection, the measured peak widths for this sample are close to (but not within) instrumental resolution. However, the patterns of the other two samples show measurable broadening due to crystallite size, particularly the middle one. Most of the peaks for sample B are wider than those in sample C.

However, notice that the 200, 400, 600, and 002 peaks have similar widths in the two samples. These peaks stand out in the pattern for B because they are sharper than the other *hkl* reflections within the same pattern. The SEM image for sample B shows that the crystallites are about 1–3 microns in length, 0.1 microns in width, and less than 0.03 microns in thickness. The peak widths that are observed in this sample suggest that the long dimension is parallel to the crystallographic c-axis ($c = 5.0$ Å), the middle dimension is parallel to the crystallographic a-axis

Fig.1.25 Scanning electron microscopy images for the **MTT** samples discussed in Fig. 1.24

(a = 21.5 Å), and the very thin dimension corresponds to the b-axis (b = 11.1 Å). These assignments are verified by electron diffraction or TEM measurements. For sample C, there are only minor differences in peak widths for the *0k0* and *h00* peaks, which suggests the crystallite dimensions are similar in the a and b crystallographic directions. The SEM image for this sample shows that the typical morphology is a bundle of tightly bound rods that are 0.2–0.5 μm in length. The ends of the rods are circular in shape and about 0.05 μm in diameter, which is consistent with the similar dimensions that are expected from our observations of the XRD peak widths.

This exercise illustrates several vital points in the application of the Scherrer equation. It is not sufficient to choose a single peak to characterize the crystal size of samples with anisotropic crystal dimensions unless it is unequivocally stated which dimension is being estimated; in fact, not stating the chosen peaks can yield very misleading or errant results. In reporting dimensions inferred from the Scherrer analysis, the investigator should mention which reflections were used, the peak shape function that was used, and what measures were taken to account for the instrumental peak broadening. These factors are often not reported, especially in the patent literature where the Scherrer analysis is frequently cited as a method to determine crystallite size. A good example may be found within a recent patent application [30] that describes a method to prepare **MTT**-type zeolites. A central issue of the patent was the determination of crystallite size using the Scherrer analysis. The patent does not explicitly mention which reflection or group of reflections was used to determine the crystallite size, but it does state that the peak width was chosen from peaks between 7 and 35° 2θ using CuKα radiation. The dimension that is often most relevant to the catalytic or adsorptive properties of a one-dimensional pore zeolite is the one that is parallel to the pore direction. In case of **MTT**-type zeolites, this corresponds to the *c*-axis, which possesses a unit cell repeat of 5.0 Å. However, there are no 00l peaks that occur between 7 and 35° 2θ; the 001 peak is systematically absent, and the 002 peak occurs at about 35.7 ° 2θ. The dimensions obtained in this patent from the Scherrer analysis are therefore most indicative of the dimensions of the crystallite within the *a–b* plane, which are often observed to be significantly smaller than the dimension parallel to the pore axis.

1.11.2 Stress

Stress in crystalline materials is another potential source of diffraction peak broadening. Microstrains induced by compressive and tensile forces can cause a distribution of *d*-spacings about the normally observed d-spacing of the material. Hence, because there is a distribution of *d*-spacings associated with each diffraction peak, there is a stress-induced broadening of the diffraction peak. The stress-induced broadening β_ε is related to the residual strain ε by

$$\beta_\varepsilon(2\theta) = 4\varepsilon \tan\theta \tag{1.14}$$

To this author's knowledge, few powder XRD studies have been performed on zeolites to examine stress-induced peak broadening. This may be because it is difficult to find several well-resolved peaks in a powder XRD pattern of a zeolite. However, there are cases in the literature where SEM images clearly show bending in crystals of as-made zeolites, so it is conceivable that large thin plates or long needles could be prone to stress. Other sources of stress include crystal defects, dislocations, and grain boundaries. Grain boundaries are dominant features in many zeolite samples. Note that, because of the $\tan\theta$ dependence of stress, it is best to use low-angle peaks if only a single peak is chosen to estimate a characteristic dimension and a Williamson-Hall analysis (vide infra) cannot be performed.

Sample-induced peak broadening due both to crystallite size and to stress is

$$\beta(2\theta) = \frac{K\lambda}{T\cos\theta} + 4\varepsilon\tan\theta \tag{1.15}$$

By multiplying each side by cos θ, we obtain $\beta(2\theta)\cos\theta = \frac{K\lambda}{T} + 4\varepsilon\sin\theta$. From the intercept and slope of $\beta(2\theta)\cos\theta$ vs. $\sin\theta$(often referred to as a Williamson-Hall plot), we can approximate the crystallite size dimension and strain. If the line

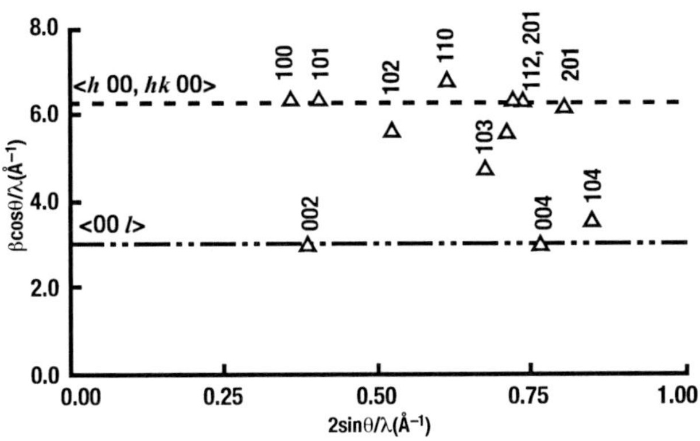

Fig. 1.26 Indexed Williamson-Hall plot for exhydroxide nitrate ZnO. Reprint permission by the Journal of Applied Crystallography

through points representing different orders of the same reflection is horizontal, then there is no measurable strain. Figure 1.26 shows a Williamson-Hall plot for a sample of ZnO. [31] In this plot, the data points for the *h00* and *hk0* reflections have approximately the same values on the *y*-axis for all order of reflections. This suggests that there is no strain in the material and that the dimensions in the *a–b* plane are the same in the a and b-directions (approximately circular). Different orders of 00 l reflections also have equivalent *y*-values, but the value is different from that measured for the *hk0* reflections. In this case, the crystal length along the *z*-axis is about twice the dimension within the *a–b* plane of the crystal. Note that the *h0l* reflections show intermediate values that are affected most by the shortest dimension when *h* and *l* are about the same.

1.11.3 Instrumental Broadening and Peak Displacement

Sources of instrumental broadening include axial divergence, effects from the finite widths of the divergence slit and receiving (detector) slit, flat specimen errors, specimen transparency errors, sample displacement errors, and the spectral distribution of wavelengths. Axial divergence, which is the divergence of an X-ray in the plane of the sample, introduces asymmetric broadening in the diffraction peaks that is most notable in the low angle region of the diffraction pattern. Axial divergence is the principal source of peak asymmetry, but it may be reduced by placing Soller slits (Fig. 1.27) before the divergence slit and after the receiving slits. Peak broadening from the divergence and receiving slits can be reduced by decreasing the slit widths, but this also significantly reduces the measured intensity. Optimal intensity and resolution are obtained when the receiving slit is close in size to the beam width. The flat specimen error occurs because the entire surface of the powder sample (often in a flat plate) is not concentric with the goniometer focusing circle.

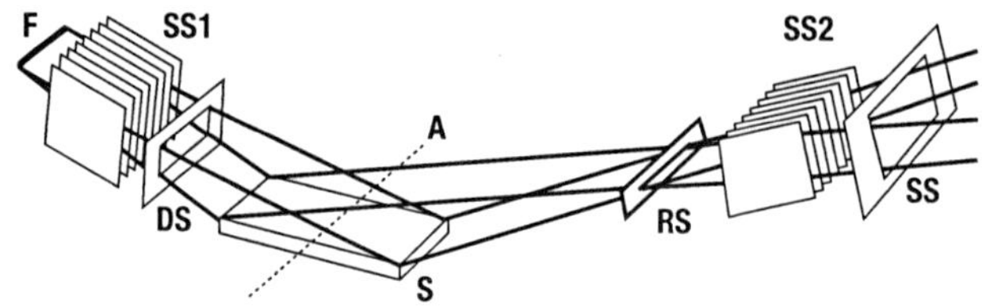

Fig. 1.27 Bragg-Brentano geometry. SS1=Soller Slits; DS=Divergence Slit; RS=Receiving Slit. Reprinted, with permission, from [2]

Table 1.2 Peak positions of selected *d*-spacings for CuKα1 and CuKα2 wavelengths

d-Spacing(Å)	CuKα1°2θ	CuKα2°2θ	Δ°2θ	Weighted average
4.439	19.986	20.037	0.051	20.003
2.254	39.967	40.072	0.105	40.002
1.199	79.950	80.193	0.243	80.031

This also introduces asymmetric broadening toward low 2θ angles. The specimen transparency error occurs because the incident beam can penetrate several layers beyond the surface of the powder sample before being diffracted. Transparency is most problematic in samples that have low absorption (see Sect. 1.12) such as organic molecules. Sample displacement, which is often the largest source of error in measured peak positions, arises from the difficulty in exactly positioning the sample surface on the focusing circle of the goniometer.

In many laboratory powder diffractometers both the $CuK_{\alpha 1}$ and $CuK_{\alpha 2}$ wavelengths are present in the X-ray beam. This presents problems in resolving peaks at high 2θ angles (>35° 2θ), where the splitting of the Kα peaks becomes pronounced. Table 1.2 shows the calculated peak positions for the $CuK_{\alpha 1}$ and $K_{\alpha 2}$ radiation for different *d*-spacings. Although most peak-fitting routines are equipped to perform $K_{\alpha 1/\alpha 2}$ fitting, it can be appreciated that as the angle increases, the difference in the α1 and α2 positions may easily fall within the differences for *hkl* reflections that have close d-spacing. This exacerbates the peak overlap problem in powder data.

Many of these problems can be reduced or eliminated by using X-rays from a synchrotron source. The synchrotron provides such an intense source of radiation that high intensity can still be obtained with a combination of narrow slits, parallel beam geometry, and diffracted beam monochromators. The flat specimen error is not present for either synchrotron or neutron devices that use cylindrical samples. Specimen transparency errors are also minimal for synchrotron sources since the sample size required is enough to fill a thin capillary. In addition, synchrotron and neutron sources offer the advantage of being highly monochromatic.

1.12 Absorption

Since heavy atoms scatter X-rays effectively, we might expect the XRD patterns for zeolites containing heavy atoms to have stronger diffracted X-ray intensity

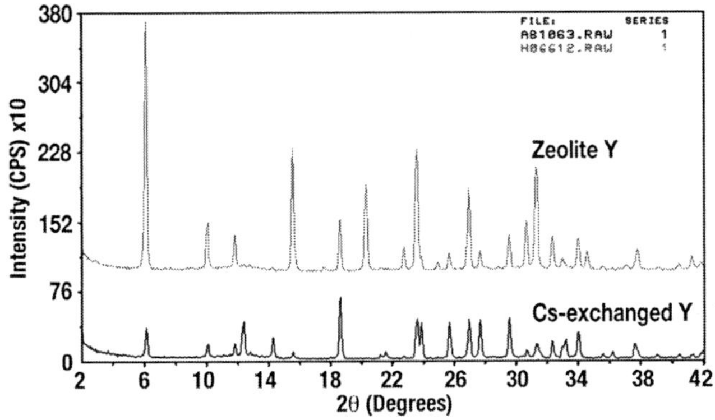

Fig. 1.28 Powder XRD patterns of zeolite Y and zeolite Y exchanged with cesium

compared to otherwise similar materials that have been isomorphously substituted with lighter atoms. However, generally just the opposite is observed. This is especially relevant for low-silica zeolites that have been ion-exchanged with heavy metal cations. Figure 1.28 compares the powder XRD patterns of zeolite Y and the same zeolite Y material that has been exchanged with cesium. Striking differences can be seen immediately both in the overall intensity of the each pattern and in the relative intensities of the individual reflections. A remarkable difference is observed for the 222 peak at about 12.3 °2θ. This peak is so weak in conventional zeolite Y or zeolite X samples that it is not even listed among the observed *hkl* reflections in the collection of simulated patterns. However, introduction of the heavy cesium atom dramatically increases the magnitude of the structure factor. An unsuspecting investigator might unwittingly conclude that this peak is due to an impurity!

To this point, there has been no discussion of the absorption of incident X-ray radiation by the specimen. The reduced intensity of the X-ray beam due to absorption is given by a Beer's law expression

$$I = I_0 \exp\left[-\left(\frac{\mu}{\rho}\right)\rho x\right] \tag{1.16}$$

where μ/ρ is referred to as the "mass absorption coefficient", ρ is the sample density, and x is the thickness of the sample through which the beam passes. The mass absorption coefficients are a function of the element and the wavelength. For a given material, the mass absorption coefficient is a weight average of the individual elements that comprise the sample. Figure 1.29 shows how μ/ρ varies with wavelength for cobalt. The wavelength of CuKα radiation is only slightly shorter than the wavelength at the absorption edge of cobalt. Note that the typical wavelengths for diffraction experiments are between 0.6 and 2.2 Å.

For single crystals or powder samples collected in transmission (Debye-Scherrer) mode, the absorption corrections vary with angle. However, in the case

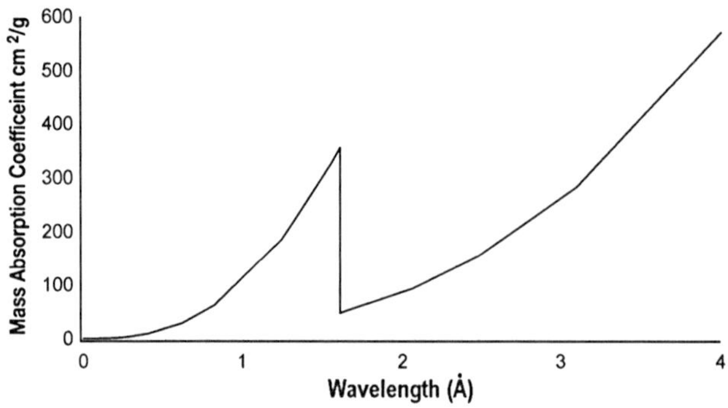

Fig. 1.29 Mass absorbtion coefficient of cobalt as a function of X-ray wavelength

for powder data collected on a flat plate in Bragg-Brentano geometry ([4], pp. 133–134), the absorption correction has a constant value of $1/2\mu$. Although the penetration depth of the beam at low angles is small, the area of the beam on the irradiated sample is larger. The effect is that the irradiated volume remains constant and therefore the absorption correction is independent of the angle.

We will compute the mass absorption coefficients for zeolite Y and cesium-exchanged zeolite Y. We will assume a unit cell composition of $Na_{56}Al_{56}Si_{136}O_{384}$ • 200 H_2O for zeolite Y and a unit cell composition of $Cs_{40}Na_{16}Al_{56}Si_{136}O_{384}$ • 200 H_2O for cesium-exchanged Y. Table1.3 shows the individual mass coefficients for each element, the mass% of each element, and the total mass absorption coeffcient for each sample. These estimates indicate that there should be an 80% decrease in the overall intensity for the cesium-exchanged material. The actual decrease will not be quite as large since the cesium atom also scatters X-rays more strongly than the other elements, but the approximation is in qualitative agreement with the experimental data.

1.13 Preferred Orientation and Graininess

For analyses that depend on peak intensities, it is usually desirable to minimize preferred orientation in powder samples. This includes quantitative determination of crystallinity, phase ratio estimation, or structure solution and refinement. Equation (1.7), for example, assumes a perfectly randomly oriented powder sample. Preferred orientation often occurs when crystallites have plate- or needle-like morphologies. In these cases, the planes of the plates or the axes of the needles tend to align parallel to the surface of a flat specimen holder. Because of this preferred alignment, there is a much greater probability that *hkl* planes parallel to the plates will be in a diffracting position than those that are perpendicular to the plates. Peak intensities will be heavily skewed toward *hkl* reflections parallel to the plates compared to what is expected from an ideal random orientation of

Table1.3 Estimates of mass absorption coefficient for zeolite Y and cesium-exchanged zeolite Y

Zeolite Y			Cesium-exchanged Y	
Element	μ/ρ (cm^2g^{-1})	Mass (%)	μ/ρ (cm^2g^{-1})	Mass (%)
Cs	325.4	0	325.4	25.6
Na	30.30	7.9	30.30	1.8
Al	50.23	9.2	50.23	7.3
Si	65.32	23.3	65.32	18.4
O	11.03	57.1	11.03	45.0
H	0.39	2.4	0.39	1.9
μ/ρ for solid sample	28.5(cm^2g^{-1})		104.5 (cm^2g^{-1})	
ρ	1.79 gcm^{-3}		2.29 gcm^{-3}	
μ	51.0 cm^{-1}		239.3 cm^{-1}	
$\mu_{cor}=\mu\times0.6$ (packing correction)	30.6 cm^{-1}		143.6 cm^{-1}	
$1/(2\mu_{cor})$	0.016 cm		0.0035 cm	

crystallites. Samples in rotating capillaries (as is often the case for synchrotron experiments) or large cylinders (neutron diffraction) are much less likely to suffer from preferred orientation effects.

In the preparation of zeolite membranes, the goal usually is to obtain preferred orientation of zeolite crystals on a film or substrate. Ideally, the researcher would like to align the crystallites so that a particular channel system is perpendicular to the membrane surface. Figure 1.30 compares the XRD pattern of a bulk UTD- 1 sample and the XRD pattern of a UTD-1 membrane prepared by the Balkus group. [32] In the pattern for the membrane, the relative intensities of the *020* and *121* reflections are much stronger than what is observed in the bulk sample. In this case, the crystallites are oriented so that 14-ring channels (which are parallel to the crystallographic *b*-axis) of the zeolite are nearly perpendicular to the substrate. Other notable XRD studies by the Tsapatsis research group nicely demonstrate preferred alignments of crystallites in **MFI**, [33, 34] zeolite A, [35] and ETS-4 [36] membranes.

Crystallites are sometimes large enough that a complete random distribution of crystal orientations is simply not possible because there are a limited number of crystals exposed to the X-ray beam. This effect is often referred to as "graininess." We have mentioned that the ideal crystal size for powder measurements is 0.1–10 μm. In the absence of preferred orientation, we expect that the greater the number of crystallites that are able to diffract, the more the sample approaches an ideal distribution. For the same sample quantity, there will be 1/1000 the number of cubic crystals in a specimen that has 10 μm edge crystals as there will be in a specimen that has 1 μm crystals. Likewise, for cubic crystals with 50 μm edges, this fraction will be $1/(1.25 \times 10^5)$.[2] In many cases, because the crystal faces are so large and because they are often anisotropic in their dimensions, these "grainy" samples also exhibit preferred orientation. It is not unusual for these effects to appear

[2]This estimation assumes the packing density of the crystals are equivalent on a per mass basis.

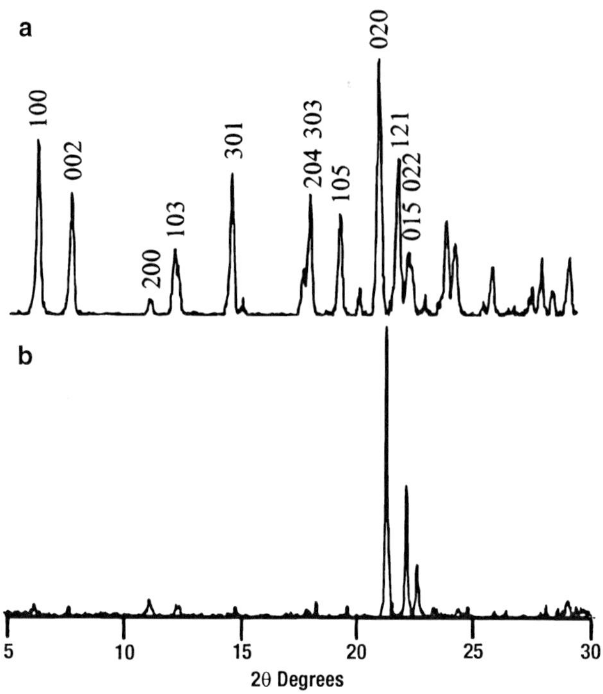

Fig. 1.30 X-ray diffraction patterns of UTD-1 (**a**) as-synthesized bulk, and (**b**) after 72 hours of hydrothermal treatment. Reprinted with permission from [32]. ©1999 American Chemical Society

in the powder XRD of zeolites prepared in fluoride media, where the crystals can sometimes have dimensions of 50–200 μm.

Figure 1.31a compares the powder XRD sample of an "ordinary" ZSM-5 with that of a ZSM-5 that possesses crystals that are 100–125 μm in length and about 20 μm in thickness (Fig. 1.31b). For the large crystal sample, it can be seen that the *h00/0k0* reflections of the orthorhombic unit cell are much more intense and that many of the other reflections are reduced in intensity. In this case, the *0k0* and *h00* planes are parallel to the wide (100 × 20 μm) cross-sections of these very large crystals. This effect is just the opposite of what is often observed in **MFI**-type membranes. In these membranes, the coffin-shaped crystals are aligned so that their lengths are nearly perpendicular to the membrane surface. Therefore, for XRD of membrane samples collected on flat plates (in reflection mode), the 00l and the h0l peaks exhibit exaggerated intensity.

For crystals that are so large, it seems that the diffraction peaks are uncharacteristically broad in the top pattern of Fig. 1.31a. The broadening of the peaks is due to the surface roughness (i.e., the distribution in the sample height of the pelleted sample) caused by these very large crystals. Reducing the crystal size through grinding could minimize this effect.

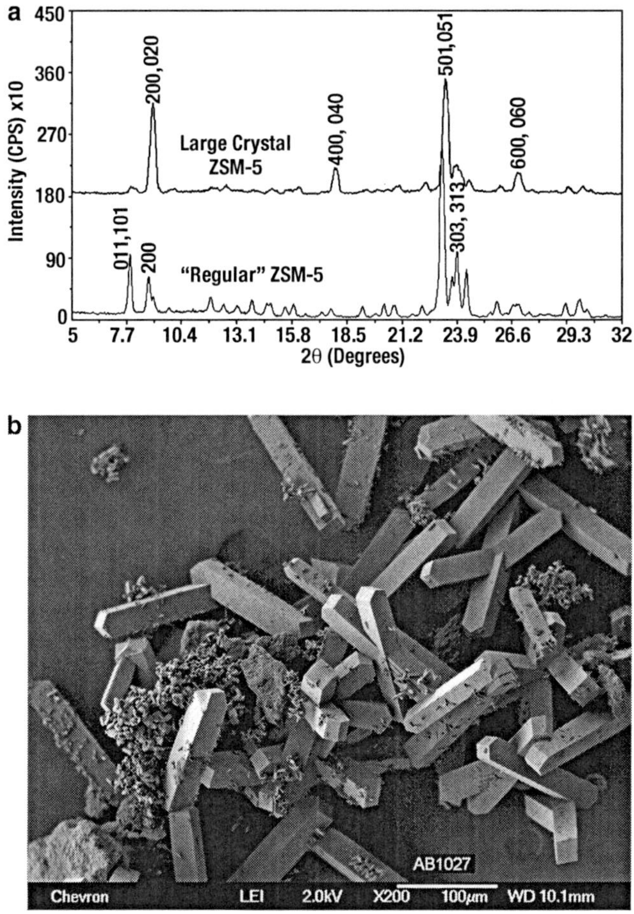

Fig. 1.31 (**a**) Comparison of the powder XRD patterns of (top) ZSM-5 with large crystals (100 x 20 x20μm) and (bottom) ZSM-5 with regular crystallite sizes. The *hkl* indices are for the monoclinic cell setting of ZSM-5 (**b**) SEM image of the large crystal sample of ZSM-5

1.14 Structure Determination

1.14.1 Unit Cell Determination

For materials with previously unreported structures, the first step of a structure solution is to determine the unit cell. Incidental effects due to the unit cell parameters may sometimes cause severe peak overlap in the powder diffraction pattern. It is usually preferred to use synchrotron X-ray diffraction data when the unit cell must be determined. The higher resolution of the synchrotron data allows discernment of peaks that might otherwise overlap due to the instrumental broadening effects of a laboratory diffractometer. Since synchrotron radiation is monochromatic, Kα splitting also is not an issue. Although low angle peaks still suffer from peak asymmetry, the gain in resolution (i.e., narrow peak widths) minimizes error in determination of peak positions. Unlike the indexing of unit cells of known materials, the inclusion of low angle peaks is vital in determining the unit cell of an unknown material.

There are several programs available for unit cell determination including Treor [37], DICVOL [38], and Ito [39]. The book by Azaroff and Buerger [40] provides a chapter that describes the Ito algorithm. From the output of these programs, the experimenter may then review the *hkl* assignments of the reflections and determine possible space groups from the systematic absences. For new investigators, before attempting to index an unknown zeolite, it is good practice to attempt indexings of known materials. Figure 1.32 compares the powder X-ray diffraction pattern of ITQ-22 collected (a) on an in-house CuKα diffractometer [41] and (b) at synchrotron

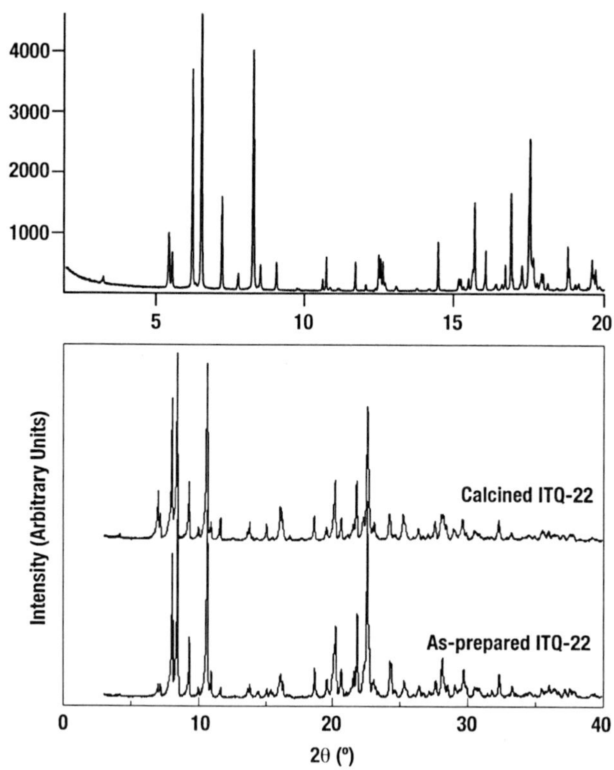

Fig.1.32 (Bottom) powder XRD patterns of as-made and calcined samples of ITQ-22 collected on an in-house CuKα diffractometer and (top) powder XRD pattern of ITQ-22 collected at Synchrotron beamline X7A at Brookhaven National Laboratory (λ = 1.1982 Å) Reprinted from [41] with permission from Nature Materials, Nature Pub. Group

beamline X7A at Brookhaven National Laboratory. The improved peak resolution in the synchrotron data is readily apparent.

The typical output from an indexing program is comprised of the proposed unit cell parameters and their standard deviations; a listing of the observed peaks provided by the user, the predicted peaks for the unit cell, and the differences in their peak positions; and figures of merit that give statistical evaluations of how well the cell matches the experimental data. In some instances the zero offset can be used as an additional refineable parameter. Table 1.4 gives an example of output for an indexing of the synchrotron powder XRD data of zeolite ITQ-22. There is superb agreement between the observed and calculated peak positions for all of the peaks; for nearly every peak, the difference is within 0.005°2θ. The figures of merit are also excellent. This agreement is due to the fact that (1) the dimensions of the crystallites results in sharp diffraction peaks, (2) the synchrotron source allows excellent resolution of the peak positions, and (3) there are no (measurable) errors due to the alignment of the synchrotron powder diffractometer. If we repeat the

Table 1.4 Output for indexing of synchrotron powder X-ray diffraction data for ITQ-22 (λ=1.1982 Å)

Orthorhombic system	a=41.961(6)	b=12.930(1)	c=12.624(2)
hkl	$2\theta^{\circ}_{obs}$	$2\theta^{\circ}_{cal}$	$\Delta 2\theta^{\circ}$
200	3.270	3.273	−.003
001	5.439	5.440	−.001
110	5.555	5.558	−.003
210	6.240	6.240	.000
400	6.546	6.548	−.002
310	7.235	7.235	.000
111	7.779	7.780	−.001
211	8.281	8.282	−.001
401	8.513	8.517	−.004
311	9.057	9.057	.000
510	9.761	9.764	−.003
020	10.633	10.634	−.001
120	10.759	10.760	−.001
002	10.889	10.893	−.004
220	11.130	11.129	.001
610	11.181	11.178	.003
511		11.182	−.001
320	11.718	11.719	−.001
121	12.066	12.064	.002
221	12.395	12.395	.000
420	12.497	12.498	−.001
212	12.562	12.563	−.001
710	12.648	12.649	−.001
402	12.715	12.719	−.004
701		12.704	.011
312	13.095	13.089	.006
711	13.784	13.778	.006

M(25)=43.2 F(25)=236

indexation using data from a well-aligned CuKα diffractometer, we are unsuccessful because the peak overlap precludes an accurate determination of the peak positions. If the correct cell were found in this case, we would expect the $\Delta 2\theta$ values to be larger in magnitude.

After the appropriate lattice setting has been established, the next step is to find the most likely space groups. In the space group determination, it is useful to write the *hkl* assignments above the observed peaks on a copy of the powder XRD pattern and then assess the output of the indexing program. The output generally lists all *hkl* reflections that are close to an observed peak (see the *610* and *511* reflections and the *402* and the *701* reflections in Table 1.4). When there is more than a single predicted reflection close to an observed peak, one must be careful in deciding whether all of the predicted reflections actually belong to the data set. The *hkl* peaks that are included in the "observed" list will guide the investigator's choice of most likely space groups. In the current example, we see that for all *h00* reflections, $h = 2n$; for *0kl*, $k = 2n$; and except for the *701* peak, $h = 2n$ for all *h0l* reflections. We note, however, that the *701* peak is 0.01° from the observed peak and that there is a much better match in the position of the 402 reflection with the observed peak. This observation, coupled with the fact that there are no other *h0l* reflections for which h is odd, suggests that the *701* reflection can likely be disregarded. From the systematic absences, the most probable space groups are *Pbam* or *Pba2*.

The volume of the unit cell is about 6850 Å^3. A typical zeolite has a tetrahedral (T) atom density of about 17 T/1000 Å^3. From this density, we estimate that there are about 116 T atoms in the unit cell. Space group *Pbam* generates eight symmetry-equivalent positions for an atom on a general position, while *Pba2* generates only four equivalent positions. For space group *Pbam*, we therefore anticipate that there are at least 116/8 ~ 15 T atoms in the asymmetric unit, and for space group *Pba2* we would anticipate about 30 T atoms in the asymmetric unit. The latter number of T sites exceeds the number of T sites found in the asymmetric unit of any known zeolite structure. We would therefore expect (or hope) *Pbam* to be the topological symmetry of the zeolite since ab initio structure solutions for a lower symmetry would be very doubtful from powder data (also the centrosymmetric space group is the more likely choice based upon statistical tests for centric distributions of intensity) ([9], p. 255). In the case of ITQ-22, the actual T atom density was ultimately determined to be 16.1 T/1000 Å^3, but four of the T sites are located on mirror planes. Therefore, there are 16 unique T atoms in the asymmetric unit.

1.14.2 Direct Methods

Since the structure factor is essentially a sum of the scattering amplitudes from each electron in the unit cell, we can reexpress (1.5) as an integral over the volume of the unit cell in terms of the electron density $\rho(x, y, z)$ as

$$F(hkl) = \int_{v} \rho(x, y, z) \exp\left[2\pi \mathrm{i}\left(h x_j + k y_j + l z_j\right)\right] \mathrm{d}v \qquad (1.17)$$

Note that this equation does not strictly account for the thermal vibration of the atoms unless we consider the electron density as the time-average of the density. This equation has the form of a Fourier transform; i.e., the structure factor is the Fourier transform of the electron density, and the electron density is the inverse Fourier transform of the structure factor. Therefore

$$\begin{aligned}\rho(x,y,z) &= \sum_h \sum_k \sum_l F(hkl)\exp\left[-2\pi\,\mathrm{i}\left(h\,x_j + k\,y_j + lz_j\right)\right] \\ &= \sum_h \sum_k \sum_l F(hkl)\exp(\mathrm{i}\phi_{hkl})\end{aligned} \tag{1.18}$$

where φ is the phase of the structure factor F.

Phase estimation is greatly simplified when the space group has a center of symmetry; that is, when for a general position (x,y,z), there is another atom at $(-x,-y,-z)$. Space groups with this symmetry relation are referred to as "centro-symmetric." We can express (1.17) as

$$F(hkl) = \sum_j f_j\left(\cos\left[2\pi\left(h\,x_j + k\,y_j + lz_j\right)\right] + \mathrm{i}\sin\left[2\pi\left(h\,x_j + k\,y_j + lz_j\right)\right]\right)$$

The sine terms for the symmetry-equivalent pairs (pairs of atoms related by the inversion center) therefore cancel and since $\cos\alpha = \cos(-\alpha)$, (1.17) becomes

$$F(hkl) = \sum_j^{N/2} 2f_j\cos\left[2\pi\left(h\,x_j + k\,y_j + lz_j\right)\right]$$

where now the sum includes only the independent atoms not related by the inversion center. In this case, since the structure factor is the sum of real numbers, the phase of the structure factor must be either 0 or π radians (values for which sine terms must be 0). The phase determination for a structure factor therefore reduces to determining whether the structure factor is positive or negative.

It can be shown, "subject to certain restrictions," that for a given set of "triplet" reflections (hkl), $(h'k'l')$, and $(h-h', k-k', l-l')$ that

$$F(hkl) \propto \sum_{h'} \sum_{k'} \sum_{l'} F(h'k'l')F(h-h',k-k',l-l')$$

Sayre [42] noted that when the magnitude of a given F(hkl) is large, the sum must tend strongly toward a positive or negative direction. Therefore if all three of the reflections (F_{hkl}, $F_{h'k'l'}$, $F_{h-h',k-k',l-l'}$, are large, then the signs "s" will in general probably be related by

$$s(hkl) \approx s(h'k'l')\cdot s(h-h',k-k',l-l') \tag{1.19}$$

If we consider the case of a reflection with indices $2h,2k,2l$, $s(2h,2k,2l) \approx s(hkl) \cdot s(hkl)$. This implies that F($2h,2k,2l$) will probably be positive if F_{hkl} and $F_{2h,2k,2l}$ are large. Direct methods use probability correlations to evaluate the likelihood of these relationships. An example of a common equation used to evaluate the probability P is

$$P = \frac{1}{2} + \frac{1}{2}\tanh\left[\frac{\sigma_3}{\sigma_2^{3/2}}\left|E_{hkl}E_{h'k'l'l}E_{h-h',k-k',l-l'}\right|\right] \qquad (1.20)$$

$$\sigma_3 = \sum_j^N n_j^3 \text{ and } \sigma_2 = \sum_j^N n_j^2, \quad \text{where } n_j = \frac{f_j}{\sum_i f_i} \text{ and } E_{hkl}^2 = \frac{\left|F_{hkl}^2\right|}{\varepsilon \sum_j^N f_j^2}$$

(ε is an integer usually equal to 1, but it may have different values for special sets of reflections). Note that n_j is simply the fraction of the scattering power for the *hkl* reflection due to atom j in the unit cell. In (1.20), 1 ≥ P ≥ 1/2 (as it should be since, at the very least, there should be an equal probability that the equation is correct or incorrect.) An important observation for this equation is that the larger the term

$$\left|E_{hkl}E_{h'k'l'}E_{h-h',k-k',l-l'}\right|$$

the closer the probability approaches 1. From this relationship a table (Table 1.5) can be created that relates the phase probability as a function of

$$|E_{av}| = \left|E_{hkl}E_{h'k'l'}E_{h-h',k-k',l-l'}\right|^{1/3} \text{ and } \frac{\sigma_3}{\sigma_2^3}$$

For noncentrosymmetric space groups, the analogue of the Sayre equation (1.19) is

$$\phi(hkl) + \phi(h'k'l') + \phi(h-h',k-k',l-l') = 0 \qquad (1.21)$$

In noncentrosymmetric space groups, because the phases can assume any value between 0 and 2π radians, the phase probabilities are expressed in the forms of distributions that give the probability as a function of the phase error relative to that

Table 1.5 Phase probabilities as a function of $|E_{av}|$ and $\sigma_3/{\sigma_2}^3$ ([9], p 259)

$\|E_{av}\|$	$\|E_{av}\|^3$	$\sigma_3/{\sigma_2}^3$					
		20	40	60	80	100	120
3.0	27.0	1.0	1.0	1.0	1.0	0.99	0.99
2.8	22.0	1.0	1.0	1.0	0.99	0.99	0.98
2.6	17.4	1.0	1.0	0.99	0.98	0.97	0.96
2.3	12.2	1.0	0.98	0.96	0.94	0.92	0.90
2.0	8.0	0.97	0.93	0.89	0.86	0.83	0.81
1.85	5.8	0.93	0.86	0.81	0.78	0.76	0.74
1.5	3.4	0.82	0.74	0.71	0.68	0.66	0.65

predicted by (1.21). When there are multiple sets of reflections that can be used to define the phase of a given *hkl*, the "tangent formula" is applied ([7], pp. 340–343):

$$\tan[\phi(h)] = \frac{\sum_{h'} G_j \, sin[\phi(h') + \phi(h - h')}{\sum_{h'} G_j \cos[\phi(h') + \phi(h - h')]}$$

where

$$G_j = 2\frac{\sigma_3}{\sigma_2^{\frac{3}{2}}} \left| E_{hkl} E_{h'k'l'} E_{h-h',k-k',l-l'} \right| \tag{1.22}$$

Similar relationships can be extended to quartet reflections. These relationships describe the fundamental principles used in direct methods. A greater level of discussion, which is beyond the scope of the current work, is required to explain how the best combinations of phases are iteratively selected and evaluated for use in structure solutions. For the student who wishes to learn more about direct methods, I recommend the books by Giacovazzo and by Stout and Jensen.

1.14.3 Patterson Methods

Patterson methods are another technique used in structure determination. If we take the Fourier transform of the $|F^2|$ coefficients (which we are able to extract from the experimental diffraction pattern) rather than the F values in (1.18), it can be shown that the peaks in the calculated map correspond to the interatomic vectors rather than to the actual atomic positions. This map is referred to as a "Patterson" map and is often represented as P(**u**), where u is an interatomic vector. For a unit cell with N atoms, there will be N^2 peaks corresponding to each interatomic vector of the unit cell. There will be a large concentration of electron density at the origin that corresponds to a vector of zero length from each atom to itself. The remaining $N(N-1)$ peaks correspond to the vectors between different atoms within the unit cell.

Each peak in the Patterson map has an integrated intensity that is proportional to the product $f_i f_j$ of the scattering factors of the two atoms separated by the vector **u**. Therefore, we can remove the massive origin peak by subtracting the sum of the squares of the scattering factor of each atom from each $|F^2_{hkl}|$. That is, we define a new

$$\left| F^{2\,\prime}_{hkl} \right| = \left| F^2_{hkl} \right| - \sum_{j=1}^{N} f_j^2$$

A Fourier transform of these redefined coefficients calculates a new map in which the origin peak does not drown out the details of the other interatomic vectors.

How can we use the Patterson map to make sense of the structure? Let us consider a scenario in which there a few atoms that are significantly heavier than the other atoms in the unit cell. If we take the case of the most simple centrosymmetric space group, P-1, then for a heavy atom at the general position (x,y,z), there is a symmetry-equivalent atom at ($-x$,$-y$,$-z$). Therefore, the interatomic vector between these heavy atoms will be located in the Patterson map at position ($2x$,$2y$,$2z$) and ($-2x$, $-2y$, $-2z$). Because these atoms are heavier than the others, the intensity for this Patterson vector will be overwhelmingly larger than the others in the map. The likely position of the heavy atom may then be deduced from the position of the highest density in the Patterson map for this simple case.

After the heaviest atoms have been identified, there are two (not necessarily independent) routes that may be taken to complete the structure solution. The next most intense peaks in the map will correspond to vectors between the heavy atoms and the lighter atoms in the unit cell. These vectors can be used to determine the likely position of these atoms relative to previously located heavy atoms. Another route would be just to use the heavy atoms as an initial model to estimate the phases of each of the reflections. An electron density map calculated from this estimate will then reveal other atoms, which in turn can be used to calculate a more accurate map. This iterative process is known as "Fourier synthesis recycling." Patterson methods are rarely used directly to determine structures of highly siliceous zeolites because they do not have enough extra-framework heavy atoms to allow exploitation of the heavy atom method, but the principles of Patterson methods are used in ab initio algorithms that are used to solve zeolite structures.

1.14.4 Structure Solution from Powder Diffraction Data

For powder data, not only is there no a priori knowledge of the phases, but many of the peak intensities are unknown due to severe overlap in the powder pattern. In turn, a high degree of overlap severely limits the phasing process because the intensities are needed for phase estimation.

There are currently two noteworthy algorithms that have been specifically tailored toward structure solutions of zeolites from powder data: FOCUS [43, 44] and ZEFSAII [45, 46]. Both routines take advantage of the crystal chemical information inherent to all zeolite structures: four-fold connectivity of the tetrahedral atoms and the distances between them. The FOCUS routine begins with a set of random phases and then assigns dominant peaks in the calculated electron density map to silicon atoms. Possible fragments of zeolite structures are then identified and used in Fourier recycling steps to calculate the next set of phases, which are in turn used to calculate another electron density map. This is repeated until there is convergence in the phase set. If a zeolite topology is found upon convergence, it is recorded. The process is performed iteratively with other sets of random phases. A histogram of possible zeolite structures is thereby created. When the correct space group is chosen, the topology found most often is usually the correct solution.

ZEFSAII is a simulated annealing technique in which a given number of silicon atoms are placed in a unit cell and annealed to find a structure that (1) is a geometrically feasible zeolite structure and (2) has a simulated diffraction pattern that shows good agreement with the experimental one. The deviations from tetrahedral distances and angles, the deviations from four-fold connectiveness, and the differences between the simulated and experimental patterns define the "energy" of the system that is minimized during the annealing and parallel tempering processes.

XLENS [47] is another program that has been used to solve some rather complex zeolite structures from powder data. It is a direct methods procedure that uses the "sum function tangent formula" to find the optimal set of phases Φ by maximizing the function S(Φ):

$$S(\Phi) = \sum_{hkl} ||E(hkl) - \langle |E| \rangle || \cdot |E(hkl, \Phi)| \tag{1.23}$$

Because $||E(hkl) - \langle |E| \rangle ||$ is used in the optimization, the weak pseudonormalized structure factors $|E_{hkl}|$ make significant contribution to the sum and are therefore included in the phasing process. This contrasts with normal phasing procedures in which only the strong structure factors are used to determine the phases. A larger amount of information can therefore be included in the phasing process. Note that the medium structure factors are not included since there is only minor contribution to the sum if $|E_{hkl}| \cong \langle |E| \rangle$. The structure solution of ITQ-22 with XLENS [41] is a notable achievement because of the large number of tetrahedral atoms (16) in the asymmetric unit.

1.14.5 Structure Refinement

There are two methods for obtaining structural information from powder diffraction data. In one approach the measured integrated intensities of the Bragg reflections are directly converted to structure factors and used to solve or refine a structure in the same way that single-crystal data is used. This works well for simple, high-symmetry structures that do not possess many overlapping diffraction peaks. However, this is usually the exception rather the rule, especially for zeolite structures, which often possess large asymmetric units.

In 1967 Rietveld developed a refinement method which uses the entire powder diffraction pattern rather than integrated peak intensities. [48] This allowed more information to be extracted from the diffraction pattern since overlapping reflections could also be included in the structure analysis. The Rietveld method is a least-squares procedure in which atomic structural parameters, background parameters, and peak profile parameters are varied until the calculated pattern best matches the experimental pattern. The requirements for a Rietveld refinement are accurate powder diffraction intensity data collected in constant step intervals of 2θ, a starting model that is reasonably close to the actual crystal structure, and a model that accurately describes the peak shape and width (as well as any systematic errors in the peak positions) as a function of 2θ. For the refinement of a zeolite structure, a reasonable starting model is usually the framework topology. Structural parameters of the framework atoms can then be refined, or extraframework atoms can be located in Fourier difference maps and subsequently refined.

The quantity that is minimized in a Rietveld refinement is

$$R = \sum_i w_i (Y_{\mathrm{io}} - Y_{\mathrm{ic}})^2$$

where Y_{io} is the observed intensity at step i (the 2θ angle), Y_{ic} is the calculated intensity at step i, and w_i is the weight given to step i which is generally equal to the square root of the observed intensity. The observed intensity at each step in a diffraction pattern consists of contributions from Bragg peaks and the background at that step. More discussion on the Rietveld method can be found in the treatise by Young. [49] Most Rietveld refinements are currently performed with the GSAS program. [50] Fig. 1.20a shows Rietveld refinements of neutron diffraction data collected for lithium-exchanged zeolite X (Si/Al = 1.0) collected at two different temperatures.

1.14.6 Background Effects in Powder Diffraction

The background in a powder diffraction pattern can arise from fluorescence from the sample, detector noise, thermal-diffuse scattering from the sample, amorphous phases that are present in the sample, incoherent scattering, and scatter of radiation from air, the diffractometer slits, or the sample containers. Background scattering from quartz capillaries is readily observed in synchrotron X-ray diffraction patterns. The background can be modeled by choosing several points that are away from the Bragg peaks and linearly interpolating between those points. Background functions with variable parameters can also be incorporated into the refinement, or a combination of fixed points can be used in conjunction with a background function.

1.14.7 Peak Profile Shapes

Because of the many combined sample and instrumental effects, diffraction peak shapes often deviate significantly from a purely Lorentzian or Gaussian form. The pure Voigt function, a convolution of Lorentzian and Gaussian functions, is a mathematically rigorous way of combining these effects. However, because of the complexity of the Voigt function, the simpler pseudo-Voigt function is most commonly used in profile fitting. The pseudo-Voigt function is a sum of a Lorentzian and Gaussian function, with a mixing parameter that varies with diffraction angle. The pseudo-Voigt function has been parameterized so that it closely approximates the Voigt function, and it contains adjustable parameters for the angular dependencies of the full-widths at half maximum (FWHM) of both the Gaussian and Lorentzian components. The pseudo-Voigt equation of Thompson-Cox-Hastings may be summarized as follows:

$$PV(\theta) = \eta(\theta)L(\theta) + (1 - \eta(\theta))\,G(\theta) \tag{1.24}$$

$$\Gamma_G(\theta) = \left(U\tan^2\theta + V\tan\theta + W + \frac{Z}{\cos^2\theta}\right)^{1/2} \tag{1.25}$$

$$\Gamma_L(\theta) = X\tan\theta + \frac{Y}{\cos\theta} \tag{1.26}$$

where G is a Gaussian function with FWHM of Γ_G, L is a Lorentzian function with FWHM of Γ_L, and η is an angle-dependent mixing term. U,V,W,X, and Y are parameters that can be adjusted during refinement of the peak profile function. Additional terms can be included to simulate anisotropic crystallite size effects or strain effects.

The peak shapes for neutron diffraction data tend to be highly Gaussian while the peak profiles of X-ray diffraction are usually highly Lorentzian. During the initial stages of a refinement for neutron diffraction data, W (Y for X-ray diffraction data) is fixed so that the calculated peak width closely matches the observed width at low angles and all other adjustable parameters are set to 0. For a given instrument, initial values of the parameters can also be obtained from previous carefully measured calibration experiments. The pseudo-Voigt function is symmetric about the calculated 2θ position for the Bragg reflection. However, both instrumental and sample effects can induce asymmetry, especially at low angles. Therefore, adjustable asymmetry corrections are often added to the profile functions included in many refinement programs. Even with the use of these corrections, the asymmetry is sometimes so pronounced at low 2θ that the lowest angle reflections are often excluded from the refinement, especially when the low angle reflections are very intense.

1.14.8 Refinement Steps in a Typical Rietveld Analysis

During the first stages of a refinement, the scale factor and background coefficients are adjusted. The unit cell parameters are next optimized, and then a zeropoint correction can also be included in the refinement. However, for well-calibrated instruments such as those present at most synchrotron and neutron sources, this correction is often so small that its effects on the refinement are negligible. One or a couple of the peak shape terms may then be adjusted. Throughout the refinement, it is important to check plots of the observed, calculated, and difference profiles of the diffraction pattern. This can quickly identify any problems in the background, calculated peak positions, or calculated peak shapes. The presence of impurity phases or phase transitions can also be identified by peaks that are not predicted Bragg reflections.

At this point atomic parameters can be refined. Soft restraints on bond distances and bond angles are often used in the beginning stages to ensure convergence to the correct structure. This author chooses to initially refine the atomic positions (of the framework atoms of a zeolite) one at a time rather than all simultaneously. Even with the use of restraints, it is important to check the bond distances and angles after the refinement cycles to make sure that the refined structure is chemically reasonable. Afterwards, the positions of all the atoms can usually be refined together without the refined structure settling into a false minimum.

Fourier difference maps can then be examined to determine the positions of extra-framework cations or organic molecules, and the occupancy factors and positions of these species can then be alternately refined. Sometimes unlocated atoms must be found in the Fourier difference maps before the parameters of other atoms can be refined in a stable manner. After the positions and/or occupancy factors of all atoms have again been alternately refined for a few cycles, the thermal displacement parameters of the atoms are then varied. The thermal vibrations of atoms are, in general, highly anisotropic and require six parameters to describe the true motion of each atom in the asymmetric unit. However, for powder diffraction data this creates a small ratio of data points (well resolved reflections) to adjustable parameters. It is therefore customary to use only isotropic thermal displacement factors in refining the atomic motion. Furthermore, framework atoms of the same element type are often constrained to have identical thermal displacement factors to reduce the number of adjustable parameters.

1.14.9 Evaluating the Quality of a Rietveld Structure Refinement

The quantities used to evaluate the progress of a Rietveld refinement and the agreement between the measured and calculated patterns are the R_p, R_{wp}, and χ values:

$$R_p = \frac{\sum_i |Y_{io} - Y_{ic}|}{\sum_i Y_{io}} \qquad R_{wp} = \sqrt{\frac{\sum_i w_i (Y_{io} - Y_{ic})^2}{\sum_i w_i Y_{io}^2}} \qquad R_e = \sqrt{\frac{n-p}{\sum_i w_i Y_{io}^2}} \qquad \chi = R_{wp}/R_e$$

From a purely statistical point of view, R_{wp} is the most significant agreement factor since its numerator is the function being minimized. Adequate refinements typically have R_{wp} values below 12%. The "expected" R index, R_e, is defined for n total observations with a model that has p parameters. The χ index, also known as the goodness of fit, is the ratio of R_{wp} to R_e. This agreement factor is a measure of how well the calculated model accounts for the experimental data. For a perfect refinement with *correctly* weighted data, χ should have a value of 1. Values much greater than 1.7 indicate an inadequate model, while values smaller than 1 could indicate the model has more parameters than are justified by the quality of the data.

Although numerical criteria are important, they should not be solely relied upon to evaluate the quality of a refinement. Difference plots should always be examined to identify any gross errors in the model. The bond lengths (or coordination distances), bond angles, and occupancy factors of the atoms in the final structure should also be chemically reasonable.

1.15 Diffuse Scattering from Materials with One-Dimensional Planar Disorder

Stacking disorder is commonly observed in zeolites and other silicate materials. Notable examples include zeolite beta [51], SSZ-26/33 [52], SSZ-31 [53], and ZSM-48. [54] The disorder often arises from stacking faults of two-dimensional layers to give one-dimensional disorder in the direction perpendicular to the layers. The stacking faults occur either from the possibility of different layer types at a given stacking position or from the possibility of translational disorder of the same layer type. The "translational" disorder will be considered for translations in directions *parallel* to the layers.

Diffraction patterns of materials with one-dimensional disorder (for simplicity, in the c-direction) manifest diffuse streaks in reciprocal space that are parallel to **c***. The powder diffraction patterns are therefore characterized by a series of Bragg peaks that are either sharp or diffuse (broadened) according to the magnitude of the translation vectors relating the planes. The following discussion attempts to explain which diffraction peaks are broadened by stacking disorder.

The structure factor F_n of a unit cell describing the nth layer of atoms for a reflection indicated by the vector **s** in reciprocal space is given by [8]

$$F_n(s) = \sum_r f_{nr} \exp(-2\pi i s \cdot u_{nr}) \tag{1.27}$$

where the sum is over all the atoms in the layer n, f_{nr} is the scattering factor for the r th atom, and u_{nr} is the vector joining the lattice node (origin) to the r th atom. The total structure factor *for all the layers* may therefore be reexpressed as

$$F(s) = \sum_n F_n(s) = \sum_n \sum_r f_{nr} \exp(-2\pi i s \cdot u_{nr})$$

The total scattering intensity for a reciprocal vector s is given by

$$I(s) = |F(s)F^*(s)|$$

where F(**s**) is the structure factor for the entire group of layers, and F^*(**s**) is its complex conjugate.

The intensity may be reformulated as

$$I(s) = \left| \sum_{n} \sum_{n'} \sum_{r} \sum_{r'} f_{nr} f_{n'r'} \exp[-2\pi i s \cdot (u_{nr} - u_{n'r'})] \right|$$

where the index n indicates the nth layer.

The effects of the stacking faults on the XRD pattern may be determined by considering the average value of $F_n F^*_{n+m}$ where F_{n+m} is the structure factor of a layer located m layers away from the nth layer. This may be expressed

$$y_m = \frac{\sum_{N} F_n F*_{n+m}}{N} \quad (1.28)$$

and the scattering power along the row in reciprocal space with index of l can be shown to be

$$I(l) \propto \left| \sum_{m=-N}^{N} \left(1 - \frac{|m|}{N}\right) y_m \exp(2\pi i l m) \right| \quad (1.29)$$

where N indicates the total number of layers. To evaluate the intensity distribution along a given row in reciprocal space, we must first determine a general expression for y_m as a function of m. We can do this by relating the stacking sequences in terms of probabilities. If we consider the case of two otherwise identical layer types that are related by a translation vector **t** (in a direction perpendicular to the stacking direction), then the structure factor of layer B is related to that of layer A by

$$F_B(s) = F_A(s) \exp(-2\pi i s \cdot t) \quad (1.30)$$

$$y_m = \frac{1}{2} F_A \left[P_m F_A^* + (1 - P_m) F_B^*\right] + \frac{1}{2} F_B \left[P_m F_B^* + (1 - P_m) F_A^*\right]$$

where P_m indicates the fraction of mth layer neighbors (or probability) of layer type A (or B) that are of layer type A (or B). For the case of identical layers related only by translations, the assignment of layer types is arbitrary so that there are an equal number of A and B layer types. This explains the factor of 1/2 that appears in front of both terms in the above equation.

If the layers A and B are related by translations of $1/2b$, then

$$F_B = F_A \exp\left[2\pi i (ha^* + kb^* + lc^*) \cdot \left(\frac{b}{2}\right)\right] = F_A \exp(k\pi i)$$

$$\begin{aligned} y_m &= \frac{1}{2} F_A \left[P_m F_A^* + (1 - P_m) F_B^*\right] + \frac{1}{2} F_B \left(P_m F_B^* + (1 - P_m) F_A^*\right) \\ &= P_m |F_A|^2 + (1 - P_m) \cos(\pi k) |F_A|^2 \end{aligned}$$

For $k = 2n$,

$$y_m = P_m|F_A|^2 + (1 - P_m)|F_A|^2 = |F_A|^2$$

So there is no effect on these reflections since ${F_A}^2$ is the average we expect in the absence of stacking faults for either a pure AAAA (or BBBB) or a pure ABAB type stacking arrangement. For $k = 2n+1$,

$$y_m = P_m|F_A|^2 - (1 - P_m)|F_A|^2 = (2P_m - 1)|F_A|^2$$

The intensities of these reflections are reduced (and the peaks are broadened) by stacking disorder. Note that for $P_m = 1$ there is no effect (as expected). To find a general expression for P_m, we note that $P_0 = 1$ (the 0th layer must be the same as itself), and $P_1 = (1-\alpha)$, where α is the probability that there is a stacking fault in a pure AAA or BBB stacking-type sequence. For a pure ABAB stacking arrangement, $P_1 = 0$. We will assume the following form for P_m:

$$P_m = \frac{1}{2} + q\rho^{|m|}$$

From the conditions for P_0 and P_1, we determine that

$$\begin{aligned} P_m &= \frac{1}{2}\left[1 + (1 - 2\alpha)^{|m|}\right] \Rightarrow y_m \\ &= \frac{1}{2}\left[1 + (1 - 2\alpha)^{|m|}\right]|F_A|^2 + \left[\frac{1}{2} - (1 - 2\alpha)^{|m|}\right]\cos(\pi k)|F_A|^2 \end{aligned}$$

We can now put this into the expression for I(l)

$$I(l) \propto \left|\sum_{m=-N}^{N}\left(1 - \frac{|m|}{N}\right)y_m \exp(2\pi i l m)\right|$$

$$= \sum_{m=-N}^{N}\left(1 - \frac{|m|}{N}\right)\left\{\frac{1}{2}\left[1 + (1 - 2\alpha)^{|m|}\right] + \left[\frac{1}{2} - (1 - 2\alpha)^{|m|}\right]\cos(\pi k)\right\}|F_A|^2 \exp(2\pi i l m)$$

For k = 2n, this sum reduces to

$$\sum_{m=-N}^{N}\left(1 - \frac{|m|}{N}\right)|F_A|^2 \exp(2\pi i l m) = \frac{\sin^2[(N+1)\pi l]}{\sin^2(\pi l)}|F_A|^2$$

This equation is the same form that we determined in the section on the Scherrer equation. The important point here is that, when $k = 2n$, the intensity is

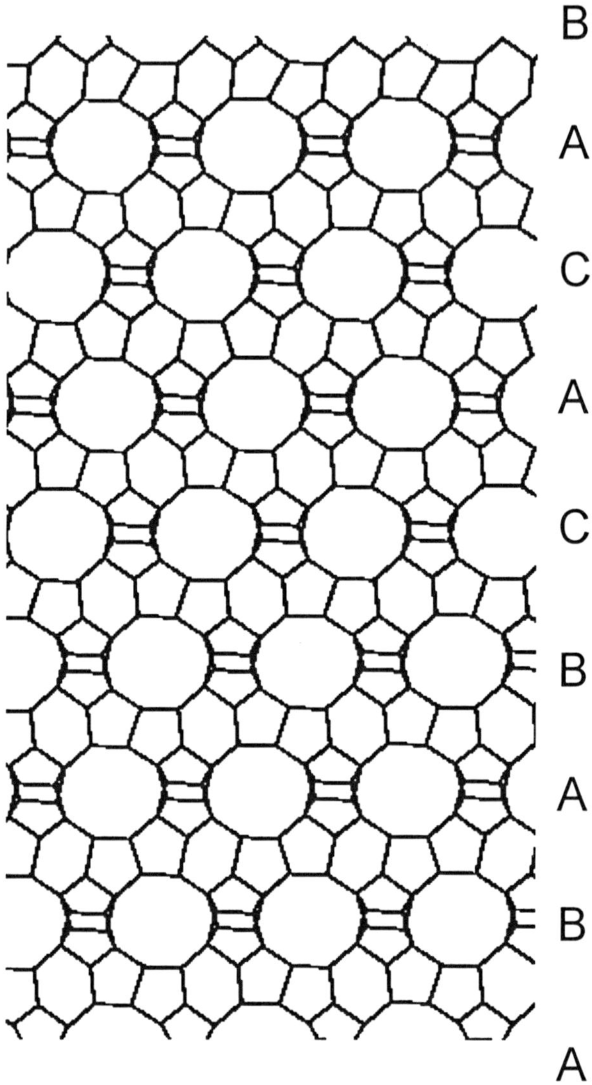

Fig.1.33 Hypothetical intergrowth structure for the SSZ-26/33 family

concentrated at nodes where there are integer values of l. As N becomes larger, this expression approaches a delta function at the integer values.

However, for $k = 2n+1$, the expression for y_m is

$$\sum_{m=-N}^{N}\left(1-\frac{|m|}{N}\right)\left[(1-2\alpha)^{|m|}\right]|F_A|^2\exp(2\pi i l m)$$

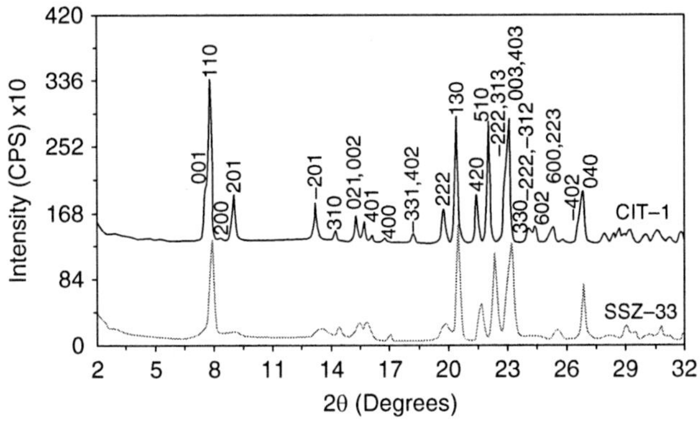

Fig.1.34 Comparison of the powder XRD patterns of CIT-1 and SSZ-33. Note that for reflections with $l \neq 3n$, the peaks are broadened

Unlike the previous case for $k = 2n$, this expression yields significant intensity at non-integer values of l. The degree of broadening is a function of the faulting probability α. The broadening is experimentally manifested in electron diffraction images in which there is streaking along the c^*-direction for odd values of k.

There are several zeolite structures in which layers are stacked to identical neighboring layers with shifts of $\pm 1/3$ of an intra-layer repeat distance. An example we will discuss is the intergrowth structure of the SSZ-26/SSZ-33 family. Figure 1.33 shows a hypothetical intergrowth of this structure. The two ordered end-members of the series are defined by stacking sequences of ABAB and ABCABC, respectively. The end-member with the ABCABC stacking arrangement is found in the ordered structure of CIT-1 (**CON**, *C2/m*; $a = 22.6$, $b = 13.4$, $c = 12.4$, $\beta = 68.9°$). In this unit cell, the layers are stacked in the a-direction, and faulting occurs by different shifts of the layers in the c-direction.

We could derive similar formula for y_m and P_m as in the previous example, but the important point here is that there will be sharp peaks for $l = 3n$ and broad peaks for $l = 3n+1$ or $3n+2$. Figure 1.34 compares the powder XRD patterns of CIT-1 and SSZ-33. The *hkl* indices have been placed above their respective peaks to illustrate how the stacking broadens selected peaks while others remain unaffected.

The program DIFFaX [55] simulates the powder diffraction patterns of materials with stacking disorder of two-dimensional layers. The required input is the atomic coordinates of the layers and stacking probabilities between different layer types. This algorithm has allowed investigators to confirm their structural models for zeolite structures that are heavily disordered.

References

1. Bish DL, Post JE (1989) Modern powder diffraction. Mineralogical Society of America, Washington, DC
2. Jenkins R, Snyder RL (1996) Introduction to X-ray Powder Diffractometry. In: Winefordner JD (ed) Chemical Analysis: A series of monographs on analytical chemistry and its applications, vol 138. Wiley, New York
3. Chung FH, Smith DK (eds) (2000) Industrial applications of X-ray diffraction. Marcel Dekker, New York
4. Cullity BD (1978) Elements of x-ray diffraction. Addison-Wesley Publishing Company, Inc., Massachusetts
5. Warren BE (1990) X-ray diffraction. Dover Publications, New York
6. Klug HP, Alexander LE (1954) X-ray diffraction procedures. Wiley, New York
7. Giacovazzo C, Monaco HL, Viterbo D, Scordari F, Gilli G, Zanotti G, Catti M (1992) Fundamentals of crystallography. Oxford University Press, New York
8. Guinier A (1994) X-ray diffraction in crystals, imperfect crystals, and amorphous bodies. Dover Publications, New York
9. Stout GH, Jensen LH (1989) X-ray structure determination: A practical guide. Wiley, New York
10. David WIF, Shankland K, McCusker LB, Baerlocher CH (eds) (2002) Structure determination from powder diffraction data. Oxford University Press, New York
11. Burns G, Glazer AM (1990) Space groups for solid state scientists. Academic Press, New York
12. Baerlocher C, Meier WM, Olson DH (eds) (2001) Atlas of zeolite framework types 5th edn. Elsevier, New York http://www.iza-structure.org/databases/
13. Treacy MMJ, Higgins JB (eds) (2001) Collection of simulated XRD powder patterns for zeolites, 4th edn. Elsevier, New York
14. Acclrss, Inc. (2000) Cerius2, v. 2.1, Product of MSI and Biosym. Accelrys seftwane, san diego
15. Carroll K, Johnson OR (1976) TEP-II: A FORTRAN thermal-ellipsoid plot program for crystal structure illustrations oak ridge, National Laboratory Report ORNL-5138
16. Marler B, Grünewald-Lüke A, Gies H (1995) Decasils, a new order-disorder family of microporous silicas. Zeolites 15:388–399
17. Hahn T (1996) International tables for crystallography. Kluwer Academic Publishers, Boston
18. Cheetham AK, Wilkinson AP (1992) Synchrotron X-ray and neutron diffraction studies in Solid-State chemistry. Angew Chem Int Ed Engl 31:1557–1570
19. Camblor MA, Corma A, Lightfoot P, Villaescusa LA, Wright PA (1997) Synthesis and Structure of ITQ-3, the First Pure Silica Polymorph with a Two-Dimensional System of Straight Eight Ring Channels. Angew Chem Int Ed 36:2659–2661
20. Wu MG, Deem MW, Elomari SA, Medrud RC, Zones SI, Maesen T, Kibby C, Chen C-Y, Chan IY (2002) Synthesis and structure determination by zefsa-ii of ssz-55: a new high-silica, large-pore zeolite. J Phys Chem B 106:264–270
21. Meier WM (1961) The crystal structure of mordenite (ptilolite). Z Kristallogr 115:439–450
22. Galli E, Quartieri S, Vezzalini G, Alberti A, Franzini M (1997) Terranovaite from Antarctica: A new 'pentasil' zeolite. Am Mineral 82:423–429
23. Dorset DL, Kennedy GJ (2004) J Phys Chem B 108:15216–15222
24. Sohn JR, Decanio SJ, Lunsford JH, O'Donnell DJ (1986) Determination of framework aluminium content in dealuminated Y-type zeolites: a comparison based on unit cell size and wavenumber of i.r. bands. Zeolites 6:225–227
25. Meyers BL, Ely SR, Kutz NA, Kaduk JA (1985) Determination of structural boron in borosilicate molecular sieves via X-ray diffraction. J Catal 91:352–355
26. Langford JI (1973) LAPOD, School of Physics and Space Research, University of Birmingham, Birmingham, UK

27. Feuerstein M, Lobo RF (1998) Characterization of Li cations in zeolite LiX by solid-state NMR spectroscopy and neutron diffraction. Chem Mater 10(8):2197–2204
28. LeBail A, Duroy H, Fourquet JL (1988) Ab-initio structure determination of LiSbWO6 by X-ray powder diffraction. Mater Res Bull 23:447–452
29. Halliday D, Resnick R, Krane KS (1992) Physics, vol 2, 4th edn. Wiley, New York, pp 989–992
30. Rouleau L, Kolenda F, Benazzi E (2002) MTT Zeolite comprising crystals and crystal aggregates with specific granulometries, and its use as a catalyst for isomerising straight chain parafins. US Patent Application 2002/0192156A1
31. Louer D, Langford JI, Ciosmak D, Niepce JC (1983) A precise determination of the shape, size and distribution of size of crystallites in zinc oxide by X-ray line-broadening analysis. J Appl Cryst 16:183–191
32. Munoz T Jr, Balkus KJ Jr (1999) Preparation of oriented UTD-1 membranes via pulsed laser ablation. J Am Chem Soc 121(1):139–146
33. Jeong H-K, Lai Z, Tsapatsis M, Hanson JC (2005) Strain of MFI crystals in membranes: An in situ synchrotron X-ray study. Micropor Mesopor Mater 84:332–337
34. Gouzinis A, Tsapatsis M (1998) On the Preferred Orientation and Microstructural Manipulation of Molecular Sieve Films Prepared by Secondary Growth. Chem Mater 10(9):2497–2504
35. Boudreau LC, Tsapatsis M (1997) A highly oriented thin film of zeolite. A Chem Mater 9(8):1705–1709
36. Jeong H-K, Krohn J, Sujaoti K, Tsapatsis MJ (2002) Oriented Molecular Sieve Membranes by heteroepitaxial growth. Am Chem Soc 124(44):12966–12968
37. Werner PE (1990) Treor90, Department of Structural Chemistry, University of Stockholm, Stockholm, Sweden
38. Boultif A, Louer DJ (1991) Appl Cryst 24:987
39. Visser JJ (1969) A fully automatic program for finding the unit tell from pwoer date. Appl Cryst 2:89
40. Azaroff LV, Buerger MJ (1958) The powder method. McGraw-Hill Book Company, New York
41. Corma A, Rey F, Valencia S, Jorda JL, Rius J (2003) A zeolite with interconnected 8-, 10- and 12-ring pores and its unique catalytic selectivity. Nature Mat 2:493–497
42. Sayre D (1952) The squaring method: a new method for phase determination. Acta Cryst. 5:60–65
43. Grosse-Kunstleve RW, McCusker LB, Baerlocher C (1997) Powder Diffraction Data and Crystal Chemical Information Combined in an Automated Structure Determination Procedure for Zeolites. J Appl Cryst 30:985–995
44. Grosse-Kunstleve RW (1996) Zeolite determination from powder data: Computer based incorporation of crystal chemical information Dissertation ETH No. 11422, Swies Federal Institute of Technology zurich
45. Falcioni M, Deem MW (1999) A biased monte carlo scheme for zeolite struchtre solution. J Chem Phys 110:1754
46. Deem MW, Newsam JM (1992) Framwork crystal struchtre solution by simulated amealing test application to known zeolite struchtres. J Am Chem Soc 114:7189
47. Rius J (1999) XLENS, a direct methods program based on the modulus sum function: Its application to powder data. Powder Diffraction 14:267–269
48. Rietveld HM (1967) Line profiles of neutron powder-diffraction peaks for structure refinement. Acta Cryst 22:151–152
49. Young RA (1996) The rietveld method. Oxford Science Publications, New York
50. Larson AC, Von Dreele RB (2000) General Structure Analysis System (GSAS), Los Alamos National Laboratory Report LAUR 86–748
51. Newsam JM, Treacy MMJ, Koetsier WT, de Gruyter CB (1988) Structural Characterization of Zeolite Beta. Proc R Soc Lond A 420:375–405
52. Lobo RF, Pan M, Chan I, Medrud RC, Zones SI, Crozier PA, Davis ME (1994) Physicochemical Characterization of Zeolites SSZ-26 and SSZ-33. J Phys Chem 98(46):12040–12052

53. Lobo RF, Tsapatsis M, Freyhardt CC, Chan I, Chen CY, Zones SI, Davis ME (1997) A model for the structure of the large-pore zeolite SSZ-31. J Am Chem Soc 119:3732–3744
54. Lobo RF, van Koningsveld H (2002) New description of the disorder in zeolite ZSM-48. J Am Chem Soc 124(44):13222–13230
55. Treacy MMJ, Newsam JM, Deem MW (1991) A general recursion method for calcuating diffracted intensikes from crystals containing planar faults. Proc R Soc London A 443:449

Chapter 2
Solid-State NMR Spectroscopy

Michael Hunger

Abstract Due to the development of new techniques and further increase of magnetic field strength available for commercial applications, solid-state NMR spectroscopy became a routine method for the characterization of zeolites. As an important advantage, solid-state NMR spectroscopy allows the investigation of the local structure of nuclei in the solids under study. The specific behavior of zeolites often depends on local effects, such as framework defects, the substitution of framework atoms, guest compounds etc. Therefore, solid-state NMR spectroscopy is a widely applied analytical method for delivering structure data, which are complementary to those of diffraction methods suitable for investigating the long-range order. While early solid-state NMR spectroscopic studies often focused on the characterization of the zeolite framework in the as-synthesized and hydrated state, an increasing number of recent works is dealing with the investigation of the framework of dehydrated and calcined zeolites and of surface sites, i.e., on the determination of their concentration, strength, and accessibility. In all these applications, the advantage of solid-state NMR spectroscopy to be a quantitative method is utilized. The present chapter demonstrates the fundamentals, various techniques, and most important applications of solid-state NMR spectroscopy making this method to an important tool of research in zeolite science.

Abbreviations

1D	One-dimensional
2D	Two-dimensional
3Q	Triple-quantum

M. Hunger
Institute of Chemical Technology, University of Stuttgart, 70550 Stuttgart, Germany
e-mail: michael.hunger@itc.uni-stuttgart.de

A.W. Chester and E.G. Derouane (eds.), *Zeolite Characterization and Catalysis*,
DOI 10.1007/978-1-4020-9678-5_2,

5Q	Quintuple-quantum
COSY	Correlation spectroscopy
CP	Cross-polarization
CRAMPS	Combined rotational and multiple-pulse spectroscopy
DOR	Double-oriented rotation
INADEQUATE	Incredible natural abundance double quantum transfer experiment
INEPT	Insensitive nuclei enhanced by polarization transfer
MAS	Magic angle spinning
MQMAS	Multiple-quantum MAS
NMR	Nuclear magnetic resonance
NOESY	Nuclear Overhauser exchange spectroscopy
REDOR	Rotational echo double resonance
SOQE	Second-order quadrupole effect
TMP	Trimethylphosphine
TMPO	Trimethylphosphine oxide
TRAPDOR	Transfer of population in double resonance

2.1 Introduction

Nuclear magnetic resonance (NMR) is an analytical tool with a high sensitivity for chemical bonds in the local structure of the resonating nuclei, such as of framework atoms, extra-framework species, surface sites, and adsorbate complexes in zeolites. Therefore, a number of research groups utilizes solid-state NMR techniques for the characterization of these industrially very important catalysts and adsorbents. Many atoms occurring in zeolites possess isotopes with a nuclear spin, which renders these isotopes NMR-active. The most frequently investigated nuclei are summarized in Fig. 2.1. Their nuclear spins and relative sensitivities (in comparison with ^{1}H) are given in parentheses. ^{11}B, ^{17}O, ^{27}Al, ^{29}Si, ^{31}P, ^{51}V, ^{67}Zn, and ^{71}Ga isotopes are accessible for NMR spectroscopy and contribute to the framework of zeolites in a broad manner. ^{1}H, ^{7}Li, ^{23}Na, and ^{133}Cs isotopes are interesting for the characterization of surface OH groups and extra-framework species, such as extra-framework cations. On the other hand, ^{1}H, ^{2}H, ^{13}C, ^{15}N, and ^{31}P isotopes allow the investigation of a wide variety of probe molecules and reactants.

Depending on the nuclear spin I responsible for the magnetic dipole moment, different line narrowing techniques are required to reach highly resolved solid-state NMR spectra of zeolites. In the case of spin $I = 1/2$, the application of the conventional magic angle spinning (MAS) technique leads to an averaging of the most important nuclear interactions, such as anisotropic chemical shielding and dipolar interactions (see Sect. 2.3.1). Isotopes with nuclear spin $I > 1/2$ are

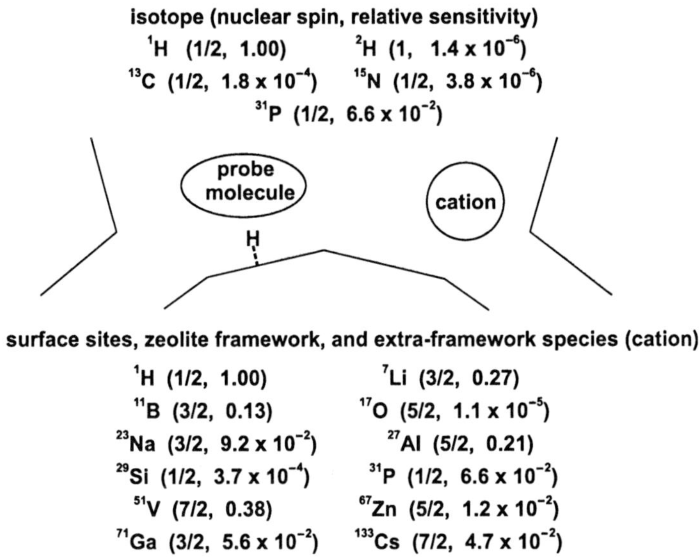

Fig. 2.1 Most suitable isotopes for solid-state NMR investigations in zeolite sciences

additionally characterized by an electric quadrupole moment. These nuclei are involved in quadrupolar interactions and often require more sophisticated solid-state NMR techniques, such as high-field MAS NMR, double-oriented rotation (DOR), and multiple-quantum MAS (MQMAS) NMR spectroscopy (see Sect. 2.3.2) making their investigation complicated. Application of these techniques makes it possible to utilize solid-state NMR spectroscopy for the study of:

- The incorporation, coordination, and local structure of framework atoms in zeolites,
- The distribution, location, and coordination of extra-framework species,
- The nature, concentration, and the chemical behavior of surface OH groups in dehydrated zeolites, and
- Adsorbate complexes suitable for the characterization of surface sites.

This chapter starts with an introduction into the fundamentals and experimental techniques of solid-state NMR spectroscopy in Sects. 2.2 and 2.3. Subsequently, the application of these spectroscopic techniques on the investigation of framework atoms and extra-framework species, such as alkali metal cations is demonstrated for the most important nuclei in Sects. 2.4 and 2.5. The advantage of solid-state NMR spectroscopy for the characterization of the nature and behavior of surface OH groups, Lewis acid sites, and base sites is outlined in Sects. 2.6 and 2.7. For further applications of solid-state NMR spectroscopy in zeolite science and heterogeneous catalysis, see [1–13].

2.2 Fundamentals of Solid-State NMR Spectroscopy

2.2.1 Introduction into Basics

The basics of NMR spectroscopy are described in a number of text books [14–18]. Therefore, this Section focuses on the mechanisms responsible for the broadening of solid-state NMR signals and their characteristic line shape function. The dominant nuclear interactions in solids are described by the total Hamiltonian

$$\widehat{H} = \widehat{H}_{\mathrm{II}} + \widehat{H}_{\mathrm{IS}} + \widehat{H}_{\mathrm{csa}} + \widehat{H}_{\mathrm{q}} \tag{2.1}$$

Here, $\widehat{H}_{\mathrm{II}}$, $\widehat{H}_{\mathrm{IS}}$, $\widehat{H}_{\mathrm{csa}}$, and $\widehat{H}_{\mathrm{q}}$ are the Hamiltonians of the homonuclear and heteronuclear magnetic dipole-dipole interaction, the anisotropic chemical shielding, and the quadrupolar interaction, respectively. In most applications of solid-state NMR spectroscopy, the scalar J-coupling can be neglected as broadening mechanism and, therefore, the corresponding Hamiltonian was omitted. Analytical expressions of the Hamiltonians in (2.1) are given [19–21]. For nuclei with spin $I = 1/2$, only the first three terms are valid. For nuclei with spin $I > 1/2$, which exhibit an electric quadrupole moment interacting with the electric field gradient at the site of the nuclei, the Hamiltonian $\widehat{H}_{\mathrm{q}}$ may cause the dominating line broadening of solid-state NMR signals. In liquids, all Hamiltonians in (2.1) are averaged due to rapid isotropic reorientation of the species under study. For solids and depending on the type of the dominating nuclear interaction, characteristic line shape functions occur in the solid-state NMR spectra. These line broadenings may cause significant problems for the separation of signals.

The center of gravity frequency, ν_{cg}, of an NMR signal occurs at

$$\nu_{\mathrm{cg}} \int_{0}^{\infty} \nu f(\nu)\,\mathrm{d}\nu \tag{2.2}$$

where $f(\nu)$ is the normalized line shape function. The value of the frequency ν_{cg} may deviate from the Larmor frequency

$$\nu_0 = \frac{\gamma_I}{2\pi} B_0 \tag{2.3}$$

which is the resonance frequency of an isolated and non-shielded nucleus with spin I characterized by the gyromagnetic ratio γ_{I}. B_0 denotes the magnetic flux density of the external magnetic field.

2.2.2 Line Broadening Mechanisms

The superposition of effects due to the various nuclear interactions complicates the interpretation and calculation of the line shape function $f(\nu)$. To describe the broadening of NMR signals caused by nuclear interactions in solid materials, it is advantageous to make use of the second moment M_2. The second moment is defined by

$$M_2 = 4\pi^2 \int_{-\infty}^{+\infty} \left(\nu - \nu_{cg}\right)^2 f(\nu)\, d\nu \tag{2.4}$$

The full width $\Delta\nu_{1/2}$ at half maximum of the solid-state NMR signal in frequency units, also called static line width, is

$$\Delta\nu_{1/2} = \frac{1}{\pi}\sqrt{2\ln 2}\,\sqrt{M_2} \tag{2.5}$$

Provided that the local structure and the dominating nuclear interactions of the spin system under study are known, the second moment can be calculated and the expected line broadening of the solid-state NMR signal can be estimated. On the other hand, characteristic line shape functions are an important source of information on the local structure of the resonating nuclei. By this way, e.g., atom-atom distances can be obtained via the evaluation of the dipolar interaction or the symmetry in the local structure of spin $I > 1/2$ nuclei can be studied since it strongly depends on the electric field gradient obtained via an quantitative evaluation of the quadrupolar interaction.

According to the total Hamiltonian given in (2.1), the total second moment of solid-state NMR signals follows to

$$M_2 = M_2{}^{\mathrm{II}} + M_2{}^{\mathrm{IS}} + M_2{}^{\mathrm{csa}} + M_2{}^{\mathrm{q,ct}} \tag{2.6}$$

with

$$M_2{}^{\mathrm{II}} = \frac{3}{5}\left(\frac{\mu_o}{4\pi}\right)^2 \gamma_I^4 \hbar^2 I(I+1)\frac{1}{N_I}\sum_{i\neq j}^{N_I} r_{ij}^{-6} \tag{2.7}$$

$$M_2{}^{\mathrm{IS}} = \frac{4}{15}\left(\frac{\mu_o}{4\pi}\right)^2 \gamma_I^2\gamma_S^2 \hbar^2 S(S+1)\frac{1}{N_I}\sum_{j=1}^{N_I}\sum_{k=1}^{N_S} r_{jk}^{-6} \tag{2.8}$$

$$M_2{}^{\mathrm{csa}} = \frac{4}{45}\gamma_I^2 B_o^2 \Delta\sigma^2\left(1 + \frac{1}{3}\eta_{\mathrm{csa}}^2\right) \tag{2.9}$$

$$M_2{}^{\mathrm{q,ct}} = \frac{92\pi^2}{7}\,\nu_{\mathrm{qs}}{}^2 \tag{2.10}$$

where $\hbar$ denotes Planck's constant h divided by 2π, μ_o is the permeability of vacuum, and I and S are the nuclear spins of the resonating and non-resonating nuclei, respectively. N_{I} and N_{S} denote the numbers of the spin I and S nuclei in the sample, γ_S is the gyromagnetic ratio of the non-resonating spins S, and r_{ij} and r_{jk} are internuclear distances. The anisotropy, $\Delta\sigma$, and the asymmetry parameter, η_{csa}, of the chemical shielding are defined by (2.13) and (2.14), respectively, while the quadrupolar frequency shift, ν_{qs}, is given in (2.15).

In the case of spin $I = 1/2$ systems and assuming that the sample does not contain paramagnetic impurities and susceptibility effects can also be neglected, the relationship between the center of gravity frequency, ν_{cg}, and the Larmor frequency, ν_0, is given by

$$\nu_{cg} = \nu_0(1 - \sigma) \tag{2.11}$$

The parameter σ is the shielding constant caused by the shielding effect of the electron shell around the resonating nuclei. This shielding constant is related to the isotropic chemical shift, δ, via $\delta = \sigma_{ref} - \sigma$, where σ_{ref} is the shielding constant of the nuclei in a reference material. The isotropic shift value δ is defined by

$$\delta = \frac{1}{3}\left(\delta_{xx} + \delta_{yy} + \delta_{zz}\right) \tag{2.12}$$

The values δ_{xx}, δ_{yy}, and δ_{zz} are the principal values of the chemical shift tensor, which are denoted in such a way that the inequality $\delta_{xx} \geq \delta_{yy} \geq \delta_{zz}$ is fulfilled. The chemical shift, δ, is given in "ppm," which means parts per million of the resonance frequency. The anisotropy, $\Delta\sigma$, and the asymmetry parameter, η_{csa}, of the chemical shift tensor in (2.9) correspond to

$$\Delta\sigma = \frac{1}{2}\left(\delta_{xx} + \delta_{yy}\right) - \delta_{zz} \tag{2.13}$$

and

$$\eta_{csa} = \frac{3}{2} \cdot \frac{\delta_{xx} - \delta_{yy}}{\Delta\sigma} \tag{2.14}$$

Solid-state NMR spectra of quadrupolar nuclei with spin $I > 1/2$ consist of the central transition (ct), $\{+1/2 \leftrightarrow -1/2\}$, and several satellite transitions, $\{\pm 1/2 \leftrightarrow \pm 3/2\}$, $\{\pm 3/2 \leftrightarrow \pm 5/2\}$, etc. (Fig. 2.2). In the case of strong quadrupolar interactions it is often impossible to excite and observe the whole spectrum, i.e., the central transition and all satellite transitions. Therefore, the second moment $M_2^{q,ct}$ in (2.10) describes the broadening of the central transition only.

Furthermore, quadrupolar interactions cause a high-field shift (to smaller chemical shift values) of the frequency of the central transition, ν_{qs}, which is called second-order quadrupolar frequency shift

$$\nu_{qs} = \nu_{cg} - \nu_0(1 - \sigma) = -\frac{I(I + 1) - 3/4}{30}\frac{\nu_q^2}{\nu_o}\left(1 + \frac{1}{3}\eta_q{}^2\right) \tag{2.15}$$

Fig. 2.2 Level scheme of nuclei with spin $I = 3/2$

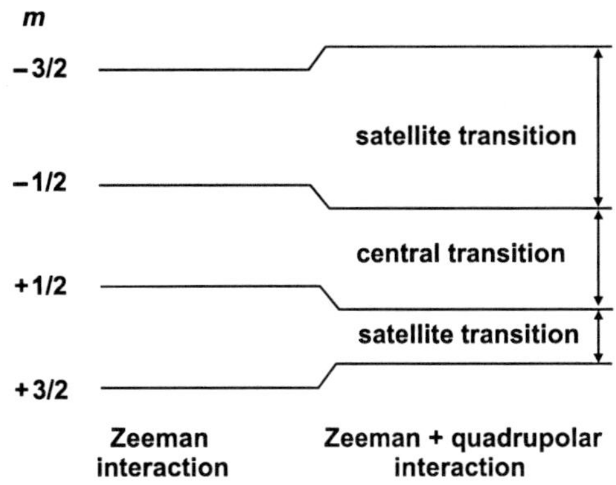

with respect to the resonance frequency, ν_{cg}, given in (2.11). Beside the quantitative evaluation of the experimental line shape function of quadrupolar nuclei, the determination of the second-order quadrupolar frequency shift, ν_{qs}, is an important method for determining the quadrupole frequency, ν_q. This quadrupole frequency, ν_q, is connected with the quadrupole coupling constant, C_q, in the following manner

$$\nu_q = \frac{3}{2I(2I-1)} C_q \tag{2.16}$$

Often, the quadrupole coupling constant

$$C_q = \frac{e^2qQ}{h} \tag{2.17}$$

is used to describe the strength of quadrupolar interaction. Here, eQ is the electric quadrupole moment of the nuclei under study and eq is the z-component, V_{zz}, of the electric field gradient. The asymmetry parameter, η_q, of the electric field gradient tensor, used in (2.15), is defined by

$$\eta_q = \frac{V_{yy} - V_{xx}}{V_{zz}} \tag{2.18}$$

The principal values V_{xx}, V_{yy}, and V_{zz} of the electric field gradient tensor in the molecular frame at the site of the resonating nucleus are denoted according to $|V_{xx}| \leq |V_{yy}| \leq |V_{zz}|$.

2.3 Solid-State NMR Techniques

2.3.1 MAS NMR Technique

The most important technique for averaging nuclear interactions in solids, i.e., for the narrowing of solid-state NMR signals, is magic angle spinning (MAS).

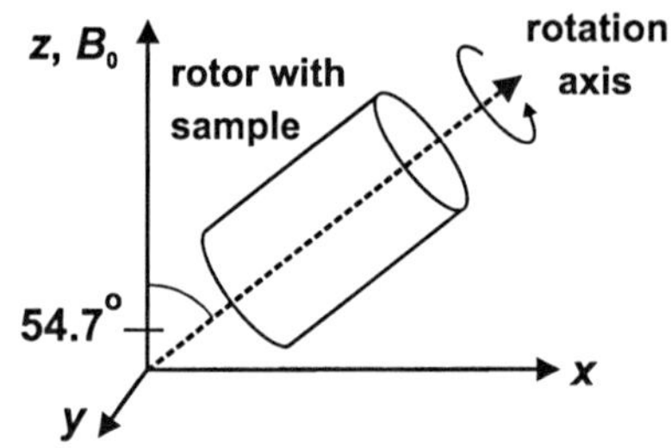

Fig. 2.3 Scheme of the MAS technique based on the rapid rotation of the sample around an axis in the magic angle of 54.7° to the external magnetic field

This experimental technique was introduced in 1958 by Andrew et al. [22]. The principle of MAS is a rapid rotation of the sample around an axis in an angle Θ to the external magnetic B_0 field (Fig. 2.3). The maximum line narrowing effect is reached for the magic angle of $\Theta_m = 54.74°$. For spinning under the angle Θ_m, the Hamiltonians of the dipolar interactions and anisotropic chemical shielding, containing the geometric term $(3\cos^2\Theta - 1)$, become zero. In the case of quadrupolar interactions, only the Hamiltonian describing the satellite transitions becomes zero, while a significant residual line broadening remains for signals due to central transitions (*vide infra*).

Averaging of Hamiltonians by MAS strongly depends on the thermal mobility of the species under study. Narrowing of solid-state NMR signals can be reached, if the correlation time of thermal mobility, τ_c, is large in comparison with the period of the macroscopic rotation, $\tau_{rot} = 1/\nu_{rot}$ [23]. This is a limiting criterion for investigating compounds by MAS NMR spectroscopy at elevated temperatures.

Nowadays, the rapid spinning of samples is performed using small turbines with gas bearing systems and rotors consisting of zirconia tubes. These MAS NMR probes reach sample spinning rates of 5–40 kHz using rotors with outer diameters of 7.0–2.5 mm. MAS is the most important technique of high-resolution solid-state NMR spectroscopy for investigations of zeolites. The first MAS NMR investigations of zeolites were performed with hydrated samples, which is a suitable approach for the study, e.g., of ^{27}Al and ^{29}Si isotopes. MAS NMR investigations of surface sites or adsorbate complexes, however, require the study of dehydrated zeolites, which makes the sample treatment more complicated. A simple approach is to dehydrate the zeolite samples in vacuum and to fill these materials into gas-tight MAS NMR rotors using a glove box purged with an inert and dry gas. Another way is to seal the dehydrated zeolite powder in small glass ampoules, which fit into MAS NMR rotors. A cheap approach is to use self-made glass ampoules. The glass tube is fixed by two chucks of a small turning-lathe, which rotates at a rate of ca. 0.5 s^{-1}. With a twofold flame burner, the glass tube is closed at the bottom. Then a waist with an inner diameter of ca. 1 mm is made at a distance of ca. 12 mm from the bottom. An alternative, but more expensive way is the application of commercial glass inserts, which are offered for all commercial MAS NMR rotor types. Approximately 50 or 200 mg of zeolite powder can be filled into a glass insert for 4 mm or 7 mm rotors, respectively. After dehydration and loading of the zeolites with probe molecules or reactants at a vacuum line, the glass insert is sealed at the

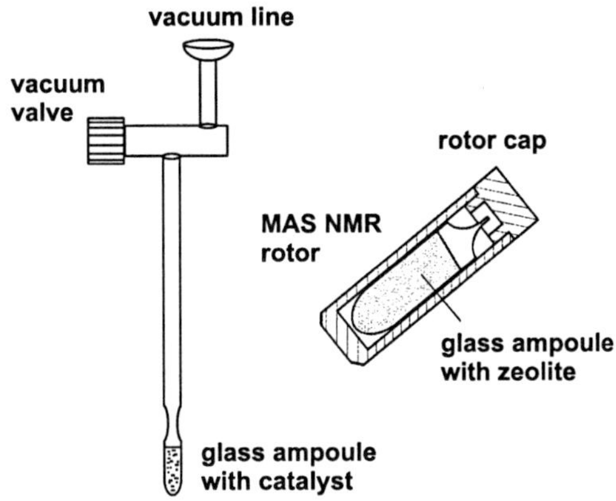

Fig. 2.4 Equipment for preparation of zeolite samples in a glass insert under vacuum and application of the sealed glass insert in an MAS NMR rotor

waist (Fig. 2.4). To prevent heating of the sample, the glass ampoule is cooled by liquid nitrogen [4]. For reviews on more sophisticated techniques for the treatment of dehydrated and activated zeolites and in situ studies of heterogeneously catalyzed reactions on by MAS NMR spectroscopy, see [7, 9–11].

If solid-state NMR signals are broadened by inhomogeneous nuclear interactions, such as heteronuclear dipolar interactions or anisotropic chemical shielding interactions [24], MAS with the sample spinning frequency, ν_{rot}, leads to NMR spectra consisting of a central line at the center of gravity frequency, ν_{cg}, and of spinning sidebands at the frequencies

$$\nu_{\mathrm{k}} = \nu_{\mathrm{cg}} + k\nu_{\mathrm{rot}} \tag{2.19}$$

Here, $k = \pm 1$, ± 2, etc. denotes the order of the spinning sidebands (Fig. 2.5).

In the case of homogeneous interactions, such as homonuclear dipolar interactions, suitable line narrowing requires a sample spinning rate, which is higher than the static line width, i.e., the full width at half maximum of the solid-state NMR signal without application of MAS ($\nu_{\mathrm{rot}} > \Delta\nu_{1/2}$) [24]. Another way is to utilize the CRAMPS (combined rotational and multiple-pulse spectroscopy) technique. In this case, the mechanical sample spinning by MAS is supported by a rapid turning of the magnetization vector via a multiple-pulse sequences leading to an additional averaging of nuclear interactions, such as strong homonuclear dipolar interactions [25, 26].

If the line broadening is caused by quadrupolar interactions, the effect of MAS is limited due to the complicated angular dependence of the corresponding Hamiltonian. Under MAS conditions, the quadrupolar broadening of the central transition is narrowed according to the second moment [27]

$$M_2{}^{\mathrm{q,MAS}} = \pi^2 \nu_{\mathrm{qs}}{}^2 \tag{2.20}$$

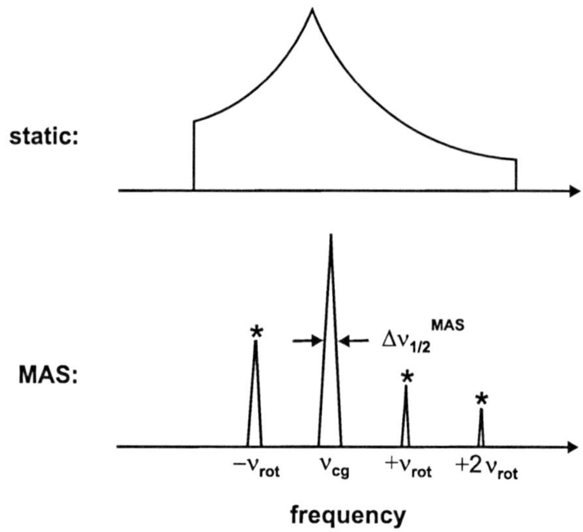

Fig. 2.5 Narrowing effect of MAS on the shape of an anisotropic solid-state NMR signal. Spinning sidebands are marked by asterisks

Comparison of (2.10) and (2.20) shows that the application of MAS for the study of the central transitions of spin $I > 1/2$ systems leads to a line narrowing by a factor $1/3.6 = 0.28$ [27]

$$\sqrt{\frac{M_2^{q,ct}}{M_2^{q,MAS}}} = \sqrt{\frac{92}{7}} \approx 3.6 \qquad (2.21)$$

In the case of strong quadrupolar interactions, therefore, MAS alone is often not sufficient to reach a line narrowing, which allows a suitable separation of NMR signals and more sophisticated solid-state NMR techniques have to be utilized.

2.3.2 DOR and MQMAS NMR Techniques

The complete suppression of the quadrupolar broadening of signals due to central transitions of spin $I > 1/2$ systems (second-order quadrupolar broadening) can be achieved by application of double-oriented rotation (DOR) [27, 28] and multiple-quantum MAS (MQMAS) NMR spectroscopy [27, 29]. In contrast to the Hamiltonians of the dipolar interaction and the anisotropic chemical shielding, the anisotropic line shape function describing the central transition of quadrupolar nuclei has two angular dependences, $(3\cos^2\Theta_1 - 1)$ and $(35\cos^4\Theta_2 - 30\cos^2\Theta_2 + 3)$ [27]. Therefore, the double-oriented rotation technique was developed to average both above-mentioned terms [28]. In the DOR device, the sample rotates simultaneously around two axes. The corresponding turbine contains an outer and an inner rotor, both with gas bearing systems. The angle Θ_1 between the external magnetic field and the rotational axis of the outer rotor corresponds to the magic angle ($\Theta_1 = 54.74°$). The angle between the rotational axes of the inner and the outer rotor is 30.56° [27, 28].

Often, the outer rotor of DOR probes has a diameter of ca. 14 mm and reaches the spinning speed of ca. 1.5 kHz, while the inner rotor has a diameter of ca. 4.5 mm and rotates with ca. 10 kHz. The "magic angle spinning" of the outer rotor averages solid-state interactions depending exclusively on the geometric term ($3\cos^2\Theta_1 - 1$) and is responsible for the occurrence of spinning sidebands. Suppression of the odd-numbered sidebands can be reached by application of rotor-synchronized pulse-sequences [30, 31]. The limiting factors of the DOR technique is the low spinning speed of the outer rotor and the small sample volume of the inner rotor in comparison with the radio frequency coil, which is arranged around the outer rotor.

To overcome the above-mentioned limitations of DOR technique, a two-dimensional experiment was introduced, which makes use of the excitation of invisible multiple-quantum transitions combined with MAS to remove the anisotropic term of the second-order quadrupolar broadening. This technique is the multiple-quantum MAS NMR spectroscopy. In typical MQMAS NMR experiments [27, 29], a pulse sequence is applied with a strong first pulse, which excites the invisible multiple-quantum coherence. The second pulse and the possibly weaker third pulse convert the coherence back to the observable single-quantum level. Going from the multiple-quantum to the single-quantum level, the sign of the phase development of the coherence is inverted. The anisotropic contribution of the second-order quadrupolar broadening is averaged out, if the times t_1 and t_2 spent on the multiple- and single-quantum level, respectively, fulfill the condition [27, 29, 32, 33]

$$t_2 = p\,\frac{36I(I+1) - 17p^2 - 10}{36I(I+1) - 27}\,t_1 = R(I,p)\,t_1 \tag{2.22}$$

Here, p is 3 and 5 for a triple- (3Q) and quintuple-quantum (5Q) coherence pathways. For 3QMAS NMR experiments of spin $I = 5/2$ systems, (2.22) is $t_2 = (19/12) \cdot t_1$ [29]. The two-dimensional MQMAS NMR spectrum is obtained after a two-dimensional Fourier transformation with respect to t_2 and t_1 leading to the F_2 and F_1 dimensions, respectively. After application of a shearing transformation, which eliminates the slope $R(I, p)$ in (2.22), the F_1 projection of the two-dimensional MQMAS NMR spectrum yields signals free of anisotropic broadening. NMR spectroscopic parameters can be obtained by simulating the line shape of slices of the F_2 projection or by evaluating the two-dimensional (2D) MQMAS NMR spectrum via the following equations for the isotropic chemical shift, δ_{cs} [29]

$$\delta_{cs} = \frac{17}{27}\delta_{F1} + \frac{10}{27}\delta_{F2} \tag{2.23}$$

and the second-order quadrupole effect (SOQE) parameter

$$\mathrm{SOQE}^2 = \frac{\delta_{\mathrm{CS}} - \delta_{\mathrm{F2}}}{k} \tag{2.24}$$

where δ_{F2} and δ_{F1} are the centers of gravity of the resonance positions along the F2 and F1 axes and

$$k = \frac{3}{10}\frac{4I(I+1)-3}{\left[4I(2I-1)\nu_0\right]^2} \cdot 10^6 \tag{2.25}$$

Finally, the SOQE parameter depends on the quadrupole coupling constant, C_{Q}, in the following manner [27, 29, 32, 33]

$$\mathrm{SOQE} = C_{\mathrm{q}}\left[1 + \left(\frac{\eta_q^2}{3}\right)\right]^{1/2} \tag{2.26}$$

For different pulse sequences suitable to excite triple- ($p = 3$) or quintuple-quantum ($p = 5$) coherences and to convert them back to the observable single-coherence, see [27, 29, 32–34].

2.3.3 *Important Pulse Sequences*

A limitation of solid-state NMR spectroscopy, especially in the case of non-abundant nuclei, such as ^{13}C and ^{15}N isotopes, is the low sensitivity of this method. An effective technique to significantly increase the sensitivity of non-abundant spins S dipolarly coupled with abundant spins I is the cross-polarization (CP) experiment [20, 35]. The principle of the cross polarization experiment is demonstrated in Fig. 2.6a. The CP experiment starts with a $\pi/2$-pulse applied in the channel of the abundant spins I. During the subsequent contact time, τ, polarization is transferred from the spins I to the dipolarly coupled spins S. The magnetic fields, $B_{1\mathrm{I}}$ and $B_{1\mathrm{S}}$, of the contact pulses, which are applied simultaneously to the spins I and S, have to fulfil the Hartmann-Hahn condition [35]

$$\alpha_{\mathrm{I}}\,\gamma_{\mathrm{I}}\,B_{1\mathrm{I}} = \alpha_{\mathrm{S}}\,\gamma_{\mathrm{S}}\,B_{1\mathrm{S}} \tag{2.27}$$

For spins $I = S = 1/2$, α_{S} and α_{I} are equal to 1. Otherwise, α_{S} is equal to $[S(S+1) - m(m-1)]^{1/2}$ if a spin $S > 1/2$ system is studied and the high-power radio frequency pulse induces transitions between the levels with the magnetic spin quantum numbers m and $m - 1$ [35]. The cross-polarization experiment leads to an enhancement of the signal intensity of the non-abundant spins S, if the abundant spins I exhibit a higher gyromagnetic ratio or a higher concentration than the spins S and the dipolar coupling between these spin systems is not averaged by thermal mobility.

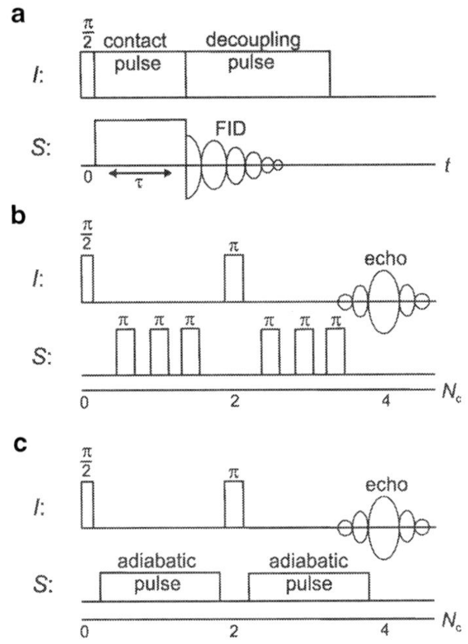

Fig. 2.6 Pulse sequences for CP (a), REDOR (b), and TRAPDOR (c) experiments

The polarization transfer during the contact pulse of the cross-polarization experiment depends on the dipolar coupling between the spins I and S, i.e., on the local structure of the compound under study. A quantitative method for investigating this dipolar coupling, i.e., for determining the atom-atom distance r_{jk}, is the REDOR (rotational echo double resonance) experiment [36, 37]. The REDOR experiment is a spin-echo double-resonance experiment in combination with MAS. A typical REDOR pulse sequence is shown in Fig. 2.6b. The rotor-synchronized echo sequence is applied to the spin I system and the echoes are detected after the time 2τ equalling the even number N_c of rotation periods. For decoupling the dipolar interaction between the spins I and S, π-pulses are applied to the spin S system at every half rotation period T_r. In the case of spins I or $S > 1/2$, the quadrupolar interactions must be weak or the experiments have to be performed in high magnetic fields (e.g., $B_0 = 17.6T$ [38]). The dipolar coupling is determined by measuring the REDOR fraction, which describes the decrease of the echo amplitude as a function of the number of rotation periods N_c [36, 37]

$$\frac{S_0 - S(N_c T_r)}{S_0} = \frac{1}{\pi^2 S(S+1)} (N_c T_r)^2 M_2{}^{\mathrm{IS}} \tag{2.28}$$

where $M_2{}^{\mathrm{IS}}$ corresponds to the second moment of the heteronuclear dipolar interaction in (2.8). This term contains the atom-atom distance r_{jk} between the dipolarly coupled spins I and S. S_0 is the echo intensity obtained without application of π-pulses to the spin *S* system. The REDOR technique was utilized, e.g., for determining the O-H distance of bridging hydroxyl groups in zeolites H-Y and H-ZSM-5 in high magnetic fields [38] and the P-B distances in BAPO-based mesoporous materials [39].

In the case of spins *S* involved in strong quadrupolar interactions, the dipolar *I*-*S* coupling can be investigated via the TRAPDOR (transfer of population in double resonance) experiment [40–42]. In the TRAPDOR experiment, again a rotor-synchronized echo sequence is applied to the spin *I* system (Fig. 2.6c). During the pulse and echo delay, the spin *S* system is continuously irradiated. Continuous irradiation of quadrupolar nuclei in combination with MAS leads to rotationally induced level transitions. These level transitions cause a dipolar dephasing, which decreases the echo intensity. Since this decrease of the echo intensity is difficult to calculate, TRAPDOR is a qualitative experiment only. On the other hand, the comparison of spectra obtained by spin-echoes recorded with and without irradiating the spin *S* system allows distinguishing spins *I* with and without dipolar coupling to spins *S*. In zeolite science, the TRAPDOR experiment is utilized for studying, e.g., heteronuclear dipolar ^{1}H–^{27}Al interactions. The ^{27}Al irradiation during the ^{1}H echo results in a dephasing of the diploar ^{1}H–^{27}Al coupling. Therefore, ^{1}H–^{27}Al spin pairs, which are coupled, do not contribute to the spin-echo of the ^{1}H nuclei. The TRAPDOR effect occurs, if the ^{27}Al irradiation frequency lies within the frequency range of $\nu_0 \pm 2\nu_q = \nu_0 \pm (3/10)\cdot C_q$ for ^{27}Al nuclei. The TRAPDOR technique was utilized, e.g., for assigning the ^{1}H MAS NMR signals of AlOH and bridging OH groups (SiOHAl) in dehydrated zeolite H-beta [43] and of silanol groups in the vicinity of framework aluminum atoms in dehydrated SBA-15 [44].

For further pulse sequences and 2-dimensional NMR spectroscopic techniques for the study of atom-atom distances, atom-atom connectivities, and electric field gradients, see [3, 5, 8, 20, 27, 33, 37, 45–50].

2.4 Investigation of the Zeolite Framework

2.4.1 ^{29}Si MAS NMR Spectroscopy

The basic units of the zeolite framework are TO_4 tetrahedra with silicon atoms at the central T-positions. Some of these T-positions may be occupied by other metal atoms, such as aluminum, boron, gallium, iron, titanium etc. Depending on the amount of metal atoms, which are incorporated in the second coordination sphere of the silicon atoms at T-positions, the SiO_4 units (Q^4) are characterized by up to five different environments denoted as Si(*n*T) with $n = 0$, 1, 2, 3, and 4. In the case of aluminum atoms incorporated at T-positions, each type of Si(*n*Al) species yields a characteristic ^{29}Si MAS NMR signal in a well defined range of chemical shifts. Fig. 2.7 gives a summary of the chemical shift ranges of these Si(*n*Al) species in zeolites [1–3, 6]. The composition of the zeolite framework influences the relative

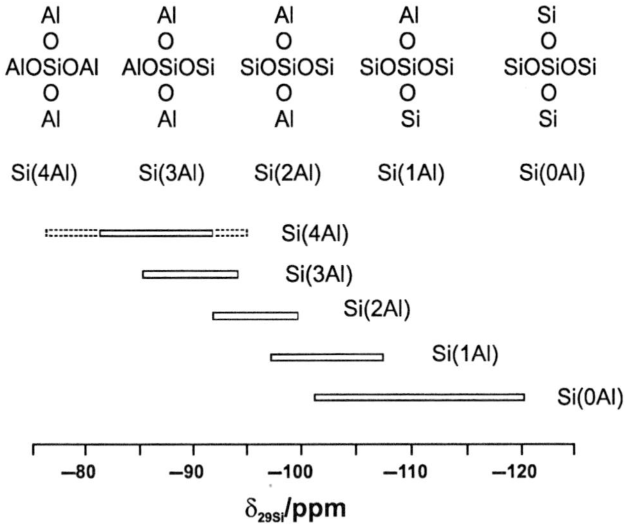

Fig. 2.7 ^{29}Si chemical shift values of Si(*n*Al) units in zeolites. The dotted lines for Si(4Al) designate the chemical shift range observed for aluminosilicate sodalites with different cage fillings

intensities of the ^{29}Si MAS NMR signals of the above-mentioned Si(*n*Al) species. Therefore, the framework n_{Si}/n_{Al} ratio of zeolites can be calculated via [1, 2]

$$n_{Si}/n_{Al} = \sum_{n=0}^{4} I_{Si(nAl)} \Big/ \sum_{n=0}^{4} 0.25 \cdot n \cdot I_{Si(nAl)} \tag{2.29}$$

Here, $I_{Si(nAl)}$ is the intensity of the Si(*n*Al) signal, i.e., of the NMR signal caused by silicon atoms with a number *n* of aluminum atoms in the second coordination sphere. Using the same method, also the chemical compositions of the framework of zincosilicates [51] and the gallium analogue of zeolite beta ([Ga]beta) were obtained [52]. The ^{29}Si MAS NMR spectra of VPI-7 and VPI-9 zeolites show signals of Si(1Zn) and Si(2Zn) units at resonance positions of −88 to −96 ppm and −78 to −81 ppm, respectively [51]. The signals of Si(1 Ga) and Si(2 Ga) units in [Ga]beta zeolites occur at −101 ppm and −95 ppm, respectively [52].

Terminal hydroxyl groups bound to silicon atoms at the outer surface or internal defect sites cause Q^3 (Si(3Si,1OH)) and Q^2 (Si(2Si,2OH)) signals depending on the number of OH groups bound to the silicon atom at T-position. It is important to note, that the ^{29}Si MAS NMR signal of Si(3Si,1OH) species ($\delta_{29Si} = -100$ to -103 ppm) occurs at similar resonance positions like Si(1Al) species ($\delta_{29Si} = -95$ to -105 ppm) [53]. The quantitative evaluation of relative intensities of Si(*n*T) signals, therefore, may require the determination of the concentration of Q^3 silicon species, which can be performed by quantitative ^{1}H MAS NMR spectroscopy of the SiOH groups in dehydrated zeolites (see Sect. 2.6).

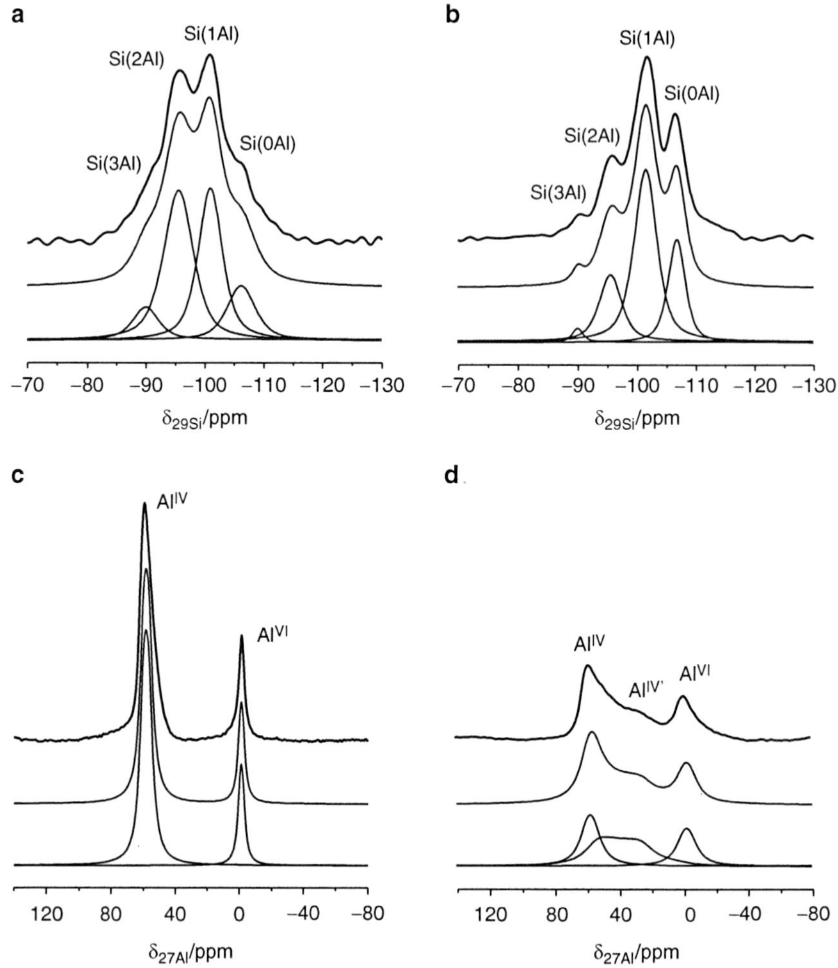

Fig. 2.8 ^{29}Si (a, b) and ^{27}Al MAS NMR spectra (c, d) of calcined and rehydrated zeolites H,Na-Y before (a, c) and after (b, d) dealumination by steaming (deH,Na-Y) [54]. Reproduced with permission from Elsevier

As an example for the application of ^{29}Si MAS NMR spectroscopy for determining the aluminum content in the framework of zeolites, Fig. 2.8., top, shows the spectra of calcined and rehydrated zeolites H,Na-Y and deH,Na-Y, the latter dealuminated by steaming [54]. In these spectra, the signals of Si(nAl) species ($n = 3, 2, 1, 0$) are assigned. Evaluating the relative intensities of Si(nAl) signals and using (2.29), framework n_{Si}/n_{Al} ratios of 2.8 and 4.0 were determined for zeolites H, Na-Y and deH, Na-Y, respectively. For these zeolites, the content of Q^3 species (Si(3Si,1OH)) to the ^{29}Si MAS NMR signal of Si(1Al) species could be neglected due to the weak ^{1}H MAS NMR signal of SiOH groups.

In contrast to X- and Y-type zeolites, a number of other zeolites are characterized by structures with crystallographically non-equivalent T-positions in chemically equivalent environments. Silicon atoms located at crystallographically non-equivalent T-positions may cause ^{29}Si MAS NMR signals of Si(nT) species with same values n at different resonance positions. For the determination of the ^{29}Si NMR shifts of silicon atoms at crystallographically non-equivalent T-positions it is useful to study all-silica materials, which consist exclusively of Si(0Al) species. Treatment, e.g., of aluminum-containing zeolites in their ammonium form with water vapor at temperatures of 873–1073 K may lead to dealumination and ultra-stabilization of the framework. The significant decrease in the ^{29}Si MAS NMR line width obtained in this way is advantageous for the investigation of the signal splitting due to silicon atoms at crystallographically non-equivalent T-positions [1].

The splitting of ^{29}Si MAS NMR signals of silicon species at crystallographically non-equivalent T-positions is explained by different bonding geometries of the corresponding SiO_4 tetrahedra. Empirical correlations and theoretical considerations indicated that the ^{29}Si NMR shifts values, $\delta_{29\mathrm{Si}}$, are linearly related to the average value of the four Si–O–T bond angles, $\overline{\alpha}$, at the silicon atom in the center of the tetrahedron. With linear regression analysis, correlations between the shift values, $\delta_{29\mathrm{Si}}$, and functions of the mean angle, $\overline{\alpha}$, were found [1, 55–58]. Studying all-silica ZSM-5 zeolites containing exclusively Si(0Al) units, the best experimentally derived correlation was obtained for $(_{29\mathrm{Si}}$ values plotted as a function of $\overline{\cos\alpha/(\cos\alpha - 1)}$ [58]. For Si(nAl) units with $n = 0$–4, theoretical investigations led to [56]

$$\delta_{29\mathrm{Si}}\Big/\mathrm{ppm} = -223.9\,\overline{\cos\alpha/(\cos\alpha - 1)} + 5\,n - 7.2 \tag{2.30}$$

Experimental studies of the ^{29}Si MAS NMR signals caused by Si(4Al) units in sodalite, cancrinite, thomsonite, and zeolites A, X, Y, and ABW gave [57]

$$\delta_{29\mathrm{Si}}/\mathrm{ppm} = -5.230 - 0.570\,\overline{\alpha} \tag{2.31}$$

In addition, the chemical shift values of the ^{29}Si MAS NMR signals of all-silica zeolites can be correlated with Si-Si distances, however, with a lower correlation coefficient in comparison with the functions of $\overline{\alpha}$ [58]. All these correlations support the structure determination of zeolites by X-ray diffraction since δ_{Si} values calculated via T–O–T bond angle and T–T distance correlations allow the experimental examination of structure data by an independent method, i.e., by ^{29}Si MAS NMR spectroscopy.

A further approach of the structure determination of zeolites by ^{29}Si MAS NMR spectroscopy is the investigation of bond connectivities between silicon atoms at crystallographically non-equivalent T-positions [59–61]. These investigations are performed, e.g., by two-dimensional COSY (correlation spectroscopy) and INADEQUATE (incredible natural abundance double quantum transfer experiment) experiments [62, 63]. The COSY experiment utilizes the J-coupling between neighboring nuclear spins. An important advantage of the INADEQUATE experiment is that it can be performed with materials containing ^{29}Si nuclei in natural abundance. Brouwer et al. [48] utilized through-space dipolar interactions between ^{29}Si nuclei in natural abundance in all-silica zeolites to provide information about bonding connectivities as well as long-range distances between silicon atoms. The combination of 1D and 2D techniques, such as cross polarization and NOESY (nuclear Overhauser exchange spectroscopy), is a suitable way for clarifying zeolite framework/adsorbate structures [50].

2.4.2 Solid-State ^{27}Al NMR Spectroscopy

According to Loewenstein's rule, the formation of Al–O–Al bondings in the framework of zeolites is forbidden and exclusively Al(4Si) units exist. Therefore, the ^{27}Al NMR spectra of tetrahedrally coordinated framework aluminum (Al^{VI}) in hydrated zeolites consist, in general, of a single signal in the chemical shift range of 55–70 ppm (referenced to 0.1 M aqueous $Al(NO_3)_3$ solution) [1]. The hydration of zeolites leads to a significant decrease of the quadrupole coupling constant of framework aluminum atoms (C_q ca. 2 MHz) accompanied by a strong decrease of the quadrupolar line broadening in comparison with solid-state ^{27}Al NMR signals of framework aluminum atoms in dehydrated samples [64]. The large quadrupole moment of ^{27}Al nuclei ($Q = 14.66 \times 10^{-30}$m^2 [65]) and the strong electric field gradient at the positions of framework aluminum atoms in dehydrated H-form zeolites cause a C_q value of up to 16 MHz [64].

Extra-framework aluminum species in hydrated zeolites obtained in result of dealumination by calcination, steaming or acid leaching are octahedrally coordinated aluminum species (Al^{VI}) causing a narrow ^{27}Al NMR signal at ca. 0 ppm. If extra-framework aluminum species exist as polymeric aluminum oxide in zeolite cages or pores, quadrupolar line broadening may occur owing to distortions of the octahedral symmetry of the AlO_6 units [66]. Broad ^{27}Al MAS NMR signals at 30–50 ppm may be caused by aluminum species in a disturbed tetrahedral coordination ($Al^{IV'}$) or by penta-coordinated species (Al^{V}) [67, 68].

For the signals of tetrahedrally coordinated framework aluminum atoms, no definite relationships between the ^{27}Al NMR shift, δ_{27Al}, and the n_{Si}/n_{Al} ratio or the silicon and aluminum order scheme of the zeolite framework exist. However, a correlation between the ^{27}Al NMR chemical shift of Al(4Si) units and the mean Al–O–Si bond angle $\overline{\alpha}$ was found [69]

$$\delta_{27Al}/\text{ppm} = 132 - 0.500\,\overline{\alpha} \tag{2.32}$$

An important prerequisite for the correlation of experimental ^{27}Al NMR shift data with structural parameters is the careful correction of the second-order quadrupolar shift according to (2.15). In result of ^{27}Al and ^{29}Si MAS NMR investigations of lithium and sodium halide aluminosilicate sodalites with high-speed MAS NMR spectroscopy, Jacobsen et al. [70] derived the following correlation between ^{27}Al and ^{29}Si NMR shift values

$$\delta_{27\mathrm{Al}}/\mathrm{ppm} = 1.03\ \delta_{\mathrm{Si}}/\mathrm{ppm} + 151.94 \tag{2.33}$$

As an example, the ^{29}Si MAS NMR signals of Si(0Al) species located at the crystallographically non-equivalent T1- and T2-positions in zeolite omega ($n_{\mathrm{Si}}/n_{\mathrm{Al}} = 4.6$) appear at -106.0 ppm and -114.4 ppm, respectively [55]. The ^{27}Al MAS NMR spectrum of this zeolite recorded at the resonance frequency of 104.3 MHz shows two signals at 61.5 ppm and 54.5 ppm due to framework Al^{IV} species at the above-mentioned T1- and T2-positions, respectively [71, 72]. The ^{29}Si MAS NMR spectra of the dealuminated zeolite MCM-22 consists of Si(0Al) signals covering a range from -104.0 ppm to -118.5 ppm [73]. The ^{27}Al MAS NMR spectra of this material recorded at the resonance frequency of 195.4 MHz show signals of framework Al^{IV} species in the chemical shift range of 49 ppm–62 ppm [73]. No second-order quadrupolar shift was observed comparing the ^{27}Al MAS NMR signals recorded at resonance frequencies of 104.2 MHz and 195.4 MHz, which indicates weak quadrupolar interactions of the Al^{IV} species in the hydrated materials under study (see (2.15)). The above-mentioned examples show a reasonable agreement with (2.33) concerning the chemical shift differences $\Delta\delta_{29\mathrm{Si}}$ and $\Delta\delta_{27\mathrm{Al}}$ of tetrahedrally coordinated framework silicon and aluminum species in hydrated zeolites omega and MCM-22, but not concerning the absolute shift values.

Generally, quantitatively reliable ^{27}Al MAS NMR spectra (spin $I = 5/2$) of hydrated zeolites containing framework as well as non-framework aluminum atoms can be obtained in high magnetic fields, with fast magic angle spinning, and using a short excitation pulse [27, 67, 74]. As an example, the ^{27}Al MAS NMR spectra of calcined and rehydrated zeolites H,Na-Y and deH,Na-Y, the latter dealuminated by steaming, are shown in Fig. 2.8c, d [54]. These spectra consist of signals at ca. 60 ppm due to tetrahedrally coordinated framework aluminum atoms (Al^{IV}) and at ca. 0 ppm caused by octahedrally coordinated aluminum species (Al^{VI}). In addition, a broad hump occurs at ca. 30 ppm in the spectrum of zeolite deH,Na-Y (Fig. 2.8d). For assigning the nature of aluminum species responsible for the broad signal at 30 ppm, ^{27}Al MQMAS NMR spectroscopy is suitable.

The ^{27}Al MQMAS NMR spectrum of zeolite deH,Na-Y in Fig. 2.9, recorded at the resonance frequency of 195.5 MHz, consists of three signals [66]. Two of them, the signals Al^{IV} and Al^{VI}, are located near the diagonal (straight line) indicating low quadrupolar interactions, which is typical for ideal (high local symmetry) tetrahedrally and octahedrally coordinated framework and extra-framework aluminum species, respectively. In contrast, the broad signal is shifted away from the diagonal hinting to a strong quadrupolar interaction. According to (2.23) and with the δ_{F1} and δ_{F2} values of 70 and 45 ppm, respectively, the chemical shift of this broad signal follows to $\delta_{\mathrm{cs}} = 61$ ppm, which is typical for tetrahedrally coordinated aluminum species. Via (2.24) and (2.25), the second-order quadrupole effect parameter, SOQE, can be calculated to 10.0 MHz. The quadrupole coupling constant, C_{q}, may be 15% larger in comparison with the SOQE parameter (see (2.26)) since the

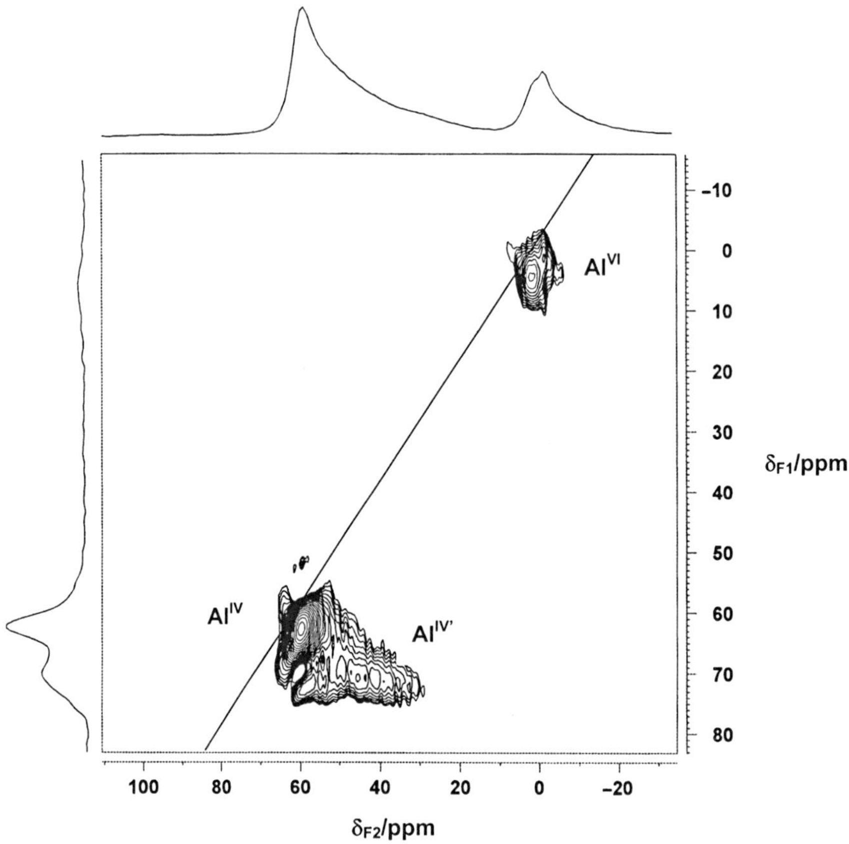

Fig. 2.9 ^{27}Al MQMAS NMR spectrum of dealuminated zeolite deH,Na-Y [66]. Reproduced with permission from Elsevier

asymmetry parameter, η_q, covers the range of 0–1. The SOQE parameter of ca. 10 MHz for tetrahedrally coordinated aluminum species in a hydrated zeolite hints to a strong strain of the local symmetry, which indicates that the corresponding $Al^{IV'}$ species are located, e.g., in the vicinity of framework defects. Quantitative evaluation of the Al^{IV} and $Al^{IV'}$ signal intensities in the one-dimensional ^{27}Al MAS NMR spectrum in Fig. 2.8d and their comparison with an external intensity standard leads to the framework n_{Si}/n_{Al} ratio of 3.8 [66].

2.5 Investigation of Extra-Framework Cations

The incorporation of aluminum atoms on T-positions in the framework of zeolites is accompanied by the occurrence of negative charges, which are often compensated by extra-framework alkali metal cations. Among the most important extra-framework species in zeolites are lithium, sodium, and cesium cations. Lithium-exchanged X- and Y-type zeolites are widely used for the separation of nitrogen and oxygen [75, 76]. On the other hand, the exchange of sodium by cesium cations leads to zeolite catalysts with basic properties [77, 78]. ^{6}Li, ^{7}Li, ^{23}Na, and ^{133}Cs MAS NMR spectroscopy are suitable methods for investigating the distribution and coordination of lithium, sodium, and cesium cations in zeolites. Although all these nuclei have a spin $I > 1/2$ (^{6}Li: spin $I = 1$; ^{7}Li: spin $I = 3/2$; ^{23}Na: spin $I = 3/2$; ^{133}Cs: spin $I = 7/2$), only the quadrupolar interaction of ^{23}Na nuclei causes a dominating line broadening effect. NMR spectroscopic parameters, which are sensitive to the cation coordination and interaction with adsorbate molecules, are the isotropic chemical shift (^{6}Li, ^{7}Li, ^{133}Cs, ^{23}Na) and the quadrupole coupling constant (^{23}Na).

2.5.1 ^{6}Li, ^{7}Li, and ^{133}Cs MAS NMR Spectroscopy

In the most cases of the investigation of lithium-exchanged zeolites by solid-state NMR spectroscopy, ^{7}Li isotopes are studied, since these nuclei have a significantly higher natural abundance in comparison with ^{6}Li nuclei (^{6}Li: 7.42%, ^{7}Li: 92.58%). Because of the weak quadrupole moment of ^{7}Li nuclei ($Q = -4.01 \times 10^{-30}\,\mathrm{m}^2$ [65]), the ^{7}Li MAS NMR spectra of lithium cations in hydrated as well as dehydrated zeolites consist of isotropic lines with a line width in the order of $\Delta\nu_{1/2} = 0.5$ ppm. In the ^{7}Li MAS NMR spectra of dehydrated zeolites Li-X and Li-A, the signals of lithium cations on SI$'$, SII, and SIII positions occurring at chemical shifts of $\delta_{7\mathrm{Li}} = -0.2$ to 0.5 ppm, -0.7 to -0.3 ppm, and -1.2 to -0.7 ppm, respectively, can be distinguished [79–84]. Therefore, ^{7}Li MAS NMR spectroscopy is suitable to determine the populations of lithium cations at crystallographically non-equivalent extra-framework positions as a function of the degree of lithium exchange [84] and as a function of the aluminum content in the zeolite framework. By variable-temperature investigations of calcined zeolite Li,Na-X it has been shown that the signal of lithium cations located on SIII positions in the supercages is influenced by the thermal mobility of these extra-framework species [82].

The chemical shifts of ^{6}Li and ^{7}Li MAS NMR signals due to lithium cations in dehydrated zeolites are sensitive to interactions of these cations with adsorbate molecules. Adsorption of nitrogen on dehydrated zeolites Li-X and Li-LSX causes a weak high-field shift ($\Delta\delta_{7\mathrm{Li}} = 0.2$–$0.3$ ppm) of the signal of lithium cations located on SIII positions in the supercages [79, 80]. There is no effect of the adsorbate molecules on lithium cations located on SI$'$ and SII positions in front of six-ring windows in the small sodalite cages and the supercages of the faujasite structure [79, 80]. Adsorption of oxygen, on the other hand, results in a strong low-field shift, $\Delta\delta_{7\mathrm{Li}}$, of the signal of lithium cations located on SIII positions by up to 7 ppm. This strong resonance shift is explained by the paramagnetic properties of oxygen atoms interacting with lithium cations. Similar results were obtained for oxygen adsorption on zeolites Li-ZSM-5 and Li-beta [85]. Accardi et al. [85] utilized the effect of oxygen adsorption on the ^{7}Li MAS NMR spectra to determine the number of accessible lithium cations. Barich et al. [86] investigated the adsorption of benzene on zeolite Li-ZSM-5 leading to a resonance shift ($\Delta\delta_{7Li} = -3.3$ ppm) of the ^{7}Li MAS NMR signal due to accessible lithium cations. By application of quantum-chemical methods, the authors showed that this shielding effect is accompanied by a chemical shift anisotropy of $\Delta\sigma = -20.4$ ppm, caused by an one-side docking of benzene molecules at the lithium cations and the influence of the aromatic π-electron system on these cations [86].

Similar to ^{6}Li and ^{7}Li nuclei, also ^{133}Cs nuclei have a weak quadrupole moment ($Q = -0.343 \times 10^{-30}$m^2 [65]). Therefore, the influence of quadrupolar interactions can be neglected in most ^{133}Cs MAS NMR studies of cesium cations in hydrated and dehydrated zeolites [87]. Due to the large electron shell of cesium atoms, strong shielding effects occur accompanied by strong spinning sidebands patterns. ^{133}Cs MAS NMR signals of cesium cations located on the different extra-framework positions of Cs,Na-X and Cs,Na-Y zeolites cover a shift range of -155 to -69 ppm in the sequence of cation positions SI$'$, SII, SII, and SII [88, 89]. Quantitative evaluation of the ^{133}Cs MAS NMR signal intensities allows the determination of cation populations at the crystallographically non-equivalent extra-framework positions [88, 89]. Impregnation of cesium-exchanged zeolites Y and X with cesium hydroxide or cesium acetate and subsequent calcination lead to the formation of cesium oxide guest compounds causing broad ^{133}Cs MAS NMR signals at $\delta_{133Cs} = -30$ ppm (CsOH/Cs,Na-Y) to 24 ppm (CsOH/Cs,Na-X) [89–91].

2.5.2 Solid-State ^{23}Na NMR Spectroscopy

Due to the large quadrupole moment of ^{23}Na nuclei ($Q = 10.4 \times 10^{-30}$m^2 [65]), the MAS NMR spectra of sodium cations in dehydrated zeolites are dominated by broad quadrupolar patterns. The overlap of the ^{23}Na MAS NMR signals of sodium cations located on crystallographically non-equivalent extra-framework positions makes the separation and assignment of the corresponding signals difficult and requires field-dependent experiments [92, 93] or application of the DOR [92, 93] and MQMAS [94–100] techniques. Utilizing the above-mentioned experimental techniques, the quadrupole coupling constant, C_q, the asymmetry parameter, η_q, and the isotropic chemical shift, δ_{23Na}, can be obtained. For the determination of the populations of sodium cations at crystallographically non-equivalent extra-framework positions, the ^{23}Na MAS NMR spectra have to be simulated and the intensity correction according to Massiot et al. [101] has to be performed.

Table 2.1 gives a survey on the NMR spectroscopic parameters of sodium cations in dehydrated zeolites. Depending on the oxygen coordination of the sodium cations in dehydrated zeolites X and Y, the quadrupole coupling constant, C_q, covers a range from ca. 1 MHz for sodium cations on SI positions in the hexagonal prisms (nearly ideal hexagonal oxygen coordination) to ca. 5 MHz for

Table 2.1 ^{23}Na quadrupole coupling constants, C_q, asymmetry parameters, η_q, and chemical shifts, δ_{23Na}, (referenced to solid NaCl) of sodium cations located on crystallographically non-equivalent extra-framework positions in dehydrated zeolites. The resonance positions are referenced to solid NaCl ($\delta_{23Na} = 0$ ppm)

Compound	Site	Cq (MHz)	η_q	δ_{23Na} (ppm)	References
Na-X and Na-LSX	SI	1.1–1.4	0	−10	[80, 100]
	SI′(1)	5.0	0	−27 to −20	[100]
	SI′(2)	3.6	0	−28	[100]
	SII	4.5–5.0	0.1	−18 to −17	[80, 100]
	SIII′(1,2)	2.2–3.0	0.5	−18 to −11	[80, 100]
	SIII′(3)	1.2–1.9	0.9	−31 to −28	[80, 100]
Na-Y	SI	1.2	0	−8.5 to −1	[95, 96, 100]
	SI′	4.8	0	−20 to −13	[95, 96, 100]
	SII	3.9	0	−15 to −3	[95, 96, 100]
Na-EMT	SI	1.0	0	1	[92]
	SI′ + SII	4.1	0	0	[92]
Na-MOR	12-ring	2.0	0	−6.8	[94]
	sidepockets	3.1	0	−16.8	[94]
Na-ZSM-5	10-ring	2.0	0	−10.8	[94]
Na-ETS-10	SI + SII + SIV	0.9 to 1.4*	–	−16 to −14	[97, 98]
	SIII + SV	1.65 to 2.1*	–	−11.2 to −9.5	[97, 98]

sodium cations on SI′ and SII positions in front of the six-ring windows inside the sodalite cages and in the supercages (threefold oxygen coordination), respectively. Sodium cations in dehydrated Na-ZSM-5, Na-MOR, and Na-ETS-10 zeolites are characterized by quadrupole coupling constants of $C_q \leq 3$ MHz.

In the same manner like ^{6}Li, ^{7}Li, and ^{133}Cs MAS NMR spectroscopy, also solid-state ^{23}Na NMR spectroscopy can be utilized for studying the interaction of adsorbate molecules with sodium cations in dehydrated zeolites [100, 102]. Adsorption of oxygen on dehydrated zeolite Li,Na-X causes a paramagnetic low-field shift of the ^{23}Na NMR signal of sodium cations located on SIII positions by $\Delta\delta_{23Na} = 2$ ppm [102]. This resonance shift is significantly weaker than the resonance shift observed for the ^{7}Li MAS NMR signal of lithium cations ($\Delta\delta_{7Li} = 7$ ppm [79, 80]) in the same zeolite, which indicates a preferred adsorption of oxygen at the smaller lithium cations [102].

2.6 Investigation of Surface OH Groups

2.6.1 Assignment of ^{1}H MAS NMR Signals

The behavior of zeolites in heterogeneously catalyzed reactions strongly depends on the nature, chemical properties, and accessibility of their surface sites, such as hydroxyl groups acting as Brønsted aicd sites. Solid-state NMR spectroscopy is a suitable method for characterizing Brønsted acid sites of solids in a direct manner or upon adsorption of probe molecules. For this purpose, zeolite samples dehydrated at elevated temperatures in vacuum are filled in gas-tight MAS NMR rotors or sealed in glass inserts (see Sect. 2.3.1). According to Table 2.2, the ^{1}H MAS NMR signals of hydroxyl groups in dehydrated solid catalysts cover a range of chemical shifts of $\delta_{1H} = 0$–16 ppm [4, 103, 104]. The lowest chemical shifts of ca. 0 ppm have been observed for metal or cation OH groups, e.g., AlOH, LaOH, and MgOH groups, in the large cages of dehydrated zeolites or at the outer surface of solid particles [105, 106]. While the ^{1}H MAS NMR signals of isolated SiOH groups occur at 1.3–2.2 ppm, silanol groups involved in strong hydrogen bonding cause signals at chemical shifts of up to 16 ppm [120, 121].

The ^{1}H MAS NMR spectra of dealuminated zeolites contain signals at $\delta_{1H} =$ 2.4–3.6 ppm due to hydroxyl protons bound to extra-framework aluminum species (AlOH) [103, 109]. This assignment was supported by $^{1}H/^{27}Al$ TRAPDOR NMR experiments [114, 116, 122, 123], which are suitable for the detection of the dipolar coupling of ^{1}H and ^{27}Al nuclei (see Sect. 2.3.3.). The resonance position of AlOH groups at 2.4–3.6 ppm is accompanied by their location in narrow channels or small cages.

Brønsted acid sites in dehydrated zeolites are caused by bridging OH groups (SiOHAl) formed at Si–O–Al bridges in the local structure of negatively charged framework aluminum atoms. Adsorption of probe molecules [4, 9, 12, 104, 124] and application of the $^{1}H/^{27}Al$ TRAPDOR NMR technique [114, 116, 123] evidenced that signals occurring at $\delta_{1H} = 3.6$–4.3 ppm are due to unperturbed bridging OH groups in dehydrated H-form zeolites. In the ^{1}H MAS NMR spectra of dehydrated zeolite H,Na-Y, the signal of bridging OH groups in the supercages occurs at $\delta_{1H} = 3.6$–4.0 ppm [4, 104]. An additional signal at $\delta_{1H} = 4.8$–5.2 ppm indicates the presence of a second type of bridging OH groups in the small sodalite cages [4, 104]. The larger chemical shift of the bridging OH groups in the sodalite cages is due to an interaction of these hydroxyl protons with neighboring framework oxygen atoms.

The detailed ^{1}H NMR shift values of unperturbed SiOHAl groups in large structural units, such as supercages of zeolites X and Y or 10- and 12-ring pores of H-ZSM-5 and H-mordenite, depend on the framework n_{Si}/n_{Al} ratio [4, 104]. The framework aluminum content affects the mean electronegativity, S^{m}, of the zeolite framework, which is calculated by the geometric average of the electronegativities S_{i} of all framework atoms ‘i’ per unit cell [127, 128]. Generally, an increasing chemical shift, δ_{1H}, is observed with decreasing framework aluminum content or increasing mean electronegativity, S^{m}, of the zeolite framework [4, 104].

In Fig. 2.10, typical ^{1}H MAS NMR spectra of dehydrated zeolites are shown [107]. The ^{1}H MAS NMR spectrum in Fig. 2.10a is that of an H-form ferrierite with $n_{Si}/n_{Al} = 9.1$ [107]. It consists of two signals at 1.8 and 4.2 ppm due to unperturbed silanol groups (SiOH) and bridging OH groups (SiOHAl), respectively, the latter acting as Brønsted acid sites. After dealumination of this zeolite by an HCl treatment [107], an additional signal appears at 2.8 ppm (Fig. 2.10b), which is caused by hydroxyl groups bound to extra-framework aluminum species (AlOH$'$) [124]. The signal at ca. 6 ppm indicates the occurrence of SiOH groups interacting with neighboring oxygen atoms.

Table 2.2 Chemical shifts, δ_{1H}, of ^{1}H MAS NMR signals caused by hydroxyl groups in dehydrated zeolites and their assignments. The resonance positions are referenced to tetramethylsilane ($\delta_{1H} = 0$ ppm)

δ_{1H} (ppm)	OH group	Assignment	References
−0.5 . . . 0.5	Metal OH	Metal or cation OH groups (e.g., AlOH, MgOH) in the large cages of zeolites or at the outer surface of solid particles	[16, 106]
1.3 . . . 2.2	SiOH	Silanol groups at the external surface of solid particles or at framework defects in zeolites	[18–114]
2.4 . . . 3.6	Metal OH′ AlOH′	Metal OH groups involved in hydrogen bonding (‘) such as extra-framework aluminum species in zeolites	[14, 110–115]
2.8 . . . 6.2	Cation OH′ CaOH′ LaOH′	Cation OH groups involved in hydrogen bonding (‘), such as OH groups located in the sodalite cages of zeolite Y	[10–109]
3.6 . . . 4.3	SiOHAl, SiO1HAl*	Bridging OH groups in large structural units such as in the supercages of zeolite Y	[14, 107–115]
4.6 . . . 5.2	SiOHAl SiO3HAl*	Bridging OH groups in small structural units such as in the sodalite cages of zeolite Y	[14, 108, 109, 111, 114, 115]
5.2 . . . 8.0	SiOHAl′	Disturbed bridging OH groups in zeolites H-ZSM-5 and H-Beta involved in hydrogen bonding (‘) with neighboring framework oxygen atoms	[112, 113, 116–119]
10 to 16	SiOH′	Internal SiOH groups involved in strong hydrogen bonding	[120, 121]

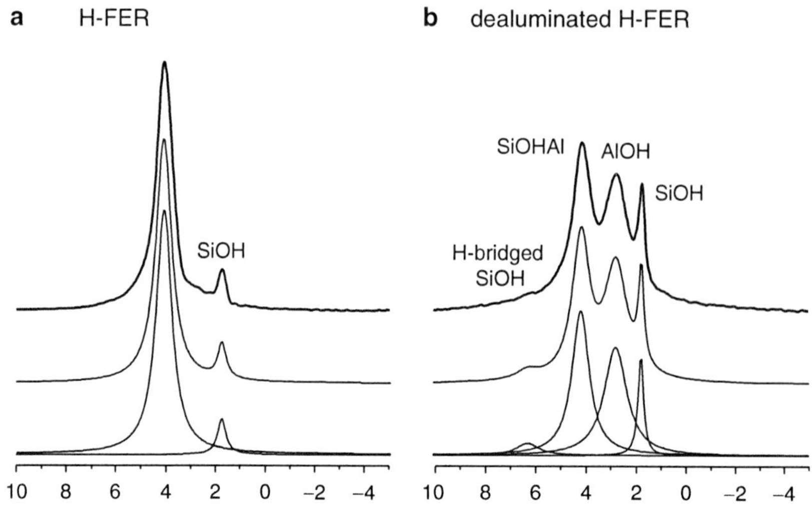

Fig. 2.10 ^{1}H MAS NMR spectra of dehydrated H-form ferrierites (H-FER) recorded before (**a**) and after (**b**) dealumination by treatment with HCl [107]. Reproduced with permission from Elsevier

^{1}H MAS NMR intensities can be utilized to calculate the concentration of hydroxyl groups in dehydrated zeolites by comparison with the intensity of an external standard, such as a well characterized H-form zeolite. For quantitative studies, the repetition time of the pulse experiments has to be large (at least factor 3) in comparison with the spin-lattice relaxation times T_1 of the different OH species, which are of the order of 1–10 s [125]. In the case of the above-mentioned H-form ferrierite, the concentration of SiOHAl groups in the parent material (Fig. 2.10a) was determined to 1.4 mmol H^+/g, while the acid leaching led to a strong decrease of this value to 0.4 mmol H^+/g (Fig. 2.10b) in result of framework dealumination [107].

2.6.2 *Hydroxyl Protons Affected by Hydrogen Bonding*

As mentioned in Sect. 2.6.1., the ^{1}H MAS NMR signals hydroxyl groups located in small structural units of zeolites, such as sodalite cages of zeolites X and Y, occur at higher chemical shift values in comparison with the same type of OH groups in large structural units, such as supercages. This finding is a hint for the presence of hydrogen bonding or electrostatic interactions of the hydroxyl protons under study with neighboring framework oxygen atoms. In Table 2.2, the corresponding hydroxyl groups are marked by an apostrophe. Comparison of structural data and ^{1}H NMR shift values of hydroxyl protons in –O–H $\cdots$ O– arrangements led to the following empirical formula describing the dependence of the low-field resonance shift, $\Delta\delta_{1H}$, of hydroxyl protons on the oxygen-oxygen distance $d_{OH\text{-}O}$ given in 'picometer' [129]

$$\Delta\delta_{1H}/\text{ppm} = 79.05 - 0.255\, d_{OH-H}/\text{pm} \tag{2.34}$$

The ^{1}H MAS NMR spectra of all-silica as-synthesized zeolites NON, DDR, MTW, AFI, and MFI obtained with organic quaternary ammonium cations, e.g., contain a signal at 10.2 ppm [120]. The resonance shift of this signal was explained by $\equiv$SiOH $\cdots$ OSi$\equiv$ hydrogen bonds between hydroxyl protons of defect silanol groups with neighboring framework oxygen atoms in an O $\cdots$ O distance of 0.27 nm. Similarly, the signal at 15.9 ppm in the ^{1}H MAS NMR spectra of RUB-18 can be explained via (2.34) by strong hydrogen bonding of silanol groups with an O $\cdots$ O distance of 0.25 nm [121]. Like for acidic zeolites X and Y, low-field shifted signals of bridging OH groups were observed in the ^{1}H MAS NMR spectra of dehydrated zeolites H-ZSM-5, H-beta, and H-MCM-22 [117, 119, 125, 126]. Also in these cases, the presence of hydrogen bonding of the hydroxyl protons of bridging OH groups with neighboring framework oxygen atoms is assumed (SiOH′Al).

2.7 Investigation of Surface Sites by Probe Molecules

2.7.1 Characterization of Brønsted Acid Sites

A suitable method for characterizing the accessibility and strength of Brønsted acid sites is the adsorption of probe molecules with different sizes and base strengths. A survey on the most important probe molecules for solid-state NMR studies of Brønsted acids sites is given in Table 2.3. The ability to protonate strongly basic probe molecules, such as pyridine or trimethylphosphine, or to form hydrogen bonds to these molecules is utilized to distinguish between acidic and non-acidic surface OH groups. A more quantitative comparison of the acid strength of acid hydroxyl groups is reached by weakly basic probe molecules, such as acetone and acetonitrile, which interact via hydrogen bonding. The adsorbate-induced low-field shifts, $\Delta\delta$, of the NMR signals caused by the interacting hydroxyl protons (^{1}H MAS NMR) or due to the functional groups of the probe molecules (^{13}C, ^{15}N or ^{31}P MAS NMR) depend on the strength and accessibility of the corresponding surface OH groups. In some cases, rapid exchange between probe molecules in the gas phase and hydrogen-bonded probe molecules may require measurements at low temperatures.

Some of the probe molecules listed in Table 2.3 are suitable for studying the location of OH groups at the outer surface of the zeolite particles or inside the pores and cages. Perfluorotributylamine (diameter $\approx$ 0.94 nm [112]), which is too large to enter the micropores of zeolites (H-ZSM-5: 0.53 $\times$ 0.55 nm; H-Y: 0.74 nm [153]), is a suitable probe molecule for studying hydroxyl groups at the outer surface of

Table 2.3 Probe molecules suitable for studying Brønsted acid sites in zeolites. ^{1}H and ^{13}C NMR shifts are referenced to tetramethylsilane ($\delta_{1H} = 0$ ppm, $\delta_{13C} = 0$ ppm), while the ^{31}P NMR signals are related to 85% H_3PO_4 at $\delta_{31P} = 0$ ppm

Probe molecule	Resonance/bservation	References
Pyridine-d_5	^{1}H: hydrogen-bonded pyridine at ca. $\delta_{1H} = 10$ ppm (SiOH) and pyridinium ions at $\delta_{1H} = 12$ to 20 ppm (SiOHAl)	[4, 104, 108, 109, 130, 131]
Acetonitrile-d_3	^{1}H: adsorbate-induced low-field shift by $\Delta\delta_{1H} = 3.6$ (H,Na-X) to 7.9 ppm (H-ZSM-5)	[108, 109, 132–134]
Perfluorotributyl amine	^{1}H: adsorbate-induced low-field shift of accessible OH groups by $\delta_{1H} = 0.25$ (SiOH) to 0.47 ppm (AlOH)	[112, 113]
Deuterated alkanes and aromatics	^{1}H: activation energy of the H/D exchange	[134–138]
^{13}C-2-acetone	^{13}C: hydrogen-bonded acetone at $\delta_{13C} = 216.8$ (H-SAPO-5) to 225.4 ppm (H-ZSM-22)	[139–142]
Trimethylphosphine (TMP)	^{31}P: $\delta_{13C} = -2$ to -3 ppm for TMP protonated by strong acid sites	[143–148]
Trimethylphosphine oxide (TMPO)	^{31}P: hydrogen-bonded TMPO at $\delta_{31P} = 53$ (H-Y) to 63 ppm (USY)	[149–151]
Triphenylphosphine (PPh_3)	^{31}P: $\delta_{31P} = 11.1$ to 14.8 ppm for PPh_3 at accessible acid sites	[152]

zeolite particles. Upon adsorption of perfluorotributylamine at SiOH and AlOH groups, low-field resonance shifts of the corresponding ^{1}H MAS NMR signals by $\Delta\delta_{1H} = 0.25$ ppm and $\Delta\delta_{1H} = 0.47$ ppm, respectively, indicate a location of these hydroxyl groups at the outer surface or in secondary mesopores [112, 113]. Often, no low-field resonance shift was found for the signals of bridging OH groups, which indicates their location inside the pores [112, 113]. Another approach is to utilize triphenylphosphine for investigating the location of bridging OH groups acting as Brønsted acid sites [152]. In this way, it was found that 6% of the Brønsted acid sites of zeolite H-MCM-22 are located at the outer particle surface.

In a number of studies, trimethylphosphine (TMP) was used as probe molecule for indentifying strong Brønsted acid sites [143–148]. The ^{31}P MAS NMR spectrum of TMP adsorbed on zeolite H,Na-Y, e.g., is dominated by a signal at -2.5 ppm (referenced to 85% H_3PO_4). This signal is due to $(CH_3)_3PH^+$ complexes arising from chemisorption of TMP at Brønsted acid sites [144]. Upon strong calcination, additional ^{31}P MAS NMR signals in the region from about -32 ppm to -67 ppm indicate the coordination of TMP at Lewis acid sites (vide infra) [144]. When TMP is used as probe molecule, it must be considered that this probe molecule is rather bulky (kinetic diameter of 0.55 nm [144]). Therefore, the characterization of surface sites in zeolites by TMP is restricted by the maximum adsorption capacity of the pore system as well as the pore diameter of the zeolite under study [145]. Similar limitations exist for the application of TMPO as probe molecule [149–151].

Upon adsorption pyridine molecules at strongly acidic bridging OH groups, a proton transfer occurs, which leads to the formation of pyridinium ions ($PyrH^+$).

Calculations of the ^{1}H NMR shifts of $PyrH^+$ protonated by bridging OH groups in 8 T zeolite clusters led to a linear correlation between the δ_{1H} values and the proton affinity of the corresponding acid site [154]. In the case of total proton transfer to the probe molecule (protonation), the smaller δ_{1H} values were obtained for the smaller proton affinities of the SiOHAl groups corresponding to the higher acid strength and *vice versa*. Experimental studies of the interaction of perdeuterated pyridine with SiOHAl groups yielded signals of $PyrH^+$ at ^{1}H NMR shifts of 16.5 ppm for zeolite H,Na-Y and at 15.5–19.0 ppm for zeolite H-ZSM-5 [4, 104, 108, 109, 130, 131].

A more frequently applied probe molecule for characterizing the strength of Brønsted acid sites in zeolites is ^{13}C-2-acetone [139–142]. Using this molecular probe, Biaglow et al. [139] investigated the acid strength of bridging OH groups in acidic zeolites by observing the ^{13}C NMR shifts of the carbonyl atom in ^{13}C-2-acetone and obtained the values of $\delta_{13C} = 216.8$ (H-SAPO-5), 219.6 (H-Y), 221.8 (H-MOR), 222.8 (H-[Ga]ZSM-5), 223.4 (H-ZSM-12), 223.6 (H-ZSM-5), and 225.4 ppm (H-ZSM-22). These chemical shift values have to be compared with the shift values of 205 and 245 ppm for carbonyl atoms of ^{13}C-2-acetone dissolved in $CDCl_3$ and in 100% sulphuric acid solution, respectively [138]. Furthermore, a scale of the Brønsted acid strength was introduced, which is based on the experimentally determined dependence of the resonance positions of carbonyl atoms in ^{13}C-2-acetone dissolved in aqueous sulphuric acid of varying concentration [155]. According to this scale, the acid strength of bridging OH groups in zeolite H-ZSM-5 ($\delta_{13C} = 223.6$ ppm [139]) corresponds to 80% H_2SO_4 in water [155]. A limitation of the application of acetone as probe molecule is its reactivity in the case of an adsorption at strongly acidic sites leading to formation of mesityl oxide.

More recently, the weak base acetonitrile found a broad application as probe molecule for the characterization of the strength of Brønsted acid sites since it is very sensitive and can be studied by ^{1}H MAS NMR spectroscopy [108, 109, 132–134]. In this way, e.g., the influence of the cation exchange and the framework n_{Si}/n_{Al} ratio of zeolites on the Brønsted acid strength was investigated via the adsorbate-induced low-field shift, $\Delta\delta_{1H}$, of the signal of acidic hydroxyl protons. For an interaction of the probe molecule via hydrogen bonding, the larger adsorbate-induced low-field shift, $\Delta\delta_{1H}$, of the ^{1}H MAS NMR signal of zeolitic hydroxyl groups corresponds to the higher acid strength and *vice versa*. As an example, Fig. 2.11 shows the ^{1}H MAS NMR spectra of lanthanum-exchanged zeolites X ($n_{Si}/n_{Al} = 1.3$) and Y ($n_{Si}/n_{Al} = 2.7$) recorded before (top) and after (bottom) adsorption of deuterated acetonitrile [108]. The signals at 1.8 and 3.6 to 3.9 ppm are due to SiOH and bridging OH groups (SiOHAl), respectively. The above-mentioned SiOHAl groups are located in the supercages, while bridging OH groups in sodalite cages cause weak shoulders at 4.6–4.8 ppm. Hydroxyl groups at lanthanum cations ($La(OH)_n$) and oxide complexes located in the sodalite cages are responsible for the signals at 5.6–6.3 ppm. The signal at 2.5 ppm occurring in the spectrum of zeolite La,Na-X/75 indicates the presence of $La(OH)_n$ groups in the supercages.

Upon adsorption of deuterated acetonitrile (Fig. 2.11, bottom), the ^{1}H MAS NMR signals of accessible bridging OH groups in the supercages of zeolites La,

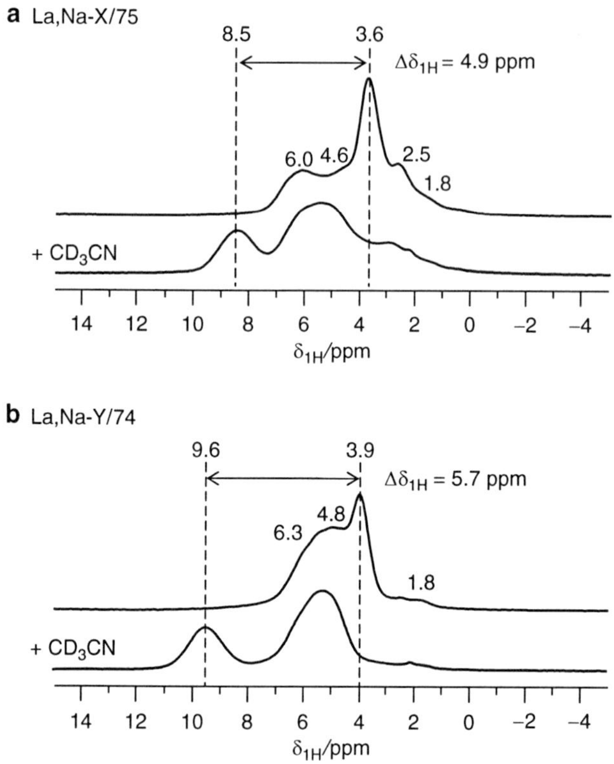

Fig. 2.11 ^{1}H MAS NMR spectra of zeolites La,Na-X/75 (**a**) and La,Na-Y/74 (**b**) recorded before (*top*) and after (*bottom*) adsorption of deuterated acetonitrile [108]. Reproduced with permission from Elsevier

Na-X/75 and La,Na-Y/74 shift by $\Delta\delta_{1H} = 4.9$ ppm and $\Delta\delta_{1H} = 5.7$ ppm, respectively, to lower magnetic field [108]. These adsorbate-induced resonance shifts have to be compared with the low-field shifts, $\Delta\delta_{1H}$, obtained for acetonitrile-loaded zeolites H,Na-X/62 (3.6 ppm [109]), Al,Na-X/32 (3.8 ppm [109]), La,Na-X/42 (3.8 ppm [109]), Al,Na-X/61 (4.4 ppm [109]), H,Na-Y/91 (5.1 ppm [134]), Al,Na-Y/34 (5.3 ppm [109]), Al,Na-Y/64 (5.3 ppm [109]), H-MOR with $n_{Si}/n_{Al} = 10$ (6.7 ppm [132]), dealuminated H,Na-Y with $n_{Si}/n_{Al} = 18$ (7.0 ppm [132]), and H-ZSM-5 with $n_{Si}/n_{Al} = 26$ (7.9 ppm [134]). The above-mentioned adsorbate-induced resonance shifts, $\Delta\delta_{1H}$, indicate that the cation type, exchange degree, and framework n_{Si}/n_{Al} ratio are important parameters for adjusting the acid strength of zeolites.

Finally, the H/D exchange rate between surface hydroxyl groups and probe molecules or reactants is studied for determining the strength of Brønsted acid sites under reaction conditions [134–138]. Deuterated benzene, e.g., was loaded on dehydrated H-form zeolites and the growth of the ^{1}H MAS NMR signal of the aromatic ring protons introduced via the H/D exchange with surface hydroxyl protons was quantitatively evaluated as a function of time. The H/D exchange rate has to be determined for at least three different temperatures. The Arrhenius plot of the corresponding H/D exchange rates gives the activation energy, E_A, of this reaction. Upon loading of benzene-d$_6$ on parent H,Na-Y, dealuminated H,Na-Y, and H-ZSM-5 zeolites, activation energies of the H/D exchange of $E_A = 107.0$, 84.9, and 60.2 kJ mol^{-1} [136], respectively, were obtained. These activation energies correlate well with the low-field shifts, $\Delta\delta_{1H}$, of the ^{1}H MAS NMR signals of bridging OH groups in the above-mentioned zeolites upon adsorption of acetonitrile (*vide supra*) [109, 132, 134].

2.7.2 Characterization of Lewis Acid Sites and Base Sites

Lewis acid sites and base sites in zeolites are difficult to investigate in a direct manner, such as by solid-state ^{27}Al or ^{17}O NMR spectroscopy. A number of recent works demonstrated that such studies are possible in general [27, 65, 74, 109, 156–159]. However, the corresponding techniques are not used for routine characterization until now. Therefore, the application of probe molecules in combination with their study by ^{1}H, ^{13}C, ^{15}N, and ^{31}P MAS NMR spectroscopy is the more suitable method. Table 2.4 gives a survey on the most important probe molecules utilized

Table 2.4 Probe molecules applied for the characterization of Lewis acid sites and base sites. ^{13}C NMR shifts are referenced to tetramethylsilane ($\delta_{13C} = 0$ ppm), while the ^{31}P NMR signals are referenced to 85% H_3PO_4 at $\delta_{31P} = 0$ ppm

Probe molecule	Resonance/Observation	References
Lewis sites		
^{13}C-2-acetone	^{13}C: $\delta_{13C} = 233$ ppm for acetone at zeolite USY	[140, 142]
^{13}C-carbon monoxide[a]	^{13}C : $\delta_{13C} \approx 770$ ppm for CO at dealuminated zeolite H-ZSM-5	[160, 161]
Trimethylphosphine (TMP)	^{31}P: $\delta_{31P} = -32$ to -67 ppm for TMP at dealuminated zeolites	[145, 146, 162]
Trimethylphosphine oxide (TMPO)	^{31}P: $\delta_{31P} = 37$ ppm for TMPO at dealuminated zeolites Y and γ-Al_2O_3	[150, 163]
Base sites:		
Trichloromethane	^{1}H: hydrogen-bonded trichloromethane at $\delta_{1H} = 7.55$ (Li-Y) to 8.23 ppm (Cs,Na-Y)	[164]
Trifluoromethane	^{1}H: hydrogen-bonded trifluoromethane at $\delta_{1H} = 6.62$ (Li-Y) to 7.6 ppm (Cs,Na-Y)	[164]
Pyrrole	^{1}H: hydrogen-bonded pyrrole at $\delta_{1H} = 8.4$ (Li-Y) to 11.5 ppm (K-X)	[165]
Chloroform	^{1}H: hydrogen-bonded chloroform at $\delta_{1H} = 7.45$ (H-Y) to 8.70 ppm (Na,Ge-X)	[166]
^{13}C-chloroform	^{13}C: hydrogen-bonded ^{13}C-chloroform at $\delta_{13C} = 77.9$ (H-Y) to 81.7 ppm (Na,Ge-Y)	[166]
^{13}C-methyl iodide	^{13}C: methoxy groups occurring at $\delta_{13C} = 58.5$ (Na-ZSM-5) to 54.0 ppm (Cs,Na-X)	[167, 168]
^{13}C-nitromethane	^{13}C: $\delta_{13C} = 102$–112 ppm for nitromethane at mixed magnesium-aluminum oxides	[170]

[a]Rapid exchange requires NMR measurements at low temperatures (ca. $T = 120$ K)

for the study of Lewis acid sites and base sites in zeolites. In the presence of Lewis acid sites, various complexes are formed by the coordination of probe molecules at these sites. Up till now, therefore, no correlation between the NMR shifts of probe molecules coordinated at Lewis acid sites and the strength of these surface sites could be found. Another situation exists for the study of base sites by probe molecules (*vide infra*). Some of the probe molecules listed in Table 2.4, such as ^{13}C-2-acetone, trimethylphosphine (TMP), and trimethylphosphine oxid (TMPO), are also utilized for characterizing Brønsted acid sites.

Upon adsorption of TMP on stongly calcined zeolites H,Na-Y, e.g., the ^{31}P MAS NMR spectrum consist of signals at −32 to −67 ppm hinting to the formation of different Lewis acid sites. With increasing calcination temperature, a significant increase of these signals occurs, which indicates that the preparation of zeolite catalysts with a high concentration of Lewis acid sites requires an optimum calcination temperature, such as 873 K for zeolite H,Na-Y [144]. The local structure of adsorbate complexes consisting of TMP coordinated at Lewis acid sites in dehydroxylated zeolite H,Na-Y was investigated by ^{31}P/^{27}Al TRAPDOR NMR experiments (see Sect. 2.3.3) [171]. The ^{31}P MAS NMR signal occurring at −47 ppm could be assigned to TMP directly bound to an aluminum Lewis acid site. Furthermore, the ^{27}Al/^{31}P INEPT (insensitive nuclei enhanced by polarization transfer) NMR technique was utilized to determine the ^{27}Al/^{31}P J-coupling [171, 172]. Upon adsorption of TMP on mixtures of $AlCl_3$ and zeolite H,Na-Y, two different TMP-$AlCl_3$ complexes were observed [172]. ^{31}P/^{27}Al J-coupling constants and internuclear distances of 299.5 Hz and 0.258 nm and of 260 Hz and 0.296 nm were determined for $AlCl_3$ molecules coordinated to one (4-fold coordinated aluminum) and two (5-fold coordinated aluminum) TMP molecules, respectively [172].

The strength of base sites in zeolites Li-Y, Na-Y, K-Y, Rb-Y, and Cs,Na-Y was investigated by adsorption of trichloro- and trifluoromethane as probe molecules [164]. Depending on the cation type, ^{1}H MAS NMR signals at resonance positions of 7.55 (Li-Y) to 8.23 ppm (Cs,Na-Y) for adsorption of trichloromethane and of 6.62 (Li-Y) to 7.6 ppm (Cs,Na-Y) for adsorption of trifluoromethane were obtained. Utilizing pyrrole as NMR probe for basic zeolites [165], the ^{1}H NMR shift of the hydrogen atoms at the ring nitrogen covers a range of 8.4 ppm for zeolite Li-Y to 11.5 ppm for zeolite K-X indicating the different base strengths of the framework oxygen atoms in these materials. Here, a stronger low-field shift corresponds to a higher base strength. Studying the formation of methoxy groups bound to zeolite oxygens in bridging positions in basic zeolites, Bosacek et al. [167, 169] found a correlation between the ^{13}C NMR shift values of surface methoxy groups and the mean electronegativity S^{m} of the zeolite framework. According to this correlation, a low ^{13}C NMR shift of methoxy groups corresponds to a high base strength of the framework oxygen atoms. The ^{13}C MAS NMR signals of surface methoxy groups bound at framework oxygens of alkali-exchanged zeolites Y and X cover a shifts range of 54.0–56.5 ppm, while a signal at 58.5 ppm occurs for methoxy groups on zeolite Na-ZSM-5 [167, 169].

References

1. Engelhardt G, Michel D (1987) High-resolution solid-state NMR of silicates and zeolites. Wiley, Chichester
2. Klinowski J (1991) Solid-state NMR studies of molecular-sieve catalysts. Chem Rev 91:1459–1479
3. Stoecker M (1994) Review on recent NMR results. In: Jansen JC, Stoecker M, Karge HG, Weitkamp J (eds) Advanced zeolite science and applications. Studies in Surface Science and Catalysis, vol 85. Elsevier, Amsterdam, pp 429–507
4. Hunger M (1997) Brønsted acid sites in zeolites characterized by multi-nuclear solid-state NMR spectroscopy. Catal Rev Sci Eng 39:345–393
5. Karge HG, Hunger M, Beyer H (1999) Characterization of zeolites – infrared and nuclear magnetic resonance spectroscopy and X-ray diffraction. In: Weitkamp J, Puppe L (eds) Catalysis and zeolites – fundamentals and application. Springer, Berlin, pp 198–326
6. Engelhardt G (2001) Solid-state NMR spectroscopy applied to zeolites. In: Van Bekkum H, Flanigen EM, Jacobs PA, Jansen JC (eds) Studies in surface science and catalysis, vol 137, Introduction to zeolite science and practice. Elsevier, Amsterdam, pp 387–418
7. Haw JF (2002) In situ NMR. In: Haw JF (ed) In-situ spectroscopy in heterogeneous catalysis. Wiley-VCH, Weinheim, pp 53–85
8. Hunger M, Brunner E (2004) Characterization I – NMR spectroscopy. In: Karge HG, Weitkamp J (eds) Molecular sieves – science and technology, vol 4. Springer-Verlag, Berlin, Heidelberg, pp 201–293
9. Hunger M, Weitkamp J (2004) In situ magnetic resonance techniques: nuclear magnetic resonance. In: Weckhuysen BM (ed) In situ spectroscopy of catalysts. American Scientific Publishers, Stevenson Ranch, pp 177–218
10. Hunger M, Wang W (2006) Characterization of solid catalysts in the functioning state by nuclear magnetic resonance spectroscopy. Adv Catal 50:149–225
11. Hunger M (2008) In situ flow MAS NMR spectroscopy: state of the art and applications in heterogeneous catalysis. Prog Nucl Magn Reson Spectrosc 53:105–127
12. Hunger M (2008) NMR spectroscopy for the characterization of surface acidity and basicity. In: Ertl G, Knoezinger K, Schueth F, Weitkamp F (eds) Handbook of heterogeneous catalysis, vol 2, chapter 3.2.4.4, 2nd edn. Wiley-VCH, Weinheim, pp 1163–1178
13. Hunger M, Wang W (2008) Solid-state NMR spectroscopy. In: Ertl G, Knoezinger K, Schueth F, Weitkamp J (eds) Handbook of heterogeneous catalysis, vol 2, chapter 3.1.3.7, 2nd edn.Wiley-VH, Weinheim, pp 912–932
14. Harris RK (1986) Nuclear magnetic resonance spectroscopy. Longman, Essex
15. Abragam A (1994) Principles of nuclear magnetism. Clarendom Press, Oxford
16. Rahman AU, Choudhary MI (1995) Solving problems with NMR spectroscopy. Academic Press, New York
17. Slichter CP (1996) Principles of magnetic resonance. Springer-Verlag, Berlin, Heidelberg
18. Freeman R (1997) Nuclear magnetic resonance. Longman, Essex
19. Haeberlen U (1976) High resolution NMR in solids. Academic Press, New York
20. Mehring M (1983) Principles of high-resolution NMR in solids. Springer, Berlin, Heidelberg
21. Stejskal EO, Memory JD (1994) High resolution NMR in the solid state. Oxford University Press, Oxford
22. Andrew ER, Bradbury A, Eades RG (1958) Nuclear magnetic resonance spectra from a crystal rotated at high speed. Nature (London) 182:1659–1659
23. Andrew ER, Jasinski A (1971) Nuclear magnetic resonance spectra of rapidly-rotated solids containing reorienting molecular groups. J Phys C:Solid State Phys 4:391–400
24. Maricq MM, Waugh JS (1979) NMR in rotating solids. J Chem Phys 70:3300–3316
25. Haeberlen U, Waugh JS (1968) Coherent averaging effects in magnetic resonance. Phys Rev 175:453–467

26. Scheler G, Haubenreisser U, Rosenberger H (1981) High-resolution ^{1}H NMR in solids with multiple-pulse sequences and magic-angle sample spinning at 270 MHz. J Magn Reson 44:134–144
27. Freude D (2000) Quadrupolar nuclei in solid-state nuclear magnetic resonance. In: Meyers RA (ed) Encylopedia of analytical chemistry. Wiley, New York, pp 12188–12225
28. Samoson A, Lippmaa E, Pines A (1988) High resolution solid-state N.M.R. – averaging of second-order effects by means of a double rotor. Mol Phys 65:1013–1018
29. Rocha J, Morais CM, Fernandez C (2004) Progress in multiple-quantum magic-angle spinning NMR spectroscopy. Topics Curr Chem 246:141–194
30. Samoson A (1993) Aproximate approach to DOR sideband suppression. Chem Phys Lett 214:456–458
31. Samoson A, Tegenfeldt J (1994) Suppression of DOR sidebands. J Magn Reson A 110: 238–244
32. Amoureux JP (1993) High-resolution solid-state NMR for spin 3/2 and 9/2: the multiple-quantum transition method. Solid State Nucl Magn Reson 2:83–88
33. Smith ME, van Eck ERH (1999) Recent advances in experimental solid state NMR methodology for half-integer spin quadrupolar nuclei. Progr NMR Spectros 34:159–201
34. Fernandez C, Delevoye L, Amoureux JP, Lang DP, Pruski M (1997) $^{27}Al\{^{1}H\}$ cross polarization triple-quantum magic angle spinning NMR. J Am Chem Soc 119:6858–6862
35. Michel D, Engelke F (1994) Cross-polarization, relaxation times and spin-diffusion in rotating solids. In: Diehl P, Fluck E, Guenther H, Kosfeld R, Seelig J, Bluemich B (eds) NMR basic principles and progress, vol 32. Springer-Verlag, Berlin, Heidelberg, pp 69–125
36. Gullion T, Schaefer J (1989) Rotational-echo double-resonance NMR. J Magn Reson 81:196–200
37. Van Eck ERH, Veeman WS (1993) Spin-density description of rotational-echo double-resonance, transferred-echo double-resonance and 2-dimensional transferred-echo double-resonance solid-state nuclear-magnetic-resonance. Solid State Nucl Magn Reson 2:307–315
38. Peng L, Huo H, Liu Y, Grey CP (2007) ^{17}O magic angle spinning NMR studies of Broensted acid sites in zeolites HY and HZSM-5. J Am Chem Soc 129:335–346
39. Huang Y, Yan Z, Richer R (2005) Characterization of BAPO- and SAPO-based mesoporous materials by solid-state NMR spectroscopy. Chem Mater 17:6545–6554
40. Van Eck ERH, Janssen R, Maas WEJR, Veeman WS (1990) A novel application of nuclear spin-echo double-resonance to aluminophosphates and aluminosilicates. Chem Phys Lett 174:428–432
41. Grey CP, Vega AJ (1995) Determination of the quadrupole coupling constant of the invisible aluminum spins in zeolite HY with $^{1}H/^{27}Al$ TRAPDOR NMR. J Am Chem Soc 117:8232–8242
42. Gullion T (1995) Measurement of dipolar interactions between spin-1/2 and quadrupolar nuclei by rotational-echo, adiabatic-passage, double-resonance NMR. Chem Phys Lett 246:325–330
43. Li X, Zhang W, Liu SB, Han X, Xu L, Bao X (2006) A high-resolution MAS NMR study on the potential catalysts Mo/HBeta for olefin metathesis: The interaction of Mo species with HBeta zeolite. J Mol Catal A: Chem 250:94–99
44. Hu W, Luo Q, Su Y, Chen L, Yue Y, Ye C, Deng F (2006) Acid sites in mesoporous Al-SBA-15 materials as revealed by solid-state NMR spectroscopy. Micropor Mesopor Mater 92:22–30
45. Van Wuellen L, Koller H, Kalwei M (2002) Modern solid state double resonance NMR strategie for the structural characterization of adsorbate complexes involved in the MTG process. Phys Chem Chem Phys 4:1665–1674
46. Amoureux JP, Trebosc J, Wiench JW, Massiot D, Pruski M (2005) Measurement of J-coupling between spin-1/2 and quadrupolar nuclei by frequency selective solid-state NMR. Solid State Magn Reson 27:228–232
47. Amoureux JP, Flambard A, Delevoye L, Montagne L (2005) A very sensitive high-resolution NMR method for quadrupolar nuclei: SPAM-DQF-STMAS. Chem Commun 27:3472–3474

48. Brouwer DH, Kristiansen PE, Fyfe CA, Levitt MH (2005) Symmetry-based ^{29}Si dipolar recoupling magic angle spinning NMR spectroscopy: A new method for investigating three-dimensional structures of zeolite frameworks. J Am Chem Soc 127:542–543
49. Fyfe CA, Brouwer DH, Lewis AR, Chezeau JM (2001) Location of the fluoride ion in tetrapropylammonium fluoride silicalite-1 determined by $^{1}H/^{19}F/^{29}Si$ triple resonance CP, REDOR, and TEDOR NMR experiments. J Am Chem Soc 123:6882–6891
50. Fyfe CA, Diaz AC, Grondey H, Lewis AR, Foerster H (2005) Solid state NMR method for the determination of 3D zeolite framework/sorbate structures: $^{1}H/^{29}Si$ CP MAS NMR study of the high-loaded form of p-xylene in ZSM-5 and determination of the unknown structure of the low-loaded form. J Am Chem Soc 127:7543–7558
51. Camblor MA, Davis ME (1994) ^{29}Si MAS NMR spectroscopy of tectozincosilicates. J Phys Chem 98:13151–1356
52. Arnold A, Steuernagel S, Hunger M, Weitkamp J (2003) Insight into the dry-gel synthesis of gallium-rich zeolite [Ga]Beta. Micropor Mesopor Mater 62:97–106
53. Engelhardt G, Lohse U, Samoson A, Magi M, Tarmak M, Lippmaa E (1982) High-resolution ^{29}Si NMR of dealuminated and ultrastable Y-zeolites. Zeolites 2:59–62
54. Jiao J, Wang W, Sulikowski B, Weitkamp J, Hunger M (2006) ^{29}Si and ^{27}Al MAS NMR characterization of non-hydrated zeolites Y upon adsorption of ammonia. Micropor Mesopor Mater 90:246–250
55. Thomas JM, Klinowski J, Ramdas S, Hunter BK, Tennakoon DTB (1983) The evaluation of non-equivalent tetrahedral sites in zeolites and related aluminosilicates. Chem Phys Lett 102:158–162
56. Radeglia R, Engelhardt G (1985) Correlation of Si-O-Si or Si-O-Al angles and ^{29}Si NMR chemical shifts in silicates and aluminosilicates - interpretation by semi-empirical quantum-chemical considerations. Chem Phys Lett 114:28–30
57. Engelhardt G, Luger S, Buhl JC, Felsche J (1989) ^{29}Si MAS NMR of aluminosilicate sodalites – correlations between chemical-shifts and structure parameters. Zeolites 9:182–186
58. Fyfe CA, Feng Y, Grondey H (1993) Evaluation of chemical shift-structure correlations from a combination of X-ray diffraction and 2D MAS NMR data for highly siliceous frameworks. Micropor Mater 1:393–400
59. Fyfe CA, Feng Y, Gies H, Grondey H, Kokotailo GT (1990) Natural-abundance two-dimensional solid-state ^{29}Si NMR investigations of three-dimensional lattice connectivities in zeolite structures. J Am Chem Soc 112:3264–3270
60. Fyfe CA, Grondey H, Feng Y, Kokotailo GT (1990) Natural-abundance two-dimensional ^{29}Si MAS NMR investigation of the three-dimensional bonding connectivities in the zeolite catalyst ZSM-5. J Am Chem Soc 112:8812–8820
61. Fyfe CA, Grondey H, Feng Y, Kokotailo GT, Ernst S, Weitkamp J (1992) Two-dimensional solid-state ^{29}Si n.m.r. investigation of the three-dimensional bonding connectivities and structure of zeolite ZSM-23. Zeolites 12:50–53
62. Ernst RR, Bodenhausen G, Wokaun A (1994) Principles of nuclear magnetic resonance in one and two dimensions. Clarendon Press, Oxford
63. Friebolin H (1991) Basic one and two dimensional NMR-spectroscopy. Wiley, Weinheim
64. Jiao J, Kanellopoulos J, Wang W, Ray SS, Foerster H, Freude D, Hunger M (2005) Characterization of framework and extra-framework aluminum species in non-hydrated zeolites Y by ^{27}Al spin-echo, high-speed MAS, and MQMAS NMR spectroscopy at $B_0 =$ 9.4 to 17.6 T. Phys Chem Chem Phys 7:3221–3226
65. Pyykko P (2001) Spectroscopic nuclear quadrupole moments. Mol Phys 99:1617–1629
66. Altwasser S, Jiao J, Steuernagel S, Weitkamp J, Hunger M (2004) Elucidating the dealumination mechanism of zeolite H-Y by solid-state NMR spectroscopy. In: Van Steen E, Callanan LH, Claeys M (eds) Recent advances in the science and technology of zeolites and related materials. Stud Surf Sci Catal, vol 154. Elsevier, Amsterdam, pp 3098–3105

67. Rocha J, Carr SW, Klinowski J (1991) ^{27}Al quadrupole nutation and ^{1}H/^{27}Al cross-polarization solid-state NMR-studies of ultrastable zeolite-Y with fast magic-angle spinning. Chem Phys Lett 187:401–408
68. Chen TH, Houthoofd K, Grobet PJ (2005) Toward the aluminum coordination in dealuminated mordenite and amorphous silica-alumina: A high resolution ^{27}Al MAS and MQMAS NMR study. Micropor Mesopor Mater 86:31–37
69. Lippmaa E, Samoson A, Magi M (1986) High-resolution ^{27}Al NMR of aluminosilicates. J Am Chem Soc 108:1730–1735
70. Jacobsen HS, Norby P, Bildsøe JHJ (1989) 1–1 Correlation between 27Al and 29Si chemical-shifts and correlations with lattice structures for some aluminosilicate sodalites. Zeolites 9:491–495
71. Klinowski J, Anderson MW, Thomas JM (1983) On the siting of aluminum in zeolite omega. J Chem Soc Chem Commun 525–526
72. Massiani P, Fajula F, Figueras F (1988) ^{29}Si and ^{27}Al MAS NMR-study of the distribution of Si and Al atoms in various forms of synthetic zeolite omega. Zeolites 8:332–337
73. Hunger M, Ernst S, Weitkamp J (1995) Multi-nuclear solid-state NMR investigation of zeolite MCM-22. Zeolites 15:188–192
74. Freude D, Haase J (1993) Quadrupole effects in solid-state nuclear magnetic resonance. In: Diehl P, Fluck E, Guenther H, Kosfeld R, Seelig J (eds) NMR, basic principles and progress, vol 29. Springer, Berlin, Heidelberg, pp 1–90
75. Reiss G (1994) Status and development of oxygen generation processes on molecular-sieve zeolites. Gas Sep Purif 8:95–99
76. Gaffney TR (1996) Porous solids for air separation. Curr Opin Solid State Mater Sci 1:69–75
77. Barthomeuf D (1994) Basicity in zeolites. In: Fraissard J, Petrakis L (eds) Acidity and basicity of solids. Kluwer Academic Publishers, Dordrecht, Boston, pp 181–197
78. Weitkamp J, Hunger M, Rymsa U (2001) Base catalysis on microporous and mesoporous materials: recent progress and perspectives. Micropor Mesopor Mater 48:255–270
79. Feuerstein M, Lobo RF (1998) Influence of oxygen and nitrogen on ^{7}Li MAS NMR spectra of zeolite LiX-1.0. Chem Commun 1647–1648
80. Feuerstein M, Engelhardt G, McDaniel PL, MacDougall JE, Gaffney TR (1998) Solid-state nuclear magnetic resonance investigation of cation siting in LiNaLSX zeolites. Micropor Mesopor Mater 26:27–35
81. Feuerstein M, Lobo RF (1998) Characterization of Li cations in zeolite LiX by solid-state NMR spectroscopy and neutron diffraction. Chem Mater 10:2197–2204
82. Feuerstein M, Lobo RF (1999) Mobility of Li cations in X zeolites studied by solid-state NMR spectroscopy. Solid State Ionics 118:135–139
83. Feuerstein M, Accardi RJ, Lobo RJ (2000) Adsorption of nitrogen and oxygen in the zeolites LiA and LiX investigated by ^{6}Li and ^{7}Li MAS NMR spectroscopy. J Phys Chem B 104:10282–10287
84. Smith LJ, Eckert H, Cheetham AK (2000) Site preferences in the mixed cation zeolite Li, Na-chabazite: a combined solid-state NMR and neutron diffraction study. J Am Chem Soc 122:1700–1708
85. Accardi RJ, Lobo RF (2000) Accessibility of lithium cations in high-silica zeolites investigated using the NMR paramagnetic shift effect of adsorbed oxygen. Micropor Mesopor Mater 40:25–34
86. Barich DH, Xu T, Zhang J, Haw JF (1998) Modeling of benzene adsorption in metal-exchanged zeolites by calculation of ^{7}Li chemical shifts. Angew Chem Int Ed 37:2530–2531
87. Malek A, Ozin GA, Macdonald PM (1996) Probing cation sites in cesium-exchanged zeolite Y via ^{133}Cs MAS NMR. J Phys Chem 100:16662–16666
88. Norby P, Poshni FI, Gualtieri AF, Hanson JC, Grey CP (1998) Cation migration in zeolites: An in situ powder diffraction and MAS NMR study of the structure of zeolite Cs(Na)-Y during dehydration. J Phys Chem B 102:839–856

89. Hunger M, Schenk U, Buchholz A (2000) Mobility of cations and guest compounds in cesium-exchanged and impregnated zeolites Y and X investigated by high-temperature MAS NMR spectroscopy. J Phys Chem B 104:12230–12236
90. Hunger M, Schenk U, Burger B, Weitkamp J (1997) Synergism of guest compounds and host framework in basic zeolites. Angew Chem Int Ed 36:2504–2506
91. Hunger M, Schenk U, Weitkamp J (1998) Mechanistic studies of the side-chain alkylation of toluene with methanol on basic zeolites Y by multi-nuclear NMR spectroscopy. J Mol Catal A: Chem 134:97–109
92. Hunger M, Engelhardt G, Koller H, Weitkamp J (1993) Characterization of sodium cations in dehydrated faujasites and zeolite EMT by ^{23}Na DOR, 2D nutation, and MAS NMR. Solid State Nucl Magn Reson 2:111–120
93. Feuerstein M, Hunger M, Engelhardt G, Amoureux JP (1996) Characterization of sodium ations in dehydrated zeolite NaX by ^{23}Na NMR spectroscopy. Solid State Nucl Magn Reson 7:95–103
94. Hunger M, Sarv P, Samoson A (1997) Two-dimensional triple-quantum ^{23}Na MAS NMR spectroscopy of sodium cations in dehydrated zeolites. Solid State Nucl Magn Reson 9:115–120
95. Lim KH, Grey CP (2000) Characterization of extra-framework cation positions in zeolites NaX and NaY with very fast ^{23}Na MAS and multiple quantum MAS NMR spectroscopy. J Am Chem Soc 122:9768–9780
96. Caldarelli S, Buchholz A, Hunger M (2001) Investigation of sodium cations in dehydrated zeolites LSX, X, and Y by ^{23}Na off-resonance RIACT triple-quantum and high speed MAS NMR spectroscopy. J Am Chem Soc 123:7118–7123
97. Ganapathy S, Das TK, Vetrivel R, Ray SS, Sen T, Sivasanker S, Delevoye L, Fernandez C, Amoureux JP (1998) Anisotropic chemical shielding, M-site ordering, and characterization of extra-framework cations in ETS-10 studied through MAS/MQ-MAS NMR and molecular modelling techniques. J Am Chem Soc 120:4752–4762
98. Anderson MW, Agger JR, Luigi DP, Baggaley AK, Rocha J (1999) Cation sites in ETS-10: ^{23}Na 3Q MAS NMR and lattice energy minimisation calculations. Phys Chem Chem Phys 1:2287–2292
99. Matijasic A, Lewis AR, Marichal C, Delmotte L, Chezeau JM, Patarin J (2000) Structural characterization of the new porous sodium silicate Mu-11 by ^{29}Si and ^{23}Na solid-state NMR. Phys Chem Chem Phys 2:2807–2813
100. Lim KH, Grey CP (2000) Characterization of extra-framework cation positions in zeolites NaX and NaY with very fast ^{23}Na MAS and multiple quantum MAS NMR spectroscopy. J Am Chem Soc 122:9768–9790
101. Massiot D, Bessada C, Coutures JP, Taulelle F (1990) A quantitative study of ^{27}Al MAS NMR in crystalline YAG. J Magn Reson 90:231–242
102. Accardi RJ, Lobo RF, Kalwei M (2001) Paramagnetic effect of oxygen in the ^{23}Na MAS NMR and ^{23}Na MQMAS NMR spectroscopy of zeolite LiNaX. J Phys Chem B 105:5883–5886
103. Freude D, Hunger M, Pfeifer H (1987) Investigation of acidic properties of zeolites by MAS NMR. Z Phys Chem (NF) 152:171–182
104. Hunger M (1996) Multinuclear solid-state NMR studies of acidic and non-acidic hydroxyl protons in zeolites. Solid State Nucl Magn Reson 6:1–29
105. Hunger M, Freude D, Pfeifer H, Prager D, Reschetilowski W (1989) Proton MAS NMR studies of hydroxyl groups in alkaline earth cation-exchanged zeolites Y. Chem Phys Lett 163:221–224
106. Mastikhin VM, Mudrakovsky IL, Nosov AV (1991) ^{1}H NMR magic angle spinning (MAS) studies of heterogeneous catalysis. Prog NMR Spectrosc 23:259–299
107. Rachwalik R, Olejniczak Z, Jiao J, Huang J, Hunger M, Sulikowski B (2007) Isomerization of α-pinene over dealuminated ferrierite-type zeolites. J Catal 252:161–170
108. Huang J, Jiang Y, Reddy Marthala VR, Ooi YS, Weitkamp J, Hunger M (2007) Concentration and acid strength of hydroxyl groups in zeolites La, Na-X and La, Na-Y with different

lanthanum exchange degrees studied by solid-state NMR spectroscopy. Micropor Mesopor Mater 104:129–136
109. Huang J, Jiang Y, Reddy Marthala VR, Thomas B, Romanova E, Hunger M (2008) Characterization and acidic properties of aluminum-exchanged zeolites X and Y. J Phys Chem C 112:3811–3818
110. Hunger M, Anderson MW, Ojo A, Pfeifer H (1993) Study of the geometry and location of the bridging OH groups in aluminosilicate and silicoaluminophosphate type zeolites using ^{1}H MAS NMR sideband analysis and CP/MAS NMR. Micropor Mater 1:17–32
111. Deng F, Yue Y, Ye C (1998) ^{1}H/^{27}Al TRAPDOR NMR studies on aluminum species in dealuminated zeolites. Solid State Nucl Magn Reson 10:151–160
112. Zhang W, Ma D, Liu X, Liu X, Bao X (1999) Perfluorotributylamine as a probe molecule for distinguishing internal and external acidic sites in zeolites by high-resolution ^{1}H MAS NMR spectroscopy. Chem Commun 12:1091–1092
113. Zhang W, Bao X, Guo X, Wang X (1999) A high-resolution solid-state NMR study on nano-structured HZSM-5 zeolite. Catal Lett 60:89–94
114. Simon A, Gougeon RD, Paillaud JL, Valtchev V, Kessler H (2001) Characterization of the acidity of Mu-14 by solid-state NMR and NH_3-STD. Phys Chem Chem Phys 3:867–872
115. Kao HM, Grey CP, Pitchumani K, Lakshminarasimhan PH, Ramamurthy V (1998) Activation conditions play a key role in the activity of zeolite CaY: NMR and product studies of Bronsted acidity. J Phys Chem A 102:5627–5638
116. Beck LW, Haw JF (1995) Multinuclear NMR-studies reveal a complex acid function for zeolite Beta. J Phys Chem 99:1076–1079
117. Brunner E, Beck K, Koch M, Heeribout L, Karge HG (1995) Verification and quantitative determination of a new type of Broensted acid sites in H-ZSM-5 by ^{1}H magic-angle spinning nuclear magnetic resonance spectroscopy. Micropor Mater 3:395–399
118. Freude D (1995) Enhanced resolution in the ^{1}H NMR spectra of zeolite H-ZSM-5 by heteronuclear dipolar dephasing spin-echo MAS. Chem Phys Lett 235:69–75
119. Hunger M, Ernst S, Steuernagel S, Weitkamp J (1996) High-field ^{1}H MAS NMR investigations of acidic and non-acidic hydroxyl groups in zeolites H-Beta, H-ZSM-5, H-ZSM-58 and H-MCM-22. Micropor Mater 6:349–353
120. Koller H, Lobo RF, Burkett SL, Davis ME (1995) $SiO^{-}\cdots HSiO$ hydrogen bonds in as-synthesized high-silica zeolites. J Phys Chem 99:12588–12596
121. Wolf I, Gies H, Fyfe CA (1999) Ordering of silicate layers by hydrogen-bonding networks: solid state NMR investigation of the perfect three-dimensional registration in the layer silicate RUB-18. J Phys Chem B 103:5933–5938
122. Ma D, Deng F, Fu R, Han X, Bao X (2001) MAS NMR studies om the dealumination of zeolite MCM-22. J Phys Chem B 105:1770–1779
123. Isobe T, Watanabe T, d'Espinose de la Caillerie JB, Legrand AP, Massiot D (2003) Solid-state ^{1}H and ^{27}Al NMR studies of amorphous aluminum hydroxides. J Coll Int Sci 261:320–324
124. Nesterenko NS, Thibault-Starzyk F, Montouillout V, Yuschenko VV, Fernandez C, Gilson JP, Fajula F, Ivanova II (2004) Accessibility of the acid sites in dealuminated small-port mordenites studied by FTIR of co-adsorbed alkylpyridines and CO. Micropor Mesopor Mater 71:157–166
125. Kennedy GJ, Afeworki M, Calabro DC, Chase CE, Smiley RJ Jr (2004) ^{1}H MAS NMR (magic-angle spinning nuclear magnetic resonance) techniques for the quantitative determination of hydrogen types in solid catalysts and supports. Appl Spec 58:698–704
126. Omegna A, Vasic M, van Bokhoven RA, Pirngruber G, Prins R (2004) Dealumination and realumination of microcrystalline zeolite beta: an XRD, FTIR and quantitative multinuclear (MQ) MAS NMR study. Phys Chem Chem Phys 6:447–452
127. Sanderson RT (1976) Chemical bonds and bond energy, 2nd edn. Academic Press, New York
128. Mortier WJ (1978) Zeolite electronegativity related to physicochemical properties. J Catal 55:138–145

129. Yesinowski JP, Eckert H, Rossman GR (1988) Characterization of hydrous species in minerals by high-speed ^{1}H MAS NMR. J Am Chem Soc 110:1367–1375
130. Hunger M, Schenk U, Breuninger M, Glaeser R, Weitkamp J (1999) Characterization of the acid sites in MCM-41-type materials by spectroscopic and catalytic techniques. Micropor Mesopor Mater 27:261–271
131. Weihe M, Hunger M, Breuninger M, Karge HG, Weitkamp J (2001) Influence of the nature of residual alkali cations on the catalytic activity of zeolites X, Y, and EMT in their Brønsted acid forms. J Catal 198:256–265
132. Jaenchen J, van Wolput JHMC, van de Ven LJM, de Haan JW, van Santen RA (1996) FTIR spectroscopic and ^{1}H MAS NMR studies of the influence of lattice chemistry and structure on Bronsted acidity in zeolites. Catal Lett 39:147–152
133. Paze C, Zecchina A, Spera S, Cosma A, Merlo E, Spano G, Girotti G (1999) Comparative IR and ^{1}H MAS NMR study of adsorption of CD_3CN on zeolite H-beta: evidence of the presence of two families of bridged Brønsted sites. Phys Chem Chem Phys 1:2627–2629
134. Huang J, Jiang Y, Reddy Marthala VR, Wang W, Sulikowski B, Hunger M (2007) In situ 1H MAS NMR investigations of the H/D exchange of alkylaromatic hydrocarbons on zeolites H-Y, La, Na-Y, and H-ZSM-5. Micropor Mesopor Mater 99:86–90
135. White JL, Beck LW, Haw JF (1992) Characterization of hydrogen bonding in zeolites by proton solid-state NMR spectroscopy. J Am Chem Soc 114:6182–6189
136. Beck LW, Xu T, Nicholas JB, Haw JF (1995) Kinetic NMR and density-functional study of benzene H/D exchange in zeolites, the most simple aromatic substitution. J Am Chem Soc 117:11594–11595
137. Ernst H, Freude D, Mildner T, Pfeifer H (1999) High temperature ^{1}H MAS NMR studies of the proton mobility in zeolites. In: Treacy MMJ, Marcus BK, Bisher ME, Higgins JB (eds) Proceedings of the 12th International Zeolite Conference. Materials Research Society, Warrendale, Pennsylvania, pp 2955–2962
138. Haw JF (2002) Zeolite acid strength and reaction mechanisms in catalysis. Phys Chem Chem Phys 4:5431–5441
139. Biaglow AI, Gorte RJ, Kokotailo GT, White D (1994) A probe of Brønsted sites acidity in zeolites–13C chemical shift of acetone. J Catal 148:779–786
140. Biaglow AI, Gorte RJ, White D (1994) 13C NMR studies of acetone in dealuminated faujasites–a probe for nonframework alumina. J Catal 150:221–224
141. Xu T, Munson EJ, Haw JF (1994) Towards a systematic chemistry of organic reactions in zeolites–in-situ NMR studies of ketones. J Am Chem Soc 116:1962–1972
142. Xu M, Arnold A, Buchholz A, Wang W, Hunger M (2002) Low-temperature modification of mesoporous MCM-41 material with sublimated aluminum chloride in vacuum. J Phys Chem B 106:12140–12143
143. Ma D, Han X, Xie S, Bao X, Hu H, Au-Yeung SCF (2002) An investigation of the roles of surface aluminum and acid sites in the zeolite MCM-22. Chem Eur J 8:162–170
144. Lunsford JH, Rothwell WP, Shen W (1985) Acid sites in zeolites–a solid-state NMR and infrared study using trimethylphosphine as a probe molecules. J Am Chem Soc 107:1540–1547
145. Lunsford JH, Tutunjian PN, Chu PJ, Yeh EB, Zalewski DJ (1989) Solid-state NMR study using trimethylphosphine as a probe of acid sites in normal and dealuminated zeolite Y. J Phys Chem 93:2590–2595
146. Lunsford JH, Sang H, Campbell SM, Liang CH, Anthony RG (1994) An NMR study of acid sites on sulphated zirconia catalysts using trimethylphosphine as a probe. Catal Lett 27:305–314
147. Luo Q, Deng F, Yuan Z, Yang J, Zhang M, Yue Y, Ye C (2003) Using trimethylphosphine as a probe molecule to study the acid states in Al-MCM-41 materials by solid-state NMR spectroscopy. J Phys Chem B 107:2435–2442
148. Zhang W, Han X, Liu X, Bao X (2003) Characterization of the acid sites in dealuminated nanosized HZSM-5 zeolite with the probe molecule trimethylphosphine. J Mol Catal A 194:107–113

149. Alonso B, Klur I, Massiot D (2002) Studies of surfaces through molecular probe adsorption and solid-state NMR. Chem Commun 804–805
150. Sutovich KJ, Peters AW, Rakiewicz EF, Wormsbecher RF, Mattingly SM, Mueller KT (1999) Simultaneous quantification of Brønsted- and Lewis-acid sites in a USY zeolite. J Catal 183:155–158
151. Karra MD, Sutovich KJ, Mueller KT (2002) NMR characterization of Brønsted acid sites in faujasitic zeolites with use of perdeuterated trimethylphosphine oxide. J Am Chem Soc 124:902–903
152. Wang Y, Zhuang J, Yang G, Zhou D, Ma D, Han X, Bao X (2004) Study on the external surface acidity of MCM-22 zeolite: Theoretical calculation and ^{31}P MAS NMR. J Phys Chem B 108:1386–1391
153. Baerlocher Ch., Meier WM, Olson DH (2001) Atlas of zeolite framework types, 5th revised edn. Elsevier, Amsterdam, p 132, 184
154. Zheng A, Zhang H, Chen L, Yue Y, Ye C, Deng F (2007) Relationship between ^{1}H chemical shifts of deuterated pyridinium ions and Brønsted acid strength of solid acids. J Phys Chem B 111:3085–3089
155. Haw JF, Nicholas JB, Xu T, Beck LW, Ferguson DB (1996) Physical organic chemistry of solid acids: lessons from in situ NMR and theoretical chemistry. Acc Chem Res 29:259–267
156. Bull LM, Cheetham AK, Anupold T, Reinhold A, Samoson A, Sauer J, Bussemer B, Lee Y, Gann S, Shore J, Pines A, Dupree R (1998) A high-resolution ^{17}O NMR study of siliceous zeolite faujasite. J Am Chem Soc 120:3510–3511
157. Pingel UT, Amoureux JP, Anupold T, Bauer F, Ernst H, Fernandez C, Freude D, Samoson A (1998) High-field ^{17}O NMR studies of the SiOAl bond in solids. Chem Phys Lett 294:345–350
158. Freude D, Loeser T, Michel D, Pingel U, Prochnow D (2001) ^{17}O NMR studies of low silicate zeolites. Solid State Nucl Magn Reson 20:46–60
159. Readman JE, Kim N, Ziliox M, Grey CP (2002) ^{17}O MQMAS NMR studies of Na-A and Ca-A. Chem Commun 23:2808–2809
160. Brunner E, Pfeifer H, Wutscherk T, Zscherpel D (1992) ^{13}C NMR investigations on the adsorption of carbon monoxide in H-ZSM-5. Z Phys Chem 178:173–183
161. Zscherpel D, Brunner E, Koch M, Pfeifer H (1995) Variable-temperature ^{13}C magic-angle spinning nuclear magnetic resonance investigations on the interaction between Lewis acid sites and carbon monoxide in H-ZSM-5 zeolites. Microporous Mater 4:141–147
162. Zhao B, Pan H, Lunsford JH (1999) Characterization of [(CH3) 3) P-H]+complexes in normal H-Y, dealuminated H-Y, and H-ZSM-5 zeolites using 31P solid-state NMR spectroscopy. Langmuir 15:2761–2765
163. Rakiewicz EF, Peters AW, Wormsbecher RF, Sutovich KJ, Mueller KT (1998) Characterization of acid sites in zeolitic and other inorganic systems using solid-state ^{31}P NMR of the probe molecule trimethylphosphine oxide. J Phys Chem B 102:2890–2896
164. Bosch E, Huber S, Weitkamp J, Knoezinger H (1999) Adsorption of trichloro- and trifluoromethane in Y-zeolites as studied by IR spectroscopy and multinuclear solid-state NMR. Phys Chem Chem Phys 1:579–584
165. Sánchez-Sánchez M, Blasco T (2000) Pyrrole as an NMR probe molecule to characterise zeolite basicity. Chem Commun 491–492
166. Sánchez-Sánchez M, Blasco T, Rey F (1999) An NMR study on the adsorption and reactivity of chloroform over alkali exchanged zeolites X and Y. Phys Chem Chem Phys 1:4529–4535
167. Bosacek V, Ernst H, Freude D, Mildner T (1997) Surface methoxy groups in zeolites studied by multinuclear MAS nmr spectroscopy. Zeolites 18:196–199
168. Schenk U, Hunger M, Weitkamp J (1999) Characterization of basic guest compounds on solid catalysts by ^{13}C CP/MAS NMR spectroscopy of surface methoxy groups. Magn Reson Chem 37:S75–S78
169. Bosacek V, Klik R, Genoni F, Spano G, Rivetti F, Figueras F (1999) Terminal and bridging methoxyls on zeolites detected by ^{13}C magic angle spinning NMR spectroscopy. Magn Reson Chem 37:S135–S141

170. Lima E, de Menorval LC, Tichit D, Lasperas M, Graffin P, Fajula F (2003) Characterization of the acid-base properties of oxide surfaces by ^{13}C CP/MAS NMR using adsorption of nitromethane. J Phys Chem B 107:4070–4073
171. Kao HM, Grey CP (1997) Determination of the $^{31}P/^{27}Al$ J-coupling constant for trimethylphosphine bound to the Lewis acid site of zeolite HY. J Am Chem Soc 119:627–628
172. Chu PJ, de Mallmann A, Lunsford JH (1991) A solid-state NMR study of aluminum Lewis acid adducts with trimethylphosphine in zeolite Y–magic angle spinning side-band analysis of dipolar and J-coupled multiple system. J Phys Chem 95:7362–7368

Chapter 3
Determination of Acid/Base Properties by Temperature Programmed Desorption (TPD) and Adsorption Calorimetry

Ljiljana Damjanović and Aline Auroux

Abstract The characterization of the acidity of zeolites and related materials is of great importance for applications of these materials, particularly in the petrochemical industry and environmental sciences. This chapter provides a comparison of the two most widely used techniques for the study of acid/base properties of zeolites, namely temperature-programmed desorption and adsorption calorimetry. The information needed to perform and interpret these experiments as well as advantages and limitations of these methods are summarized. The curves and data obtained for the two most studied zeolites Y and ZSM-5 are discussed, which can be particularly useful for young researchers in this field.

3.1 Introduction

The study of the acidity of zeolites is motivated by their applications as solid acid catalysts in industrially important reactions, particularly in petroleum refinery processes, as well as gas sensing.

In order to design zeolites and related materials with high activity and selectivity, a detailed characterization of the active sites, including their nature, number, strength, and strength distribution, is required. A question of basic interest is to determine the correlation between these properties and the promotion of catalytic activity. Therefore, the investigation of acidic properties of zeolites originating from both Brönsted and Lewis types of active sites is a very important subject. However, the

A. Auroux (✉)
Institut de Recherches sur la Catalyse et l'Environnement de Lyon, UMR 5256 CNRS/Université Lyon1, 2 avenue Albert Einstein, 69626, Villeurbanne, France
e-mail: aline.auroux@ircelyon.univ-lyon1.fr

L. Damjanović
Faculty of Physical Chemistry, University of Belgrade, Studentski trg 12-16, 11158, Belgrade, Serbia
e-mail: ljiljana@ffh.bg.ac.rs

A.W. Chester and E.G. Derouane (eds.), *Zeolite Characterization and Catalysis*,
DOI 10.1007/978-1-4020-9678-5_3, © Springer Science+Business Media B.V 2009

determination of the acidity of zeolites is often difficult because of the possible presence of both types of acid centers and the possible existence of a heterogeneous distribution of acid strengths. Even if less studied, the basicity of zeolites also leads to important applications.

Various methods have been successfully applied to study the active sites present on zeolites. Most of these methods are based on the adsorption or desorption of gas-phase probe molecules, which are chosen on the basis of their reactivity and molecular size. The conventional methods, temperature-programmed desorption (TPD) [1, 2] or adsorption microcalorimetry [3–6] of probe molecules give information regarding the strength and distribution of active sites. Spectroscopic methods, such as infrared (IR), nuclear magnetic resonance (NMR) or X-ray photoelectron spectroscopy (XPS) have been widely used to study the nature of acid sites found on zeolites [7–9]. It has been shown that combining different experimental techniques to investigate the interaction of gaseous probe molecules with a solid surface in order to obtain complementary results about active sites is highly informative [10].

The versatility of the acidic properties of zeolites and related materials is caused by the existence of crystallographically inequivalent sittings and different environments of the framework atoms.

Brönsted acid sites in zeolites of aluminosilicate type are assigned to bridging hydroxyl groups between Si and Al atoms of the framework. These sites can be formally represented as Si-O(H)-Al. The Si-O and Al-O distances, and the Si-O-Al angle are the most important structural parameters influencing the acidity of the OH groups located in the internal cavities. The number of Brönsted acid sites is equal to the number of framework $[MO_4]^-$ tetrahedra (where M is a trivalent cation such as Al^{3+}, Fe^{3+}, Ga^{3+}, As^{3+}, etc.) in the silicate network of zeolites, while the strength of these sites depends on (1) the structure of the zeolite, (2) the ratio of silicon to trivalent metal ions, and (3) the nature of trivalent metal ion in the framework [11, 12].

The negative charge of the framework can be balanced not only by protons (in the form of Brönsted groups) but also by monovalent alkali cations (Li^+, Na^+, K^+, etc.), divalent alkaline earth cations (Ca^{2+}, Mg^{2+}, etc.), transition monovalent and divalent metal ions (Cu^+, Cu^{2+}, Zn^{2+}, Rh^+, Co^{2+}, etc.) and, more rarely, by trivalent elements (Al^{3+} in extraframework position) as well.

Lewis acid sites are essentially electron acceptor centers and they can be cations or different aluminum species located in defect centers; the latter ones are known as so-called true Lewis sites. Some authors associate Lewis acidity in zeolites with framework trigonal Al atoms, originated from the former Brönsted sites as a result of thermal dehydroxylation [13]. Conversely, other research groups have hypothesized that Lewis sites are made up of extraframework Al^{III} species, leached from the zeolites framework during severe chemical or thermal treatments which induce dealumination [14–16].

Beside the effect on acidity associated with the local structure of the sites, long distance interactions (such as electrostatic interactions involving the second and third shell atoms of the framework and other ions present in the structure) also play a role. These collective effects are analogous to those which, in homogeneous phase, are known as solvent effects. On the basis of these considerations it is evident that a detailed description of the three-dimensional structure of zeolites and of the location of the acid sites in the micropores is of primary importance [17].

The two techniques most commonly used for the study of acid/base properties of zeolites and related materials are adsorption calorimetry and TPD. To the best of our knowledge, these two methods are rarely compared from the point of view of obtained data, limitations, usefulness and scientific interest.

In TPD experiments, temperature increases linearly and the concentration of desorbed gas is recorded as a function of temperature, whereas calorimetry involves the adsorption of gases onto the sample's surface while it is kept at a constant temperature and a heat-flow detector emits a signal proportional to the amount of heat transferred per unit time. The peak maxima temperatures in the TPD spectra are influenced by the acid site strength, the number of acid sites, the zeolite structure and the heating rate [18]. Additional information can be obtained concerning the nature of the acid sites and their strength by coupling IR spectroscopy or XPS with calorimetry. In particular, adsorption microcalorimetry gives access to the number, strength, and strength distribution of the acid sites in a single experiment.

This chapter collects information on how to analyze the acidity of zeolites by both techniques, adsorption calorimetry and TPD. In particular we compare the respective curves obtained for the two most studied zeolites, Y and ZSM-5. Of course, the applications of these techniques extend just as well to any other micro or mesoporous material.

3.2 Experimental Techniques

3.2.1 Temperature Programmed Desorption (TPD)

3.2.1.1 Description of the TPD Technique

The temperature-programmed desorption (TPD) of a basic molecule such as ammonia (NH_3-TPD) is one of the most commonly used methods for measuring the surface acidity of zeolites and related materials. It is widely used because of its simplicity and it is an inexpensive method [19].

As the sample is heated, pre-adsorbed gas is progressively desorbed, and sometimes decomposes. With increased temperature, the desorption rate increases, goes through a maximum, and finally goes back to zero as the surface is depleted of adsorbate. A desorption spectrum is a record of the concentration of desorbed gas as a function of temperature [20].

TPD of adsorbed basic probe molecules from an acidic surface may yield a profile with several desorption peaks. The temperature dependence of desorption is related to the strength of acid sites which may involve either physical or chemical adsorption. In general, in TPD spectra the peak areas show the amount of acid sites, while the temperatures of the peak maxima are correlated to the acid strength distribution of surface sites releasing the probe molecules. This method can only be used to roughly distinguish the various acid site strengths, since very often it assigns a single population of acid sites with the same strength to a broad TPD peak. In addition, the positions of the desorption peaks may be greatly affected by experimental conditions.

Curve deconvolution techniques have proved to be of great help in extracting this information from the usually poorly resolved desorption spectra [1, 21–23]. The area below TPD thermograms is proportional to the total number of sites of the zeolite where the basic probe molecule is adsorbed whereas the area under each of the mono-energetic curves is proportional to the relative amount of acid sites with that energy. While the acid strength distribution of active sites and their amount can be obtained from TPD spectra, adsorption enthalpies and desorption activation energies can hardly be extracted from a TPD analysis [24], due to the onset of diffusional limitations [25]. However, TPD is well suited for measuring adsorbate coverages and initial reaction products [24].

TPD consists of the following steps: (1) sample pretreatment, (2) preadsorption of the adsorbate, (3) evacuation or flushing with the carrier gas to remove physically adsorbed gas, (4) programmed desorption of the residual chemisorbed gas into the stream of a carrier gas, (5) detection of the desorbed gas in the carrier, and (6) trapping and analysis of the desorbed gas. Generally, studying the mechanism of bonding between adsorbates and the adsorbing surface requires ultra-high vacuum, as is the case for example, in XPS experiments. In the case of TPD, ultra-high vacuum is not necessary, and the experimental conditions may approach more closely those under which the catalyst is used in general practice [26, 27]. The carrier gas is usually helium at atmospheric pressure, but argon and nitrogen have also been used. Thermal conductivity cells can be employed as detectors, but the ideal detector is a mass spectrometer which measures the composition of the effluent stream as a function of the sample temperature.

For TPD a mass spectrometer is recommended as the detector because the qualitative and quantitative analysis can be done simultaneously, in particular in the case of NH_3 as a probe molecule. For the identification of desorbed ammonia, the co-presence of parent ($m/e = 17$) and fragment ($m/e = 16$) peaks should be confirmed in the mass spectrum, because the parent peak is strongly influenced by water, and the fragment peak is affected by oxygen-containing substances such as carbon dioxide [28]. It is also recommended to use $m/e = 15$ in order to avoid the interference of water fragmentation masses [29]. The fragment with $m/e = 16$ is usually used for the quantification of ammonia. If all the secondary fragments of the probe in the mass spectrum show exactly the same behavior as the molecular ion, it is a confirmation that they were generated only by the probe molecule, and that this particular probe does not undergo significant decomposition, either on the zeolite surface or after desorption, under the conditions of TPD measurements. The concentration of ammonia in the gas phase is thus plotted as a function of the temperature [28].

The transient nature of the TPD technique, in which both temperature and the surface coverage vary with time, has the advantage of providing information which is not available from steady-state kinetic measurements. In the temperature-programmed desorption the temperature is usually continuously varied and the rates of desorption are simultaneously influenced by the manner in which the process depends both on time and temperature.

To demonstrate how important is influence of experimental parameters on TPD results, Farneth and Gorte [2] have measured the desorption of ammonia from an HZSM-5 in two experimental configurations, one using a vacuum system and one using acarrier gas. One measurement was carried out on a 20 mg sample in vacuum with a heating rate of 20 K min^{-1}. The other one was carried out on a 44 mg sample in vacuum with a heating rate of 5 K min^{-1}. Both measurements were performed in a microbalance. A difference of 150° was observed between the respective maxima of the peaks and 40 kJ mol^{-1} between the desorption activation energies. This study showed the difficulty to determine intrinsic desorption rates from TPD on microporous materials. By measuring a series of TPD curves with different carrier gas flow rates, Sharma et al. [25] observed a shift in the peak temperature which could be used to calculate the average heat of desorption of ammonia from HZSM-5. The value they obtained was in good agreement with numbers obtained from microcalorimetry (145 kJ mol^{-1}).

The activation energies and entropies of desorption of gases adsorbed on the surfaces of solids can be determined by measuring the rates of desorption at several constant temperatures [20].

The parameters that have to be optimized in TPD analysis are: (1) flow rate of carrier gas, (2) reactant gas/inert gas ratio, (3) sample volume/mass, (4) the sample particle size, (5) the geometry of the reaction vessel, (6) the heating rate, (7) the signal intensity, and (8) the system pressure [30].

When using the TPD technique, the influence of experimental conditions such as sample mass, particle size and pelletizing pressure on the number of peaks and peak temperatures should be carefully examined in order to obtain reproducible and reliable results. The peak temperature in the TPD spectrum is not directly related to the acid strength. The peak maximum temperature increases with increasing the W (sample weight)/F (flow rate of carrier gas) ratio on a given sample [31]. On various zeolites, it has been clarified that readsorption of ammonia freely occurs under usual conditions of the NH_3-TPD, and the apparent desorption rate is controlled by the equilibrium [31, 32]. So, the peak temperature is affected by the amount of acid sites and experimental conditions. This is recognized as a problem which limits the amount of information obtained by NH_3-TPD [33]. Gorte commented on this method and strongly opposed the use of TPD technique to distinguish the acidic strength [33], based on experimental evidence showing that the peak temperature for ammonia desorption from the Brönsted sites in H-MFI zeolite might be shifted by more than 150° for the same sample depending on the conditions used for the measurement [2] and that peaks separated by 100° in TPD might be caused by changes in the particle size of the catalyst, rather than by different types of sites

[24]. Recently, attempts have been made to overcome this limitation of the TPD method [28].

Under controlled equilibrium with free readsorption, one can determine two of the thermodynamic parameters, namely, the standard entropy and enthalpy changes of desorption (the latter is the heat of adsorption of ammonia), based on the acid site amount and dependence of the peak maximum temperature on the W/F ratio observed by multiple measurements of the NH_3-TPD on one sample. Subsequently it has been found that the entropy change was almost constant on various zeolites, and a theory was derived [32] to determine the heat of ammonia adsorption as an index of the acid strength from a single measurement of NH_3-TPD. Moreover, a curve-fitting method for the determination of the acid strength distribution by analyzing the shape of observed desorption curve was developed [34]. Thus, the number and strength of the acid sites on a solid surface can be determined by the thus improved NH_3-TPD technique [34]. However, the direct measurement of heats of adsorption by calorimetry remains the most accurate technique for determining the strength of sites.

As noted previously, deriving kinetic parameters from TPD data is a demanding task. It is therefore necessary to use simplified model cases. The simplest case to be considered is that of a homogeneous surface where desorption of the first order occurs in the adsorbed sample, while diffusion and readsorption can be neglected. Using the linear heating rate, the desorption rate can be expressed as follows [26]:

$$\frac{d\theta}{dT} = -\frac{\theta A}{\beta}\exp(\frac{E_d}{RT}) \tag{3.1}$$

where θ is the surface coverage, A is the pre-exponential factor and E_d is the activation energy of desorption (independent of θ), while β is the heating rate.

Equation (3.1) yields [26]:

$$2\ln T_M - \ln\beta = \frac{E_d}{RT_M} + \ln\left(\frac{E_d}{AR}\right) \tag{3.2}$$

where T_M is the temperature of the peak maximum.

Equation (3.2) can be used to obtain E_d experimentally by varying β (the rate of temperature increase), measuring corresponding T_M and plotting:

$$\ln\left(\frac{T_M^2}{\beta}\right) = f(1/T_M) \tag{3.3}$$

This requires an appreciable range of β values and considerable experimental precision. In this case the TPD spectra should be symmetrical curves. However, symmetrical curves with only one peak rarely occur in TPD experiments.

The other limiting case is when readsorption of the sample occurs freely and its rate is only limited by the availability of the unoccupied adsorption sites (1–θ).

In this case the enthalpy change of desorption, ΔH, can be obtained from the following equation [26]:

$$2\ln T_M - \ln\beta = \frac{\Delta H}{RT_M} + \ln\left(\frac{(1-\theta_M)^2 W\Delta H}{FA^*R}\right) \tag{3.4}$$

where θ_M is the coverage at peak maximum, F is the flow rate of the carrier gas, W is the sample weight and A^* stands for exp($\Delta S/R$) (since the equilibrium constant of desorption is $K = \exp\left(\frac{\Delta S}{R}\right)\exp\left(\frac{\Delta H}{R}\right) = A * \exp\left(\frac{\Delta H}{R}\right)$). ΔH can be obtained when plotting $\ln\left(T_M^2/\beta\right) = f\left(1/T_M\right)$. Thus, ΔH may be estimated from series of TPD runs where the position of the maximum T_M is shifted either by varying the heating rate, β, or the sample weight, W, or in the case of experiments carried out under an inert gas stream, by varying the flow rate of the carrier gas, F [35].

In general, when the experiments are carefully conducted and the samples pretreated identically, similar values of acid sites can be obtained from both calorimetric and TPD measurements. Similar values of heat of adsorption can also be obtained. For example, when a HZSM-5 and a H-mordenite sample were examined by both methods: calorimetry showed that NH_3 adsorbs with ΔH_{ad} of 150 kJ mol^{-1} on HZSM-5 and 157 kJ mol^{-1} on H-mordenite while the value of ΔH obtained from TPD was 153 kJ mol^{-1} for H-mordenite, and the desorption temperature was higher than for HZSM-5 [25]. However, the results for pyridine adsorption were quite different for the two techniques, due to the fact that desorption of pyridine was limited by diffusion within the zeolite crystallites. Thus, although calorimetry showed that pyridine adsorbed more strongly on H-mordenite than on HZSM-5, the desorption temperature of pyridine was higher on the latter sample because of more severe diffusion limitation [36].

The various factors that affect the shape and position of the spectra are: (1) surface heterogeneity (multiple adsorption sites with different energies), (2) readsorption, (3) mass transfer effects (diffusional resistances), (4) subsurface diffusion and adsorption, and (5) desorption kinetics [30].

Surface heterogeneity, i.e., when a surface contains more than one type of adsorption sites, plays an important role in the analysis. If the heats of adsorption on these sites differ significantly, the TPD spectrum will contain multiple peaks, one for each type of adsorption site. When this happens, the heat of adsorption on each site can be determined separately. However, multiple peaks in a TPD spectrum do not always indicate the presence of multiple adsorption sites on the catalyst surface. Extra peaks can also result from lateral interactions between adsorbate molecules (induced heterogeneity) and diffusion of adsorbate in the subsurface region. As noted above, quantitative analysis of TPD spectra, i.e., the determination of reliable kinetic parameters, is quite a demanding task because of the magnitude of experimental errors [30]. Different kinetic models have been developed to account for surface heterogeneity. Various types of mathematical models for evaluation of the kinetic parameters for adsorption-desorption during TPD analysis are given in [30].

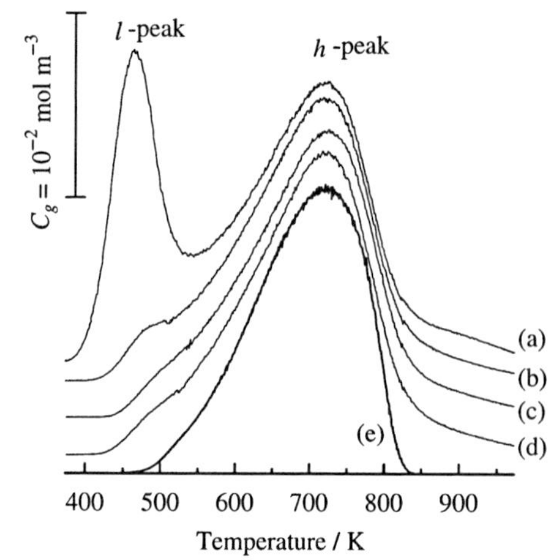

Fig. 3.1 Ammonia TPD spectra of H-mordenite (Si/Al = 7.5, Na/Al = 0.04, ion exchanged from MOR used by the Catalysts Society of Japan as reference catalyst, JRC-Z-M15) measured by the conventional method (**a**), and with water vapor treatment method with 1 (**b**), 2 (**c**) and 4 (**d**) cycles of introduction of water and evacuation. The curve (**e**) was simulated by the curve fitting method assuming an acid site amount of 1.7 mol kg^{-1} and a desorption heat of 145 kJ mol^{-1} [28]

Readsorption and diffusion of probe molecules in porous materials are two factors which are difficult to monitor or assess, and complicate the desorption process. It seems that adsorption at higher temperatures in order to ensure adsorption-desorption equilibrium, followed by TPD, gives more reliable data.

A typical TPD profile from an acid type zeolite has two desorption peaks, termed *l*- and *h*-peaks (low and high temperatures, respectively), as shown in Fig. 3.1a ([28]).

On H-form zeolite samples with high crystallinity, the number of ammonia molecules desorbed in the *h*-peak always agrees with the number of aluminum atoms in the zeolite. On the other hand, the *l*-peak is found even on a Na form zeolite, and its intensity is not related with catalytic activity for acid-catalyzed reactions. This finding suggests that the *l*-peak corresponds to ammonia adsorbed on non-acidic sites, while the *h*-peak corresponds to ammonia adsorbed on acid sites. Several studies have shown that the species responsible for the *l*-peak are readily replaced by water, whereas the *h*-peak was never influenced by the co-feed of water vapor in the carrier gas during TPD measurements [37–39]. The intensity of the *l*-peak depends on the evacuation conditions after the adsorption of ammonia: the longer the evacuation time, the smaller the *l*-peak. The intensity at a constant evacuation time, which must indicate the strength of adsorption, is higher on the Na form of the zeolite than on the H form. These observations suggest that the *l*-peak consists of ammonia adsorbed on an ion exchange site which has already adsorbed a cation [40]. In the Na form, *l*-peak ammonia is presumably adsorbed on Na^+, whereas in the H form, it should be adsorbed on NH_4^+ formed by prior adsorption of ammonia on the Brönsted acid site. The driving force of this type of adsorption should be the electrostatic interaction between the ion exchange site and N–H bonds. The interaction of water with such a site (so-called hydrogen bond) is stronger than that of ammonia because of the higher polarity of O-H compared to N–H, resulting in the replacement of this type of ammonia by water. According to

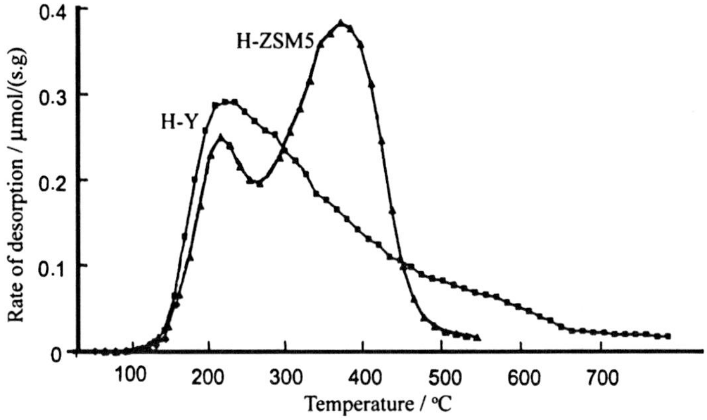

Fig. 3.2 NH_3-TPD profiles of zeolites ZSM-5 and Y [41]

this explanation, it is obvious that the *l*-peak on the H form zeolite does not give any information on the acidic properties of the solid. On the other hand, ammonia remaining after water vapor treatment must be bonded to the acid site with an acid-base interaction, because the basicity of ammonia is stronger than that of water [28].

NH_3-TPD profiles of zeolites ZSM-5 and Y are shown in Fig. 3.2, illustrating the presence of acid sites with different strengths on these zeolites.

One of the drawbacks of the TPD method is that it can distinguish sites by sorption strength only, but not Lewis from Brönsted sites, i.e., no information on the structure of the acid site is available from this method. One way to overcome this limitation is to employ the IRMS–TPD method, in which IR and mass spectrometry (MS) are used on line together to measure the behavior of adsorbed ammonia species as well as that of desorbed ammonia, respectively. Using this method, the structure of acid sites is determined clearly, and it is possible to measure the amount and strength individually for each type of acid sites. Investigations of the variations of IR band intensities with respect to temperature in the context of IR spectroscopy coupled with ammonia TPD on H–Y and dealuminated H–Y zeolites have been reported in the literature [42, 43]. The joint use of IR and MS to simultaneously study adsorbed and desorbed ammonia has been reported by Niwa et al. [44]. Using this method, a quantitative study is possible, because the number of acid sites is determined from the amount of desorbed ammonia measured by MS. Furthermore, the unique feature of this method to study acid sites is that the interaction of the OH band with ammonia species is clearly revealed.

Since in different types of zeolites the OH groups are located in pores, channels, and cages of different size and shape, a confinement effect should be considered when comparing the acid properties of zeolites on the basis of adsorption of probe molecules [45–47]. This confinement effect (originating from the physical interaction between the adsorbed substrate and the zeolite framework), also detectable in the effective desorption energy of NH_3-TPD, should result from the nonspecific interaction of ammonia with the zeolite framework. The effect also exists in certain zeolites if the OH groups are localized in different segments of the pore system. A detailed interpretation of NH_3-TPD therefore demands an assessment of this nonspecific contribution to the overall interaction. Using appropriate analytical models the strength of the nonspecific interaction can be calculated and used to correct the effective desorption energy values derived from NH_3-TPD experiments [48]. For a detailed understanding of the interaction of bases with acid sites in zeolites it is necessary to carefully account for confinement effects arising from van der Waals interactions and also to consider solvation effects due to the action of the zeolite framework as a solid solvent [47].

The combination of TPD and thermogravimetric analysis (TGA) of probe molecules is also widely used to determine the number of acid sites. The key aspect of this TPD-TGA technique is that TPD-MS analysis provides information about the nature of acid sites according to the different products evolved during TPD, while TGA gives quantitative amounts for the corresponding types of acid sites. For example, Gorte et al. have suggested the use of a combination of TPD and TGA of the adsorption of amines (e.g., isopropylamine) on zeolites to determine the number of surface acid sites [33]. As long as the chosen probe amine is small enough to enter the cavities of the zeolite, a 1:1 complex reaction (one surface proton decomposes one adsorbed amine molecule) is always observed. Biaglow et al. [49] demonstrated this stoichiometric relationship by employing the TPD-TGA-MS technique for the adsorption of isopropylamine on a H-MFI sample with a bulk Si/Al ratio of 26 (0.63 mmol Al/g sample). Fig. 3.3 [33] shows two stages of weight loss for desorption of isopropylamine in the TGA profile, and two stages of product evolution in the TPD profiles as recorded by a mass spectrometer (MS). The TPD-TGA results in Fig. 3.3 clearly show that 0.65 mmol g^{-1} of isopropylamine are desorbed from the surface at temperatures lower than 500 K, and 0.50 mmol g^{-1} of isopropylamine are decomposed to propylene and ammonia above 550 K. Biaglow et al. [49] attributed the latter population (0.50 mmol g^{-1}) to Brönsted acid sites and the other surface species (0.65 mmol g^{-1}) to other types of adsorbed complexes. The lower concentration of Brönsted acid sites measured by TPD-TGA (0.50 mmol g^{-1}) in comparison with the bulk Al content (0.63 mmol g^{-1}) indicated the presence of extra-framework Al (EFAL) in the sample.

TPD experiments can also be performed using a coupled thermogravimetry-differential scanning calorimetry system, which provides information on the acid strength of zeolites as well as the strength distribution [50, 51].

In contrast to the traditional TPD techniques, which are carried out with a constant heating rate, it is possible to use isothermal steps followed by steps of heating at a constant rate in order to quantitatively determine, for example, ammonia desorption from acid sites of different types and strengths. This method is called stepwise temperature programmed desorption (STPD) [52–54]. As shown in Fig. 3.1a above, typical TPD spectra of NH_3 desorbed from zeolites present one large peak which seems to consist of two overlapping peaks. The application of well-selected isothermal steps can separate these two regions. Zhang et al. [55] have performed STPD measurements on MOR zeolites with different Si/Al ratios (Fig. 3.4). They applied five isothermal holding periods at 180, 250, 350, 440 and

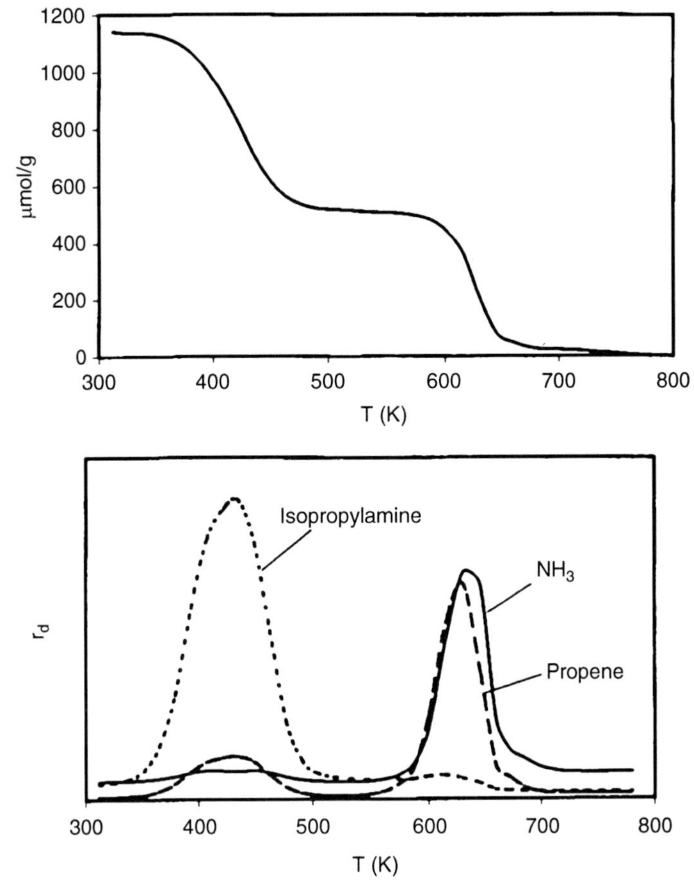

Fig. 3.3 TPD-TGA results for isopropylamine in an H-MFI (Si/Al = 26) sample. The desorption of propylene and ammonia above 550 K corresponds to decomposition of the amine at the Brönsted acid sites [33]

540°C. The addition of more isothermal periods did not result in any changes in the ammonia desorption profile, while a decrease in the number of isothermal periods or a variation in the heating steps resulted in overlapping peaks. The largest desorption peaks occurred at high temperatures, indicating strong acidity of the mordenite samples. Combining STPD data with FT-IR results made it possible to distinguish among different types of acid sites.

A new experimental technique called quasi-equilibrated temperature programmed desorption and adsorption (QE-TPDA) has been recently developed, using hydrocarbons (*n*-hexane) as probe molecule [56]. The investigated sample is activated by heating in a flow of pure He, then adsorption is performed at room temperature, by switching the flow through the reactor to a He/probe molecule mixture. After adsorption, QE-TPDA experiments are carried out by heating the sample with preadsorbed probe molecules in the same flow of He/probe

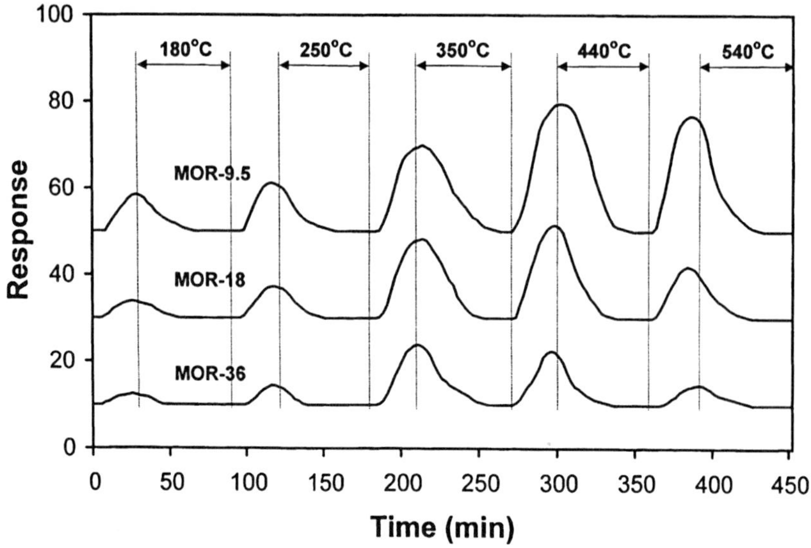

Fig. 3.4 NH_3-STPD profiles from mordenite samples; Si/Al = 9.5 for MOR-9.5, 18 for MOR-18, 36 for MOR-36 [55]

molecule mixture, applying a complex temperature programme consisting of several heating and cooling ramps [56]. This method allows the construction of the adsorption isosteres from the experimental thermodesorption profiles and the determination of isosteric adsorption heats. TPD profiles are obtained while changing temperature and concentration, and can be quantified using a model assuming the equilibrium control of desorption. The main advantages of the QE-TPDA method are the relatively short time of measurements, the simple and relatively inexpensive equipment, the very small amount of sample required, and the possibility of performing cyclic measurements combined with modifications of the studied sample. This method seems well suited for monitoring changes in the pore structure of various micro- and mesoporous materials, combined with studies of their catalytic activity or thermal and hydrothermal stability.

3.2.1.2 Limitations of the TPD Technique

Despite the simplicity of TPD technique and usefulness of the data obtained, the main limitations of this technique have to be pointed out.

- It cannot distinguish different types of acid sites (i.e., Lewis or Brönsted), so no structural information can be obtained.
- It is mostly a qualitative technique; quantitative studies are a very demanding task and many attempts have been made to overcome this problem, one of them being the combination of TPD-MS with TGA.

- It is difficult to separate the kinetic, diffusion and readsorption effects influencing TPD profiles obtained under typical flow setups operating under atmospheric pressures [24].
- TPD gives an average value of acid strength rather than a distribution.
- It is strongly affected by mass or heat transfer and diffusion.
- Desorption from weak sites is hindered by adsorbates on strong sites.
- During desorption, a readsorption may occur.
- Peaks overlap, so TPD can be used only to roughly distinguish the various acid site strengths; application of the curve deconvolution technique is highly recommended.

It is generally the case that adsorption microcalorimetry provides a much better description of the surface acid site strength distribution than TPD [18].

3.2.2 Adsorption Calorimetry

3.2.2.1 Description of the Calorimetry Technique

Values of the enthalpy of adsorption, determined either from the variation of adsorption with temperature (isosteric enthalpy of adsorption) or by direct calorimetric measurements, provide a valuable insight into the mechanism of adsorption. When taken together with data of adsorption isotherms, they provide information which could not be extracted from either set of data alone.

Heats of adsorption and other thermodynamic parameters can be obtained either by direct calorimetric determination, $-\Delta H = f(n_a)_T$ (n_a = adsorbed amount), or by using the Clausius-Clapeyron equation and the data from isosteric measurements. Isosteric heats of adsorption are calculated from the temperature dependence of the adsorption isotherms, i.e., from the isosteres. Indeed, q_{st} can be computed from the experimental isosteres for each average temperature according to the equation

$$q_{st} = -RT^2\left(\frac{\partial(\ln P)}{\partial T}\right)_{n_r} \tag{3.5}$$

where T is the absolute temperature, R the gas constant and n_r the number of reversibly adsorbed molecules.

However, adsorption in micropores is often reported to be irreversible, and this effect is mostly responsible for the frequently encountered uselessness of adsorption heats calculated from isosteres.

The average errors in evaluation of the differential heats of adsorption, as estimated by Stach et al. [57], are 1–2% only for the direct measurement and around 5% for the isosteric measurements.

The direct measurement of heats is also more accurate than the computation of energies from TPD data, which quantifies average activation energies of desorption.

Whereas calorimetry is capable of yielding the strong acid sites distribution in detail, the TPD of ammonia usually yields only an average value [58], except when appropriate kinetic models are employed.

In the direct calorimetric determination, $(-\Delta H = f(n_a)_T)$, the amount adsorbed (n_a) is calculated either from the variations of the gas pressure in a known volume (volumetric determination) or from variations of the weight of the catalyst sample in a static or continuous-flow apparatus (gravimetric determination), or from variations of the intensity of a mass spectrometer signal [59]. In a static adsorption system, the gas is brought into contact with the catalyst sample in successive doses, whereas in a dynamic apparatus the catalyst is swept by a continuous flow. Comparative calorimetric studies of the acidity of zeolites by static (calorimetry linked to volumetry) and temperature-programmed (differential scanning calorimetry linked to thermogravimetry) methods using ammonia adsorption and desorption have appeared in the literature [50].

Presently, calorimetry linked to the volumetric technique is still the most commonly used method [60]. Prior to adsorption, the catalyst is outgassed at the desired temperature and under high vacuum (~0.1 mPa) in the calorimetric cell. After cooling down to the adsorption temperature and establishing the thermal equilibrium of the calorimeter, a dose of gaseous probe molecule is brought into contact with the catalyst sample, and both the pressure and heat signal are monitored until equilibrium is reached. Then, successive new doses are added and the new equilibrium pressures are recorded together with the corresponding evolved heats. The data obtained directly from adsorption calorimetry measurements can be expressed in different ways (Fig. 3.5) as follows.

(a) The raw data obtained for each dose of probe molecule, i.e., the evolution of the pressure above the sample (P) and the exothermic heat evolved signal (Q) as a function of time (Fig. 3.5a).
(b) The amount of gas adsorbed at constant temperature plotted as a function of the equilibrium pressure (adsorption isotherm I).

In order to accurately determine the chemisorbed amount from the overall adsorption isotherm, the sample can be further outgassed at the same temperature to remove the physically adsorbed amount, after which a new adsorption procedure is carried out to obtain isotherm II. The difference between the first and second isotherm gives the extent of irreversible adsorption (V_{irr}) at a given temperature (Fig. 3.5b), and can be considered as a measurement of the amount of strong sites in the catalyst. However, in the first approximation, the magnitude of the heat of adsorption can be considered as a simple criterion to distinguish between physical and chemical adsorption.

(c) The corresponding calorimetric isotherms (Q_{int} vs. P) (Fig. 3.5c).
(d) The integral heats (Q_{int}) as a function of the adsorbed quantities (n_a) (Fig. 3.5d).

This representation leads to the detection of coverage ranges with constant heat of adsorption, for which the evolved heat is a linear function of the coverage.

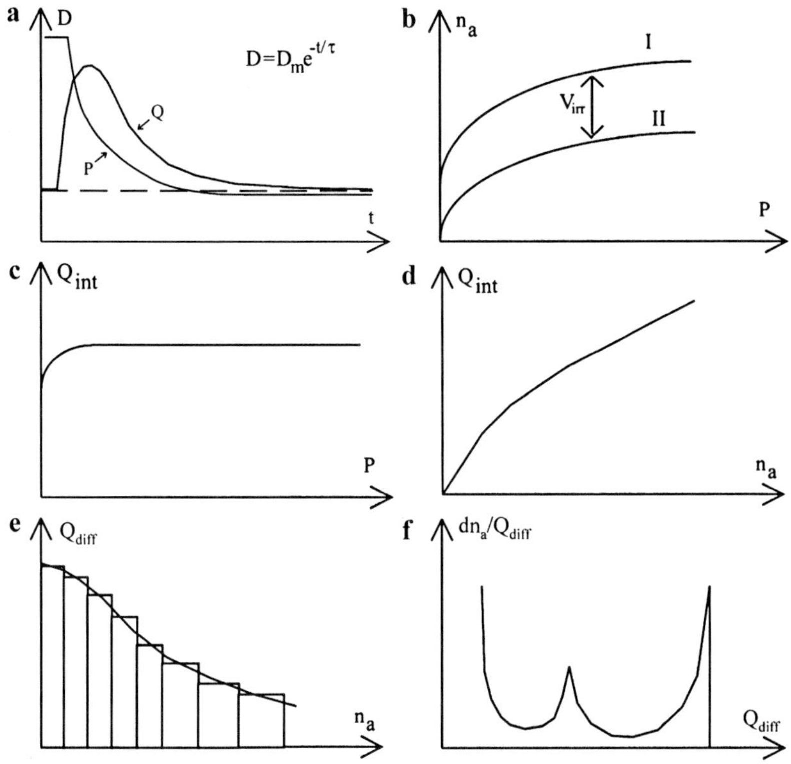

Fig. 3.5 Calorimetric and volumetric data obtained from adsorption calorimetry measurements [61]

(e) The differential heat $Q_{\text{diff}} = \partial Q_{\text{int}}/\partial n_a$ (molar adsorption heat for each dose of adsorbate) as a function of n_a (Fig. 3.5e).

The ratio of the amount of heat evolved for each increment to the number of moles adsorbed (in the same period) is equal to the average value of the differential enthalpy of adsorption in the interval of the adsorbed quantity considered. The curve showing the differential heat variations in relation to the adsorbed amount is traditionally represented by histograms. However, for simplification, the histogram steps are often replaced by a continuous curve connecting the centers of the steps. The curves of differential heat of adsorption vs. coverage are generally distinguished by the following features [6] (Fig. 3.6):

1. An initial region of high heat of adsorption, representing adsorption on the strongest sites which are ascribed mainly to Lewis sites. The initial drop in the curve of Q_{diff} vs. coverage, even in the case of adsorption on apparently homogeneous surfaces, could be ascribed to residual surface heterogeneities. Thus, an exponentially decreasing adsorption energy distribution for 1–5% of the surface exhibiting strong heterogeneity is often observed, especially in the presence of extra-framework aluminum species.

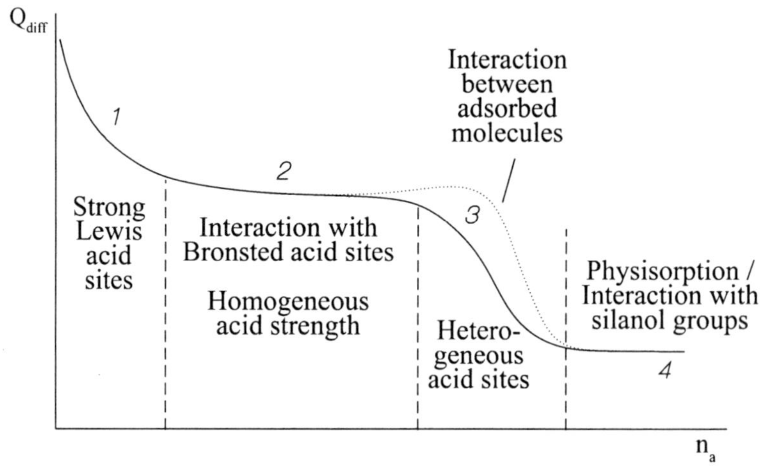

Fig. 3.6 Regions in a typical curve of differential heat of adsorption versus adsorbed amount [6]

2. One or more regions of intermediate strength sites which are predominantly Brönsted acid sites. There is always a fairly good relationship between framework aluminum content and population of Brönsted acid sites, provided the framework aluminum is totally accessible to the probe. A region of constant heat in this domain is characteristic of a set of acid sites of homogeneous strength.
3. A region where heats decrease more or less steeply depending on the heterogeneity of the sites. Sometimes interactions between already adsorbed molecules can create a small heat increase with gas loading.
4. At high coverages the heat of adsorption approaches a nearly constant value characteristic of hydrogen-bonding between the probe and the sample or physisorption of the probe. This constant value depends on the nature of the probe (i.e., of its enthalpy of vaporization at the adsorption temperature). However, there is also evidence in the literature for weak Lewis acidity giving rise to low heats of adsorption [62]. The heat then falls to a value close to that of the heat of liquefaction of the adsorbate.

(f) Acidity spectra.

In some cases the variation of the adsorption heats with progressive coverage corresponds to step-shaped curves. Such a behavior may be associated with the discrete surface heterogeneity due to the existence of several energetic levels [63]. In such cases, to describe the change in the adsorption heats with coverage, another approach is to plot acidity spectra (Fig. 3.5f): assuming that the variation in the adsorption heats coincides with energy distributions, one may wish to measure the number of sites with the same energy, i.e., sites that give rise to the same differential heat. This is achieved upon plotting $-\mathrm{d}n/\mathrm{d}Q_{\mathrm{diff}}$ as a function of Q_{diff}. The area below the curve included between Q_{diff} and $Q_{\mathrm{diff}} + \mathrm{d}Q_{\mathrm{diff}}$ represents the population of sites of identical strength estimated via Q_{diff}.

(g) Plot of the variation of the thermokinetic parameter as a function of the adsorbed amount of probe.

Heat conduction microcalorimetric output consists of power versus time and hence can undergo analysis to produce not only thermodynamic but also kinetic data. The kinetics of heat release during adsorption can be monitored by the change in the thermokinetic parameter τ [64, 65]. The calorimetric signal decreases exponentially with the adsorption time after the maximum of each adsorption peak. This can be approximated by $D = D_m \exp(-t/\tau)$, where D and D_m are the deviation at time t and the maximum deviation of the calorimetric signal, respectively (see Fig. 3.5a). In this expression, the thermokinetic parameter τ, known also as time constant, can thus be calculated as the reciprocal of the slope of the straight line obtained upon plotting log D as a function of time [64]. This thermokinetic parameter is indeed not constant and varies with coverage. One can then plot the variations of the thermokinetic parameter with the amount of adsorbed probe.

For most zeolites, when NH_3 is used as probe molecule at a given temperature, the time needed to establish thermal equilibrium after each dose at first increases with increasing adsorbed amount, passes through a maximum, then decreases rapidly and finally reaches a value close to the time constant of the calorimeter [6].

In fact, the establishment of the adsorption equilibrium is monitored both through the change in the heat signal and through the variations of pressure in the system. The time required to establish equilibrium depends on the quantity of adsorbed probe, on the temperature and on the inertia of the calorimeter. At low temperatures, a slower adsorption is observed in covering the strong adsorption centers than at higher temperature. The long time to establish equilibrium is apparently related to a redistribution of the adsorbed probe on the centers that are energetically more favorable [66].

A calorimetric study of ammonia adsorption on zeolites has confirmed that, at 303 K, the adsorption process involves a slow redistribution of adsorbed species, so that a longer time is needed to achieve thermal equilibrium [67]. The rate of heat liberation depends strongly on the amount of ammonia adsorbed. At higher temperature this phenomenon cannot be observed. The data related to the thermokinetic behavior of ammonia adsorbed on faujasite, L and mordenite type zeolites have been discussed in the literature [67]. The position of the maximum on the kinetic curves depends on the chemical composition of the zeolite and on the degree of filling of the cavities of the zeolite by the molecules adsorbed. The slow liberation of the heat is associated with a redistribution of the adsorbed molecules. When the adsorption temperature is increased to 373 K and then to 473 K, there is a decrease in the time required to establish equilibrium and a decrease in the region of coverages in which slow adsorption is observed [66]. At 573 and 673 K, equilibrium is established rapidly at all coverages. The time to establish thermal equilibrium in this case is determined solely by the inertia of the calorimeter. When using probes with a molecular diameter much larger than that of NH_3, other diffusional problems also occur.

(h) Estimate of the entropy of adsorption from the adsorption equilibrium constants obtained from adsorption isotherms and heat of adsorption data obtained microcalorimetrically.

The differential molar entropies can be plotted as a function of the coverage. Adsorption is always exothermic and takes place with a decrease in both free energy ($\Delta G < 0$) and entropy ($\Delta S < 0$). With respect to the adsorbate, the gas-solid interaction results in a decrease in entropy of the system. The cooperative orientation of surface-adsorbate bonds provides a further entropy decrease. The integral molar entropy of adsorption S^a and the differential molar entropy $S^a{}_{\text{diff}}$ are related by the formula $S^a{}_{\text{diff}} = \partial(n_a S^a)/\partial_{na}$ for the particular adsorbed amount n_a. The quantity S^a can be calculated from

$$\Delta_a S = S^a - S^{g,P} = \frac{Q_{\text{int}}}{T.n_a} + \frac{R}{n_a}\int_0^P n_a d(\ln P) \qquad (3.6)$$

where S^a is the molar entropy of the adsorbed phase, $S^{g,P}$ the molar entropy of the gaseous phase (available from tables), and n_a the adsorbed amount. After integration of the plot of n_a vs. $\ln(P)$ between 0 and P from the adsorption isotherm at the temperature T, and of the plot of Q_{diff} vs. n_a between the same boundaries to obtain Q_{int}/n_a, the value of S^a can be obtained, and then that of $S^a{}_{\text{diff}}$.

As indicated above, the extent of adsorption can also be measured directly by weighing the catalyst sample together with the adsorbed probe molecule by means of a microbalance set inside the calorimeter. The sample is outgassed as previously described and then contacted with successive doses of probe gas. The adsorbed amount is calculated from the weight gain for each dose [68].

Calorimetry is also a suitable technique for co-adsorption studies. For example, it has been shown that the enthalpy change of ammonia adsorption on the remaining sites of an H-mordenite after methanol preadsorption is not affected by the irreversibly bound methanol species, but that methanol preadsorption has the effect of reducing the number of strong adsorption sites [69].

Another application of the calorimetric technique deals with simultaneous measurements of loading, composition and heats of adsorption from gaseous mixtures. The components of a binary mixture are dosed alternatively, so that the individual differential heats of adsorption can be determined from two successive measurements [70, 71].

An alternative method to the static adsorption method is flow adsorption microcalorimetry, which involves the use of a carrier gas passing continuously through the adsorption cell [72–76].

In the pulse flow method, the procedure consists in injection of a precise and well-defined gas volume (probe molecule + carrier gas) into the stream which flows through the catalyst bed held on the fritted glass of a specially designed calorimetric cell. For each pulse, the calorimetric signal is recorded and the amount of gas which has not been retained by the catalyst is measured by a gas chromatograph (or mass spectrometer) connected on-line to the calorimetric cell. The major disadvantage of this technique is that the weakly chemisorbed portion of the probe gas is not held by the catalyst and gives rise to an endothermic peak of desorption which follows immediately the exothermic peak of adsorption, and thus necessitates peak deconvolution.

As an illustration, let us mention a study by Brown et al. [72], who investigated by flow adsorption calorimetry the relative accessibilities of acid sites in zeolite Y and K_{10} clay catalysts in both acid and Na^+ forms, reaching the conclusion that, under flow conditions and nitrogen at atmospheric pressure as carrier gas, the sites first covered by ammonia are not necessarily those with the highest heats of adsorption.

Another example is a calorimetric study by Aguayo et al. [74] of the adsorption of different bases (ammonia, *n*-butylamine, *tert*-butylamine, pyridine) in gas phase on a FCC (H-Y-type) catalyst in flow regime through an isothermal bed of solid; the authors have determined the conditions under which the adsorption is exclusively chemisorption, as well as those under which the adsorption is not limited by the internal diffusion of the base.

Flow microcalorimetry (FMC) has been compared with batch volumetry methods, i.e., two-cycle adsorption (TCA) and temperature-programmed desorption (TPD) for the determination of the surface acidity of various porous materials in powder form, using ammonia chemisorption in all cases [77]. Flow-calorimetric measurements of the transfer of matter and thermal effects accompanying one adsorption-desorption cycle of pure ammonia in helium carrier gas at 373 K were carried out on powders of porous materials such as H-Y zeolites with varying Si/Al ratios (HY 30, HY 20, HY 9, HY 6) and surfactant-templated aluminosilicates (MSA 20, MSA 10, MSA 5).

The comparison of the flow-microcalorimetry method with two-cycle ammonia adsorption and thermoprogrammed ammonia desorption revealed some differences in the number of surface acid sites (Fig. 3.7), which could be ascribed to variations in experimental conditions used during desorption or pre-treatment stages in each of these methods. Therefore much attention should be paid to this aspect when comparing the results of surface acidity tests reported in the literature. The same experiments were performed on some selected samples pre-saturated with water vapor, and the non-negligible influence of vicinal water on the results of surface acidity tests was demonstrated [77].

Finally, let us mention that calorimetric measurements can also be used to monitor adsorption phenomena in liquid phase (in a solvent). For example, the so-called *cal-ad method* [78–81] has been used to measure the adsorption heats evolved upon addition of dilute solutions of pyridine in *n*-hexane to a solid acid (TS, H-ZSM-5) in a slurry with *n*-hexane. The amount of free base in solution is measured separately with a UV-Vis spectrophotometer [78, 79]. Contrarily to earlier calorimetric results for H-MFI in vacuum, the authors [78] suggested the presence of two energetically different types of sites, with only the strong sites ($\approx$176 kJ mol^{-1}) capable of transferring a proton to the adsorbed base. More recently, de Macedo et al. have characterized Na and proton exchanged mordenite zeolites by the cal-ad method (calorimetric and adsorption studies of pyridine

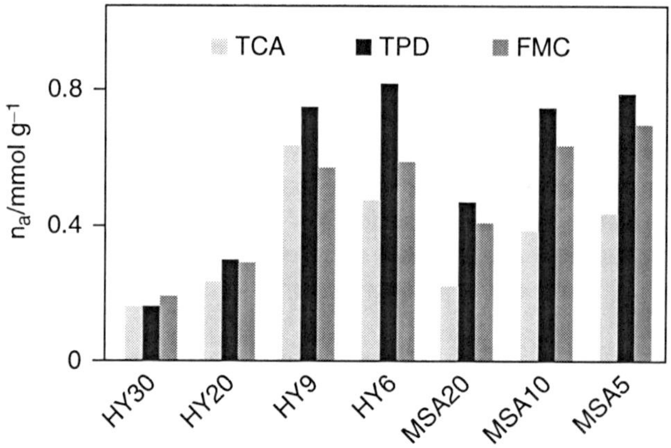

Fig. 3.7 Schematic diagram of the difference in number of surface acid sites, determined by two-cycle adsorption (TCA), thermoprogrammed desorption (TPD) and flow adsorption microcalorimetry (FMC), on various zeolites and mesoporous aluminosilicates [77]

interaction with the solid in cyclohexane slurry) [82]. A two-site adsorption process has been proposed to describe the acidic sites of H-MOR, with 0.081 mmol g^{-1} of Brönsted sites ($\Delta H_1 = -82.9$ kJ mol^{-1}) and 0.422 mmol g^{-1} of Lewis sites ($\Delta H_2 = -36.7$ kJ mol^{-1}). However, Savitz et al. [80, 81] using vapor-phase measurements for the co-adsorption of pyridine and *n*-hexane in H-MFI claimed that equilibrium is not achieved for pyridine titration measurements in a solvent at room temperature and, therefore, that the ability of the cal-ad method to distinguish Brönsted acid sites of widely different strengths in H-MFI is unproven [80, 81]. Finally, it is also worth mentioning that calorimetric measurements of immersion into water and *n*-butylamine have been used to estimate the hydrophilic and acid-base characters of a series of Zn(II) exchanged Na-X zeolites [83].

3.2.2.2 Limitations of the Calorimetry Technique

Although it is quite a powerful and informative technique, some limitations of calorimetry have to be mentioned, the most important one being the time-consuming nature of experiments.

- Wall adsorption, although significant mainly in metallic volumetric systems, appears not to result in major problems if a suitably careful calibration is performed. Rapid transfer of heat from the sample to the cell wall is recommended [84].
- Careful heat flow calibrations have to be performed. Chemical calibrations present many disadvantages: they rely on prior results, with no general agreement and no control of rate, and are generally available only at a single temperature. On the contrary, electrical calibrations (Joule effect) provide many advantages and they are easy to perform at any temperature [6].

- It appears that it is often difficult to determine the nature of the adsorbed species, or even to distinguish between the different kinds of adsorbed species, from the calorimetric data. In many cases this technique fails to distinguish between cations and protonic sites and so to discriminate Lewis and Brönsted sites, due to the insufficient selectivity of the adsorption if no complementary techniques are used. As no exact information can be obtained regarding the nature of the acid centers from the calorimetric measurements, suitable IR, MAS NMR, and/or XPS [10] investigations are necessary to identify these sites. However due to the complex nature of the acid strength distribution, it is currently still not possible to establish a detailed correlation between sites of different nature and their strength.
- Another limitation is that the rate of heat liberation depends strongly on the amount of preadsorbed ammonia [85]. In view of the high sensitivity of calorimeters, heat liberation can be observed for a significantly longer period, but the true time necessary to establish equilibrium cannot be determined accurately because the heat liberation curve approaches the base line asymptotically. However the use of a highly sensitive and accurate capacitance manometer for the measurement of the equilibrium pressure, and of a sufficiently sensitive calorimeter with a low time constant, limits the error on this determination.
- The adsorption of the gas should not be limited by diffusion, neither within the adsorbent layer (external diffusion) nor in the pores (internal diffusion). Should diffusion limitations occur, then adsorption on active but less accessible sites may only occur after better exposed but less active sites have interacted. Diffusion may thus cause the "smoothing out" of significant details in the energy spectrum, and the differential heat curves determined under the influence of diffusion phenomena may indicate less surface heterogeneity than actually exists on the adsorbent surface [86].

The validity of energy distributions derived from heats of adsorption has been examined in the literature [87]. According to Biaglow et al. [88], there is a possibility that the mobility of molecules larger than NH_3, such as pyridine, can be limited even at weaker sites. A way to get around this is to increase the mobility of the probe molecule by increasing the calorimeter temperature, but this might be limited by the thermal stability of the probe in the presence of the sample.

- Microcalorimetry can give erroneous results if adsorption equilibration is too slow, a particularly serious problem if one works at low temperatures. If, e.g., a fraction of the adsorbate molecules do not reach the aluminum sites of the framework during the time of the experiment, the measured heats will not be indicative of the strong Brönsted acid sites [89]. The temperature has to be chosen according to a systematic study of the temperature dependence of the heat curve corresponding to probe molecule chemisorption. At low temperature, selective sorption on the stronger acid sites is not observed; instead, random distribution of the gaseous molecules occurs simultaneously on acid and non-acid sites. It follows that temperatures high enough to avoid steric hindrance or diffusional limitations should be used for the acidity determinations [90, 91].

Generally speaking, the adsorption temperature should not be too low, in order to allow the detection of differences among the sites; otherwise, under certain circumstances the measured evolved heat can be just an average value. Another important issue is that one must ensure that chemisorption predominates over physisorption.

- The absence of a plateau of constant heat in the differential heat curve can be the result of molecular interactions between molecules adsorbed at neighboring sites rather than a true indication of differences between sites [88]. This matter can be checked by varying the probe size or the site density.

3.3 Probe Molecules

Appropriate probe molecules are required depending on the type of information desired: about the structure and topology of the zeolite using pore filling or about acid/base properties of the zeolite using adsorption of basic/acidic molecules.

The micropores of zeolites have diameters comparable to molecular dimensions, which are responsible for the phenomenon of shape selectivity in catalytic reactions. For example, branched hydrocarbons and substituted aromatic molecules are separated using zeolitic sorbents and catalytically converted in large-scale technical processes relying on the principles of shape selectivity. One of the ultimate goals of fundamental research in catalysis is to design tailor-made catalysts. In order to achieve this goal it is of utmost importance to obtain precise information about the structure of zeolites, and in particular about the topology of pores and cavities, as well as the active sites. Adsorption-desorption of properly chosen probe molecules monitored by TPD or microcalorimetry can provide valuable insight into molecular arrangements in these complex solids.

Even though, as a general rule, adsorption calorimetry makes it possible to estimate the total number of adsorption sites and potentially catalytically active centres; the values obtained depend on the nature and size of the probe molecule. Appropriate probe molecules for adsorption microcalorimetry should be stable over time and with temperature. The desired properties of the probe depend on the materials to be studied: in the case of zeolites of the aluminosilicate type, the negatively charged aluminum atoms in the framework generate an electrostatic field, making zeolites capable of interacting strongly with polar molecules. On the other hand, non-polar molecules can be adsorbed on zeolites via van der Waals interactions as well as pore filling [92]. Furthermore, the probe molecules should be small enough to readily penetrate into the intracrystalline space of most zeolites. In general, the probe adsorbed on the zeolite surface should also have sufficient mobility to equilibrate with active sites at the given temperature. For example, the heats of adsorption of CO_2 and NH_3 are extremely sensitive to changes in the chemistry and geometry of the surface of the void spaces and channels in zeolites, because the significantly large quadrupole (CO_2) and dipole (NH_3) moments cause a considerable contribution of specific interactions to the total energy of adsorption of these molecules [93]. Moreover, repulsive interactions between adsorbate molecules could arise from the ionic repulsion of two positively charged adsorbates, or from changes in a protonic site due to adsorption at adjacent sites [88].

3.3.1 Probing the Pore Structure of Zeolites

Zeolites present different configurations of porous systems. The porous framework can be uni-, bi- or tridimensional, and channels may or may not be interconnected. These architectural features of the zeolite have important consequences for the modes and rates of diffusion of molecules.

Zeolites can be divided roughly into three main classes in terms of pore openings:

1. large-pore zeolites (6.5–8 Å): faujasite, mordenite, ... , which can admit large molecules like alkylaromatics;
2. medium-pore zeolites (5–6.5 Å): offretite, ZSM-5, ferrierite, ... , which can admit alkanes and possibly simple aromatics;
3. small-pore zeolites (3.5–5 Å): Rho, ZK-5, ... , which admit only normal alkanes.

Physical techniques such as XRD (X-ray diffraction), neutron and electron diffractions are the most adequate tools to determine the crystalline structure of zeolites. However, despite the advances of these techniques, the determination of zeolite structures is not a simple task. Valuable insight into the pore topology of molecular sieves can also be obtained by studying the adsorption and reaction of well-selected probe molecules. The potential of adsorption studies combined with microcalorimetry goes beyond the determination of the pore dimensions, and it is believed that these methods can give a reliable idea of the pore topology even in more complicated cases involving structures formed by pores with different dimensions. This type of studies can be very helpful even in cases where the zeolite structure has been determined by crystallographic techniques. Indeed, knowledge of the crystallographic structure only provides a limited amount of information about micropore accessibility, existence of diffusion restrictions, amounts of organic molecules adsorbed in the micropores, etc., since these characteristics depend on the pore shape and the particular packing of the molecules [94]. It is important to realize that the crystallographic pore diameter does not represent a suitable threshold molecular diameter, since molecules 20% larger than this diameter are readily admitted, especially at higher temperatures. For example, naphthalene (9.1 $\times$ 6.8 $\times$ 3.4 Å) is readily taken up at 100°C by H-ZSM-5 zeolite (pores 5.1 $\times$ 5.5 and 5.3 $\times$ 5.6 Å). While adsorption and reactivity studies cannot give details on the crystalline structure, they can be of special interest since they indicate how the chosen molecules "see" that particular structure [95].

Adsorption experiments with molecules of different sizes can be also used to estimate the relative extents of the external and internal surface areas [96] and can be successfully followed by adsorption microcalorimetry. Each adsorbate will enter into micropores according to their relative sizes; the amount that fills them up is determined from the abrupt change in differential heat of adsorption that occurs at the completion of micropore filling. Also, the shape and position of the curve of differential heat vs. coverage yields information on the energetics of the adsorbate/surface interaction. The total amount adsorbed inside the micropores depends not only on the micropore volume, but also on how the adsorbate molecules can be accommodated inside the microporous space. This is related to the shape and size of the molecules and to the presence and dimensions of the channels, cavities, side pockets, etc. [97]

The pore structure of zeolites hinders or restricts the access of reactants to some of the potentially active sites. With relatively bulky reactants, some surface sites of suitable acid strength may remain inactive during the reaction under investigation; the same sites, however, could be active toward less bulky reactants.

It is well known that under identical conditions of adsorption on molecular sieves of the same chemical composition but different pore systems, the structure of the pore system and the diameters of the pores of the microporous adsorbents are responsible for the phenomenon of selective adsorption. The intensity of the electrostatic field is also a determining factor. In acid-catalyzed reactions, zeolites often show shape selectivity because of their unique pore structure. Outer surface acidity, however, diminishes this important property, and can even cause pore blocking by coke formation [6].

Zeolites with connected channels of 10 and 12-membered rings (MR) offer an interesting pore arrangement for different applications, e.g., for catalysis since they may combine high conversion and shape selectivity features. Large channels of 12 MR allow large molecules to diffuse and react, while products formed may diffuse with different abilities in the two pore systems available. Preferential diffusion paths may exist depending on the relative sizes of the reactants and products.

In addition to the specific interaction with the acidic centers, each probe molecule is expected to interact through dispersive forces with the walls of the pores, giving rise to the confinement effect [98, 99]. N_2 and CO single out mainly Lewis sites and are also sensitive to confinement effects due to dispersive forces which are governed by the micropores. Polar molecules such as CH_3CN and NH_3 are not preferentially adsorbed on Lewis sites with respect to the Brönsted ones, by virtue of their ability to generate either protonated species or strong H-bonded adducts.

Adsorption of *n*-alkanes on zeolites and other porous solids is important not only in the context of hydrocarbon processing in the petroleum refining industry, but also in studies of porous materials. The micropores of zeolites are similar in size to a range of hydrocarbon molecules, and a number of shape-selective reactions have been utilized in practical applications. However, the relationship between pore size and restrictions of reactions through steric hindrances has yet to be adequately understood [100].

Ruthven et al. [101] recommended adsorption of *n*-hexane for determination of the micropore volume of zeolites. Mittelmeijer-Hazeleger et al. [102] proposed temperature programmed desorption of *n*-nonane as a comparative method for probing the size of micropores in zeolites (4A, 5A and 13X) and carbon molecular sieves.

Makowski et al. [103] have shown that quasi-equilibrated temperature programmed desorption and adsorption (QE-TPDA) of *n*-nonane on zeolites H-ZSM-5 and H-Y, as well as on mesoporous MCM-41 and SBA-15, allows the determination of the micropore and mesopore volume as well as the mesopore size. Complex adsorption/desorption profiles with high temperature desorption peaks related to adsorption in the micropores and low temperature peaks attributed to capillary condensation in the mesopores were observed. Thus, QE-TPDA may be a complementary method to the standard technique of low temperature N_2 adsorption.

Microcalorimetric studies of *n*-hexane, toluene, *m*-xylene and 1,3,5-trimethylbenzene adsorptions on bidimensional 10-MR (ZSM-11), unidirectional uniform 12-MR (SSZ-24), constrained unidirectional (ZSM-12), 12- and 8-MR (mordenite), and 12- and 10-MR (CIT-1) zeolites have been perfomed by Guil et al. [94], who suggested that the following set of measurements should be performed in order to determine the pore topology:

1. Adsorption of four probe molecules on the investigated material, monitored by adsorption microcalorimetry, to test whether the adsorbate enters into the micropores or not. This allows the determination of the approximate micropore diameter.
2. Comparison of total uptake of each adsorbate on micropore network shows the existence of micropore sets of different diameters, since bigger molecules will enter only the wider channels.
3. The variation of adsorption heats with uptake, besides defining the micropore capacity for each adsorbate, provides information on the surface chemistry.
4. Packing molar densities of the adsorbates give information on the particular micropore shape and the existence of cavities, lobes, etc.
5. Toluene and/or *n*-hexane post-adsorption experiments after preadsorption of different adsorbates give information on the connectivities or crossings between sets of channels.

Adsorption and microcalorimetric study [104] of molecules with different kinetic diameters, such as toluene, meta- and orthoxylene, and 1,2,4-trimethylbenzene, together with consecutive adsorption of these molecules, confirmed the presence of 10-MR channels and cavities of larger dimensions in MCM-22, all of which can be reached from the external surface. The dimensions of the two types of void species have been estimated from the kinetic diameters of the adsorbates: diameters of ~5.9 and 7.0 Å have been calculated for the micropores and cavities, respectively.

Corma et al. [95] have shown that the pore topology of zeolite IM-5 can be determined by combining catalytic test reactions and adsorption microcalorimetry of hydrocarbons with different sizes and shapes (*n*-hexane, toluene, *m*-xylene, and 1,3,5 trimethylbenzene). The pore diameter should be close to but somewhat smaller than 5.5 Å. It has been found that the channel network consists of a system of unidirectional 10-MR with lobes or side pockets, or crossing 10-MR pores with a pore diameter somewhat smaller than those in ZSM-5. This structure offers interesting shape selectivity features for catalytic reactions.

A calorimetric study [105] of adsorption of light alkanes on the acidic zeolites theta-1 (H-TON), ferrierite (H-FER) and ZK-5 (H-KFI) has shown that sorption of alkanes on the Brönsted acid sites is energetically favored, and the uptake at low alkane partial pressures is directly correlated to the concentration of acid sites. With H-FER, the accessibility of the two intersecting channel systems depends on the chain length of the alkanes. Although propane and *n*-butane adsorb in both channels, the sorption of *n*-pentane is hindered in the 8-MR channels and sorption of *n*-hexane only occurs in the 10-MR channels. In the presence of cavities (such as found in H-KFI), pronounced intermolecular interactions between sorbed alkanes were observed. In general, the heat of adsorption of alkanes increases with decreasing pore size. At pore size smaller than 0.5 nm repulsion forces start to be important, their impact being most prominent for larger isoalkanes.

Eder et al. [106] have reported that the sorption of alkanes in molecular sieves seems to be governed by a multitude of influences. In addition to a general increase of the strength of interaction with the size of the *n*-alkane and a decrease in the pore size of the molecular sieve, factors such as ordering, intermolecular interactions, and interaction with the acid sites are important. The influence of ordering on the shape of the adsorption isotherm and the maximum uptake is illustrated with the sorption of *n*-hexane and *n*-heptane in MFI zeolite, when the chain length of the alkane is similar to the length of zig-zag channels (see Fig. 3.8).

Molecules sorbed in the zig-zag channels of MFI hinder the sorption in the straight channels by partially blocking the intersections. To reach pore filling, these molecules have to be confined to their position, so higher partial pressures

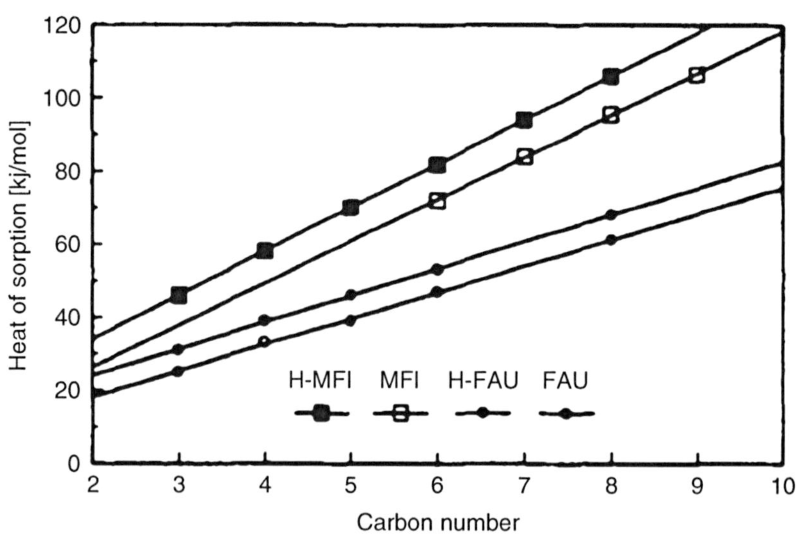

Fig. 3.8 Heats of *n*-alkane sorption on acidic and neutral MFI and FAU zeolites [106]

are necessary and a kink or even a plateau is observed in the isotherms. At elevated temperatures (e.g., *n*-heptane sorption at 373 K) the thermal motion of the molecules is too high to realize such confinement. In contrast, sorbate-sorbate interactions lead to an increase of the heat of adsorption with increasing loading in large pore molecular sieves such as faujasites.

Alkane sorption experiments have revealed [107] that the 10-MR channel system of MCM-22 persists unperturbed in MCM-36. The major fraction of the acid sites, and thus the favored sorption sites, are located in the 10-MR channels of these layers. Only about 10% of acid sites exist on the outer surface and are accessible through the mesopores. Consequently, alkane sorption in MCM-22 and MCM-36 at low equilibrium pressures is dominated by the 10-membered ring channels in the layers.

Yang et al. [99] have studied the environment provided by the zeolite channels in high-silica zeolites having the MFI (ZSM-5), CHA (Chabazite), MOR (Mordenite), TON (ZSM-22), MTW (ZSM-12), FER (Ferrierite), and FAU (Faujasite) structures. Calorimetric measurements using CH_4 and O_2 at ~210 K showed that structure affects the adsorption properties in a manner which depends on the pore dimensions. The differential heats for CH_4 at low coverage were 28 kJ mol^{-1} in FER, 27 kJ mol^{-1} in TON, 25 kJ mol^{-1} in MOR, 21 kJ mol^{-1} in MFI, 20.5 kJ mol^{-1} in MTW, 19.5 kJ mol^{-1} in CHA, and 14 kJ mol^{-1} in FAU.

The thermodynamic parameters of the adsorption of hydrocarbons, and especially of normal paraffins, olefins and cyclic hydrocarbons, have been measured on highly dealuminated Na-Y zeolites, and for comparison on silicalite $(SiO_2)_{96}$, by Stach et al. [57, 108]. The dealuminated Na-Y named US-Ex $(Na_2(AlO_2)_2(SiO_2)_{190})$ presented mesopores with diameters between 3 and 10 nm besides the unchanged micropore structure of the parent Na-Y material. Dealuminated zeolites showed less energetic heterogeneity in the adsorption of hydrocarbons than Na-X and Na-Y zeolites. A comparison of the equilibrium data for hydrocarbons adsorbed on highly dealuminated zeolites revealed a strong influence of the diameters of the pore systems on the thermodynamic parameters. While the electrostatic fields within the pore systems of the zeolites influenced the heats of adsorption of hydrocarbons, no such influence on the entropy values was found [57]. In the case of *n*-paraffins, the heat increased with the number *n* of carbon atoms of the chain of the adsorbed *n*-paraffins. Moreover, the enthalpy values of *n*-paraffins adsorbed in faujasite increased with the number of adsorbed molecules, attained a maximum and then decreased [57].

The isosteric heats of adsorption of *n*-hexane and 3-methylpentane in zeolites Na-Y and H-Y have been reported by Pires et al. [109]. The difference between the heats of adsorption of *n*-hexane on the Na^+ (56 kJ mol^{-1}) and the H^+ forms (58 kJ mol^{-1}) of Y zeolites was within the range of experimental error, but a significant difference between the heats of adsorption of 3-methylpentane on Na-Y (51 kJ mol^{-1}) and H-Y (57 kJ mol^{-1}) was observed. These results were compared with those obtained on a ZSM-20. The adsorption of both molecules with a higher energetic effect on Y zeolite relative to ZSM-20 reflected structural effects, in agreement with the existence of larger supercages in ZSM-20 than in Y zeolite, and hence a higher confinement effect for the sorbate molecules in zeolite Y [109].

Differential molar heats of adsorption have been determined calorimetrically by Thamm et al. [110] for benzene and cyclopentane on H-ZSM-5 type zeolites with different aluminum contents. The results showed that the structural OH groups represent energetically preferred sorption sites for benzene and presumably also for cyclopentane. The occurrence of inflection points in the sorption isotherms as well as of pronounced maxima in the sorption heat curves for benzene and cyclopentane on the Al-poor H-ZSM-5 samples indicated that the influence of sorbate-sorbate interactions on the state of the sorbed molecules rose with decreasing Al content. Heats of adsorption of benzene and ethylbenzene were also measured by Nieβen et al. [111] on a H-ZSM-5 sample (Si/Al = 33.6) and were 64.5 and 86.5 ± 5 kJ mol^{-1}, respectively.

Differential microcalorimetry has also been used to study the adsorption of tetrachloroethene at 298 K on three ZSM-5 zeolites with Si/Al ratios of 26.5, 339 and 500. The diameter of this non-polar molecule is very close to the pore opening of the adsorbents. The differential molar enthalpies continuously increased with increasing coverage up to a value of 4 molecules per unit cell, and then showed from this value on a stepwise increase of about 10 kJ mol^{-1}. Subsequently, the heat curves were almost horizontal, before abruptly decreasing to the value of the heat of vaporization of tetrachloroethene when approaching micropore saturation [112].

Of course, the strict boundary between probing the pore size and probing the acidity is not so easy to define.

A microcalorimetric study of the acidic properties of Al containing MCM-41 molecular sieves in comparison with relevant FAU and MFI zeolites has been performed by *n*-hexane, benzene, acetronitrile, and water adsorption [113]. Where as the heats of adsorption of acetonitrile are mainly influenced by specific interaction on the acidic sites in the meso- and microporous molecular sieves, the adsorption heat of a non-polar molecule like *n*-hexane is determined by the pore size or density. However, the pore size effect on the adsorption heats of acetonitrile in acidic molecular sieves has to be taken into account while employing those heats as a measure of acidic strength.

Calorimetric data obtained by Yang et al. [99] for acetonitrile in the acidic forms of MFI, CHA, MOR, TON, MTW, FER and FAU zeolites suggests that adsorption heats are independent of the structure within experimental error, with differential heats for the 1:1 adsorption complexes all equal to 100 ± 10 kJ mol^{-1}. A possible explanation lies in the additional orientation-dependent interaction resulting from the hydrogen bonding. The size and structure of zeolite cavities significantly change heats of adsorption of physically adsorbed species, so that calorimetric measurements can provide useful insights into "confinement" effects for molecules that do not interact strongly with the acid sites. However, for molecules which form strong hydrogen bonds that orient the molecule with respect to the acid sites, heats of adsorption are not a simple sum of the specific hydrogen-bond interaction and the heat of physisorption.

3.3.2 *Probing the Acidity of Zeolites*

Different molecules are used as probes for the acidity of zeolites, and a variety of experimental techniques are employed to characterizate acid-base sites. Spectroscopic methods such as IR, NMR and XPS are often used, as well as TPD and microcalorimetry. As stated previously, the latter two techniques allow determining not only the density but also the strength of acidic centers.

Though it is very reliable, the calorimetric method depends on the choice of the basic probe molecule utilized to neutralize the acid sites. When a zeolite or related material is studied, a first problem lies in the transport limitations encountered by a given probe molecule during diffusion to the acid site. This is usually avoided by choosing the smallest gaseous basic probe molecule, which is NH_3.

Ammonia, being a small molecule, can penetrate all pores, whereas larger probes such as pyrrole can penetrate the large pores of H-Y but not the smaller pores of ferrierite. Large-pore zeolites display lower adsorption heats than medium or small-pore zeolites. Their number of acid sites per unit cell is also lower. However, the differential heats as a function of coverage reveal a uniform distribution of the strong acid sites in wide-pore zeolites, while it is much more heterogeneous in other zeolites. For example, comparing zeolites with variable offretite-erionite character, only crystals free from stacking faults such as H-offretite adsorb pyridine. The presence of erionite domains in offretite drastically reduces pyridine, but not ammonia chemisorption. On pure erionite, pyridine adsorption is not observed. This is why both ammonia and pyridine should always be used to ascertain that site accessibility is not restricted by the size of the probe molecules that represents the size of a reactant molecule [4].

As already mentioned, differential heats of adsorption for ammonia and pyridine vary with the zeolite structure [114]. These differences are not large and cannot be correlated to catalytic activities [115]. However, the differences do suggest that the specific interactions between the protonated molecule and the lattice are different. For example, the heats of formation of complexes with ammonia and a whole series of amines are uniformly higher by 15 kJ mol^{-1} on H-MOR compared to H-ZSM-5 [116]. On the other hand, the heats for complexes with substituted pyridines are, if anything, higher on H-ZSM-5. While calorimetric techniques obviously cannot provide direct structural information about the complex, these results suggest that the zeolite cavity does influence the overall results. In summary, analysis of the differential heats allows inferences to be made about the nature of the specific bonding in the adsorption complexes. Referencing the heats of formation of adsorbed complexes formed by strong bases to proton affinities (PA) allows one to abstract reasonable estimates for the specific interactions [117].

3.3.2.1 Acidic and Basic Probe Molecules

Two main types of probe molecules are most widely used for the characterization of acidic properties of zeolites [118]. Probe molecules such as NH_3, pyridine and amines form a chemical bond with the protons of hydroxyl groups, thus giving information about the concentration of acid sites of zeolites. On the other hand, aromatics, olefins, CO and H_2S can be used to acquire information on the strength and the accessibility of the acid sites to the probe molecules. The criteria for the selection of probe molecules have been summarized in a review by Knözinger [119].

Ammonia is among the smallest strongly basic molecules, and its diffusion is hardly affected by the porous structure, if at all. This makes it the most commonly used probe in calorimetry for testing the acid sites. Ammonia is adsorbed as an ammonium ion, and the corresponding heat of adsorption depends both on the proton mobility and on the affinity of ammonia for the proton.

Figure 3.9 shows the differential heats of ammonia adsorption versus coverage on samples of different types of zeolites, all of which presented only very few extra-framework species. The Si/Al ratio, the activation temperature, and the adsorption temperature are given for each sample in the figure caption.

Ammonia (pK_a = 9.24, proton affinity in gas phase = 857.7 kJ mol^{-1}) and pyridine (pKa = 5.19, proton affinity in gas phase = 922.2 kJ mol^{-1}), being strong

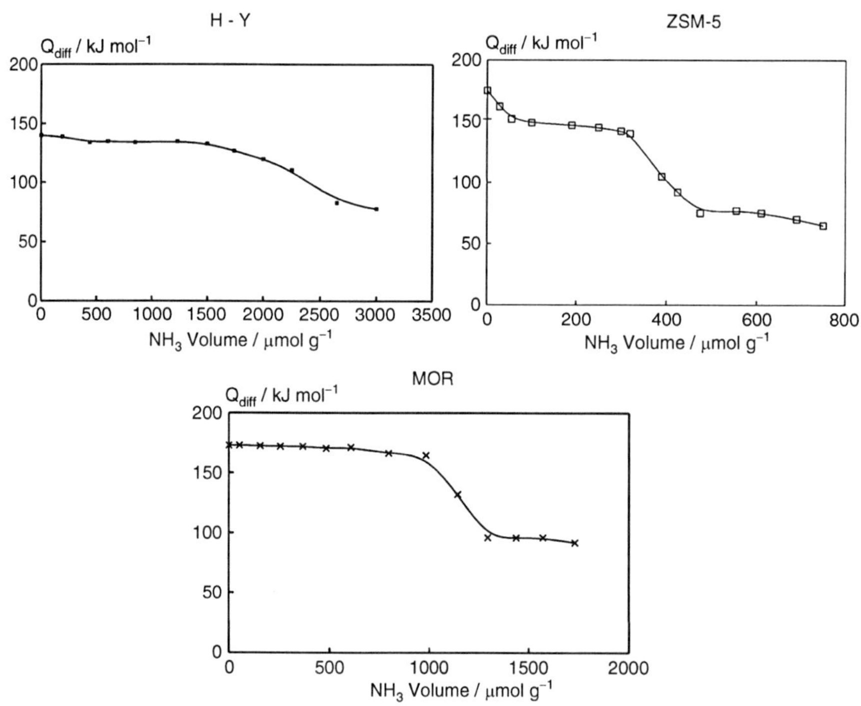

Fig. 3.9 Differential heat of adsorption vs. NH_3 uptake for a H-Y zeolite (Si/Al = 2.4, 80% exchange, pretreatment at 673 K, adsorption at 423 K), a ZSM-5 zeolite (Si/Al = 33.4, pretreatment at 673 K, adsorption at 423 K) and a mordenite (Si/Al = 16.2, pretreatment at 673 K, adsorption at 423 K) [6]

bases, are the favored molecules to probe the overall acidity of zeolites, since both Lewis and Brönsted acid sites retain these molecules. In order to distinguish qualitatively and unambiguously between these two types of sites it is necessary to use IR spectroscopy or to analyze the shifts in the binding energies of N_{1s} core-level lines of adsorbed N-probe molecules observed by XPS. Due to its small molecular dimensions (3.70 Å × 3.99 Å × 3.11 Å) [120], ammonia is a suitable probe for all OH groups accessible through pores, channels or windows greater than about 4 Å. The use of substituted pyridines, such as 2,6-dimethylpyridine (lutidine), has also been considered in order to probe specifically the Brönsted sites, because of their basic character and the steric effect of the methyl groups [121].

Carbon monoxide meets many criteria that have promoted its extensive use as a surface probe for the study of Lewis as well as Brönsted sites particularly by IR and NMR spectroscopy [7, 119, 122]. The CO molecule has an asymmetric charge and is easily polarized. Therefore it is sensitive to the electrostatic fields surrounding cations. As a result of the polarization by the metal cation of the zeolitic lattice, upon adsorption its stretching frequency increases relative to the gas phase. There are many results showing a correlation between the carbonyl frequency and the polarizing power of the cation which, being an electron acceptor, acts as Lewis acid site in the zeolite framework. Besides the strength of the Lewis acid sites, the adsorption of CO also provides information about their location, aluminum distribution in the surroundings, etc [123].

The use of different probe molecules has greatly enlarged the amount of available information on zeolite acidity, since different molecules can probe different types of acid sites. Weakly interacting probe molecules are much more specific than strongly interacting ones and therefore can provide more detailed information about acid sites. Recently, a growing amount of attention is being paid to studying the acid-base properties of zeolites using weakly interacting probe molecules like N_2O [124–127], benzene [128], CO [129–134] or hexane [106, 135–138], N_2 and H_2 [139, 140].

Acetonitrile is also an interesting molecule for probing acid sites in zeolites [91, 113, 141, 142]. It is a weak base, so no protons are abstracted and actual hydroxyl groups can be observed. It also allows the investigation of both Lewis and Brönsted acidities. While it is normally considered to be a weak base, it actually has a moderately high proton affinity (798 kJ mol^{-1}, compared to 857 kJ mol^{-1} for ammonia and 773 kJ mol^{-1} for methanol). Other nitriles and alcohols have also been used to probe the acid sites of zeolites [143, 144].

For basicity measurements, the number of acidic probes able to cover a wide range of strength is rather small [145]. Moreover, a difficulty stems from the fact that some acidic probe molecules may interact simultaneously with cations (such as Na^+). The ideal probe molecule should be specific to basic sites and should not be amphoteric. It should not interact with several types of basic sites or give rise to chemical reactions [145]. For instance CO_2 ($pK_a = 6.37$) is commonly chosen to characterize the basicity of zeolites, but it may either be adsorbed on the cations or physisorbed, or may react with hydroxyls and framework oxide ions to give carbonated species. However, the energetic aspect of the adsorption of CO_2 on various molecular sieves in a large domain of temperature and pressure can provide interesting information on the nature of the adsorbate-adsorbent interactions [93].

Li et al. [146] used CO_2 adsorption calorimetry and stepwise TPD of CO_2 for characterization of zeolites X modified by alkali metal oxides occluded in their pores. Results from CO_2 adsorption microcalorimetry and STPD for CsOx/CsX (13.5 Cs per unit cell), CsOx/KX (16.3 Cs per unit cell) and KOx/KX (15.2 K per unit cell) are shown in Fig. 3.10. From Fig. 3.10a it is evident that the CsOx occluded in supercages of CsX was a slightly stronger base than CsOx occluded in KX or KOx occluded in KX. The values of CO_2 adsorption capacity from STPD were slightly higher than those from adsorption microcalorimetry. This small difference may arise from the fact that the amount of CO_2 adsorbed with $-\Delta H_{ads}$ less than 60 kJ mol^{-1} was neglected [146]. Nevertheless, results from CO_2 STPD

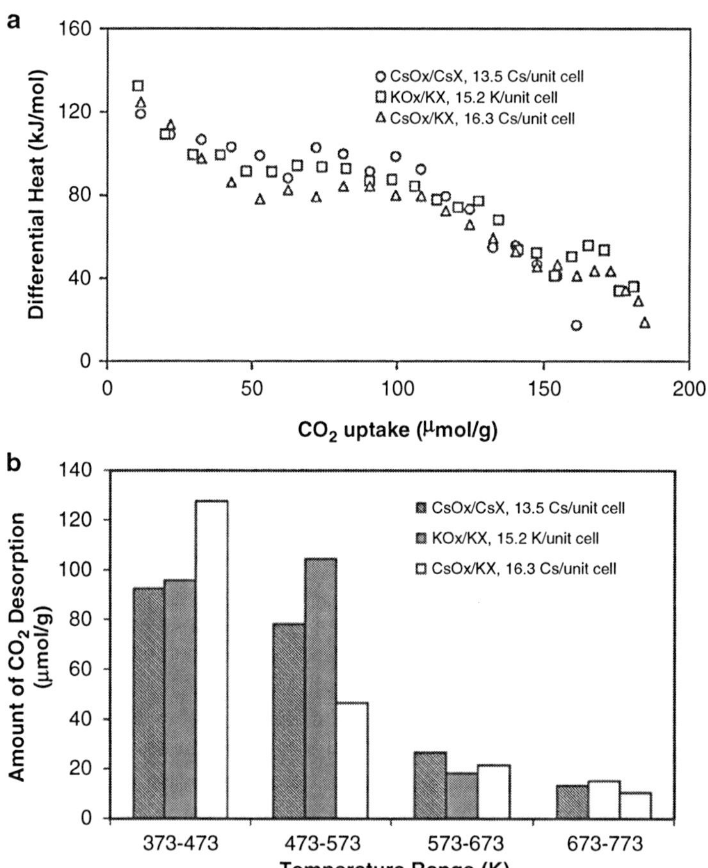

Fig. 3.10 (**a**) Differential heats of adsorption as a function of CO_2 uptake (pretreatment at 773 K, adsorption at 373 K) and (**b**) evolution of the STPD spectra of different alkali-modified zeolites [146]

indicated that about four CO_2 molecules adsorbed per unit cell at 373 K. Most of the CO_2 desorbed between 373 and 573 K and little desorbed above 673 K.

The same problems, as in the case of CO_2, may arise when using SO_2 as an acidic probe, despite the fact that SO_2 ($pK_a = 1.89$) is more acidic than CO_2 and, thus, more likely to probe the total basicity of the surface. Carboxylic acids such as acetic acid can also be used, but dimers can be formed, particularly at high coverage. Pyrrole may also be used, particularly at low adsorption temperature, but has sometimes shown some amphoteric character. The model of pyrrole chemisorbed on a basic site consists of a bonding between the framework oxygen (Lewis basic site) and the H atom of the NH group of the pyrrole molecule.

3.3.2.2 Influence of the Probe on the Adsorption Heats

Even though the interaction of basic probe molecules with surface acid sites is by no means simple, there are some general correlations that may be taken as a theoretical basis for measurements of surface acid site strengths by microcalorimetric adsorption of basic molecules.

Due to their strong basicity, ammonia and pyridine are adsorbed very strongly even on the weakest acid sites. As a consequence, their interactions with surface acid sites are relatively unspecific. Therefore, a significant number of studies using these probe molecules report homogeneous acid site strength [123, 147].

In any case, the size and strength of the probe molecule are important parameters and have to be carefully considered when performing adsorption studies.

As shown in the case of many zeolites (ZSM-5, mordenite, H-Y) [116, 121], it is difficult to compare adsorption of ammonia and pyridine, because of their difference in basicity and specific interactions of each of these molecules with the host zeolite. The adsorption heats should indeed be correlated not to the basicity of the probe molecules in liquid phase or in aqueous solutions (where NH_3 is more basic than pyridine by about 20 kJ mol^{-1}, since the pK_a values are 9.3 and 5.2, respectively) but to their basicity in gas phase, which can be expressed in terms of the PA. In gas phase, pyridine is a stronger base than ammonia: the PA value of pyridine is ca. 922.2 kJ mol^{-1}, while that of ammonia is only ca. 857.7 kJ mol^{-1}. The other parameters which have to be taken into account are size (the kinetic diameters of pyridine and ammonia are 0.533 and 0.375 nm, respectively) and the secondary interactions between these adsorbates and the zeolite structure (hydrogen bonds may lead to the formation of a monodentate with pyridine and a bidentate with ammonia), as shown by Parillo et al. [116, 148]. These differences are reflected in the adsorption heats, which are usually 20–30 kJ mol^{-1} higher in the case of pyridine than in the case of ammonia; a similar difference is observed for the chemisorption limit [114, 121].

Adsorption enthalpies have been measured for ammonia, pyridine, and isopropylamine as a function of coverage at 480 K by Parrillo et al. [114] on several high silica zeolites, including H-ZSM-5, H-ZSM-12, H-mordenite (H-M, Si/Al = 15) and H-Y. Except for isopropylamine in H-ZSM-5, each adsorbate exhibited a constant heat of adsorption up to a coverage of one molecule per Brönsted acid site, followed by a sharp drop in the measured heats. For isopropylamine in H-ZSM-5, the results were found to depend on the Si/Al ratio and synthesis procedures, and it appears that hydrogen bonding between the adsorbate molecules affected the measured heats. For ammonia, the adsorption enthalpies were 150 kJ mol^{-1} for H-ZSM-5, H-ZSM-12 and H-Y, and 160 kJ mol^{-1} for H-M. Adsorption enthalpies for pyridine were between ~190 and 210 kJ mol^{-1} for H-ZSM-5, H-ZSM-12 and H-M, but only ~180 kJ mol^{-1} for H-Y. Heats of adsorption for isopropylamine tended to be slightly higher than those for pyridine on each sample (between 200 and 240 kJ mol^{-1} for H-ZSM-5 [149]), but appeared to depend on site concentrations. As the various structures exhibited almost the same adsorption energies, the authors suggested that the isolated Brönsted acid sites in high silica materials are almost identical [114].

A comparison of the adsorption energies of the complexes formed by ammonia, pyridine and isopropylamine with a simple potential-energy model, which assumes that the heat of adsorption linearly scales with gas phase proton affinities, suggests that proton transfer dominates the interaction between the adsorbate and acid site [149].

In another study by Parrillo et al. [148] on a H-ZSM-5 sample, the average heats of adsorption in the low coverage regime were determined for the following probe molecules: ammonia (145 kJ mol^{-1}), methylamine (185 kJ mol^{-1}), ethylamine (195 kJ mol^{-1}), isopropylamine (205 kJ mol^{-1}), *n*-butylamine (220 kJ mol^{-1}), dimethylamine (205 kJ mol^{-1}), trimethylamine (205 kJ mol^{-1}) and pyridine (200 kJ mol^{-1}). Again a good correlation was found between these heats and gas-phase proton affinities. The heats of adsorption increase in increments that are identical to the gas-phase proton affinity differences, except for relatively small deviations with trimethylamine and *n*-butylamine. The deviations for trimethylamine and *n*-butylamine imply that there are some specific interactions between these amines and the zeolite [148]. The deviation for trimethylamine could be explained by a decrease in the ability of this molecule to form hydrogen bonds, while the deviation for *n*-butylamine could be due to interactions between the alkyl groups and the zeolite walls. In contrast to the excellent correlation between heats of adsorption and gas-phase acidity scales, the correlation with solution-phase acidities (pK_a) is very poor [148].

The microcalorimetric technique has also been applied to investigate the adsorption of acetonitrile, dimethylether, water, pyrrole and ammonia on ferrierite, a small pore zeolite [150]. This investigation showed that the results of the determination of the site strength distribution are dependent on the strong or weak basicity of the probe, on the acidity and porosity of the acid solid and on the adsorption temperature. A similar calorimetric investigation of the acidity of dealuminated Y-type zeolites was performed using the same various basic probes [151].

The data concerning the adsorption of different bases indicate that adsorbed bases interact with zeolites more strongly than hydrocarbons of similar structure and molecular weight [63]. A comparison of adsorption heats of various bases such as ammonia, pyridine, *n*-butylamine with benzene on A, X, Y and mordenite zeolites modified by ion exchange and aluminum extraction was carried out by Klyachko et al. [63]. Surprisingly, the heats of pyridine adsorption were found to be virtually the same on sodium and hydrogen zeolites. Furthermore, the sorbed amounts of large molecules such as pyridine and *n*-butylamine were severely limited by the finite void volume of zeolites. In the case of ammonia, things were more complicated because of penetration of NH_3 into the small cages.

An important observation is that it is possible to quantitatively evaluate the location of acid sites by using different probe molecules, e.g., on one hand ammonia, which is able to reach all acid sites, and on the other hand pyridine, cyclohexane, benzene, etc, probes that can only attach to sites in the main channels.

Finally, it is worth mentioning that water can be used as a probe molecule to provide knowledge of hydration enthalpies, which is of considerable interest when assessing the hydrophilic or hydrophobic character of zeolites [152].

For example, adsorption of water vapor on X and Y zeolites exchanged with barium (Ba-X, Ba-Y) has been studied by thermogravimetry and calorimetry [153]. The faujasite zeolite was found to exhibit a stronger adsorption affinity for water molecules when exchanged with barium than with sodium or potassium.

The following Tables (Tables 3.1 and 3.2) are summarizing some results of calorimetric measurements on zeolites ZSM-5 and Y. More detailed infomation can be found in a review by Cardona-Martinez et al. [170]. Besides the various probe molecules used in adsorption calorimetry, Tables 3.1 and 3.2 present the Si/Al ratio, the exchange level, the pretreatment temperature of the zeolites and the adsorption temperature. The corresponding adsorption enthalpies are given at nearly zero coverage (initial heat of adsorption) and at the plateau or half coverage (average heat of adsorption).

In conclusion, the choice of a suitable probe molecule depends on the physico-chemical characteristics of the studied samples, such as the pore size distribution, aluminum content, etc.

3.4 Comparison of Adsorption Calorimetry and TPD Methods for the Determination of Acidic Properties of ZSM-5 and Y Zeolites: A Case Study

3.4.1 Influence of the Adsorption or Desorption Temperature of the Probe

Achieving thermodynamic equilibrium between the sorbed probe molecule and the acid sites on the time scale of microcalorimetric measurements is a prerequisite for application of adsorption microcalorimetry to the study of solid-acid catalysts.

Table 3.1 Calorimetric measurements on ZSM-5 zeolites

Probe molecule	Si/Al ratio	$T_{activation}$(K)	$T_{adsorption}$ (K)	Initial Q_{diff} (kJ mol^{-1})	Average Q_{diff} (kJ mol^{-1})	Refs.
Ammonia	34	723	473	170	140	[25]
	35	723	473	170	150	[69]
	31	750	480	155	150	[89]
	23	750	480	150	145	[148]
	15	673	303	125	70	[154]
Pyridine	33	723	473	195	140	[155]
	34	673	473	235	160	[147]
	31	750	480	215	210	[89]
Diethylether	25.5	700	360	135	130	[144]
Monomethylamine	35	723	473	230	185	[69]
	23	750	480	175	185	[148]
Dimethylamine	35	723	473	300	250	[69]
	23	750	480	205	210	[148]
Trimethylamine	35	723	473	160	140	[69]
	23	750	480	205	205	[148]
Ethylamine	23	750	480	180	195	[148]
Isopropylamine	31	750	480	230	240	[114]
n-butylamine	23	750	480	220	225	[148]
Methanol	35	723	373	116	80	[69]
	25.5	750	400	120	110	[143]
Water	35	723	373	100	60	[69]
	25.5	723	400	105	90	[143]
Acetonitrile	25.5	723	400	110	110	[143]
	25.5	750	400	110	110	[142]
	30	n.r.	400	110	110	[99]
Ar	15	673	303	28	20	[156]
N_2	15	673	303	38	25	[156]
CO	15	673	303	50	33	[156]
CH_3CN	15	673	303	60	50	[154]
Acetone	25.5	700	360	130	130	[144]
p-xylene	26	773	352	110	85	[157]
	40	773	352	130	85	[157]

n.r. not reported

Microcalorimetry gives an accurate assessment of the heat of adsorption if the temperature is sufficiently high to achieve good surface mobility [155]; however, information about the equilibrium constant of adsorption may be difficult to obtain if the adsorption temperature is not sufficiently high to achieve equilibrium between the adsorbed and gaseous probe molecules.

The choice of adsorption temperature, however, is constrained and involves a compromise between using high temperatures to reduce equilibration times, and using lower temperatures to achieve high coverage of the acid sites at reasonable pressures. Lower temperatures are also used to ensure that the basic molecule does not decompose on the surface. In general, bases that adsorb more strongly must be studied at higher temperatures [159].

The effect of temperature on ammonia adsorption by H-ZSM-5 (Si/Al = 27) samples has been investigated by microcalorimetry, varying the adsorption

Table 3.2 Calorimetric measurements on H-Y zeolites

Probe molecule	Si/Al ratio	Exchange level(%)	$T_{activation}$(K)	$T_{adsorption}$ (K)	Initial Q_{diff} (kJ mol^{-1})	Average Q_{diff} (kJ mol^{-1})	Refs.
Ammonia	2.54	98.6	675	425	200	140	[158]
	2.4	82	723	423	130	115	[159]
	2.4	80	753	573	110	110	[160]
	2.5	83	673	473	125	100	[161]
	2.4	80	673	423	140	140	[50]
	2.43	85	675	416	125	125	[162]
	2.56	79	673	353	150	130	[163]
	2.4	100	673	423	133	120	[164]
	2.4	85	673	423	114	110	[165]
	2.42	85.7	673	423	113	110	[166]
Pyridine	2.4	84	623	473	300	140	[147]
	2.4	39	623	473	190	120	[147]
Trichloroethylene	2.6	0	623	298	55	65	[167]
Propyne	2.8	100	773	303	280	170	[168]
Pyrrole	2.4	90	623	296	175	150	[151]
Dimethylether	2.4	90	623	296	150	110	[151]
p-xylene	2.4	0	773	298	85	90	[169]
m-xylene	2.4	0	773	298	100	100	[169]

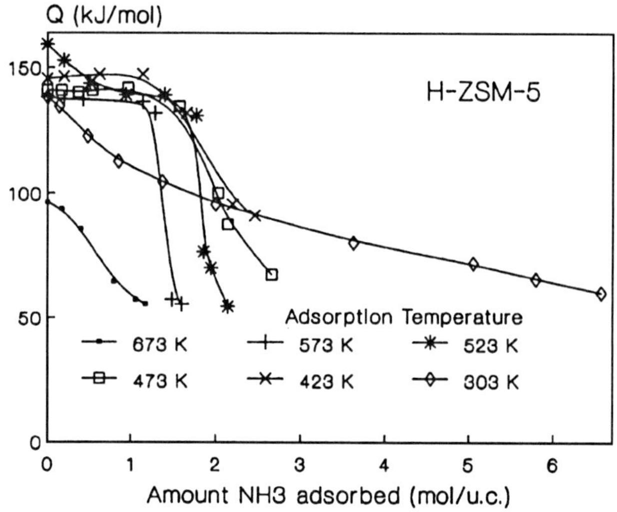

Fig. 3.11 Differential heats of adsorption vs ammonia uptake on a ZSM-5 (Si/Al = 2.5) zeolite at various temperatures of adsorption [4]

temperature from 423 up to 673 K, Fig. 3.11 [4]. The initial heats of adsorption were independent of temperature, at least up to 573 K. When the adsorption temperature increased, there was a competition between the formation of ammonium ions on Brönsted sites and their decomposition. The total number of titrated sites decreased with increasing adsorption temperature. Room temperature adsorption is a slow phenomenon and it did not allow differentiating strong and weak acid sites, and above 573 K the decomposition of ammonium ions became predominant. It appeared that an adsorption temperature between 423 and 573 K is appropriate for these calorimetric experiments.

Heats of adsorption of ammonia and pyridine on H-Y zeolites (Si/Al = 2.5) pretreated at 673 K were measured by Tsutsumi et al. [161] using a microcalorimeter in the temperature range of 313–673 K. The heats of adsorption appeared to depend on temperature in two different ways. One was that heats of adsorption on H-Y did change with coverage to a greater extent at high temperature (above 473 K) than at low temperature, and more for pyridine than for ammonia. The difference in the shape of the heat curves was attributed to the difference in the selectivity of adsorption at different temperatures, that is, adsorption at higher temperatures occurred preferentially on stronger acidic sites, while adsorption on weaker or non-acidic sites also took place in the low temperature range. The other kind of temperature dependence was a slight decrease in heats of adsorption with a temperature rise, which was observed in cases of ammonia adsorption on H-Y zeolites above 473 K and on Na-Y above 313 K, while the shapes of heat curves were similar irrespective of the adsorption temperature. Such a decrease may be attributed to the fact that the temperature dependence of the heat of adsorption is thermodynamically defined by the difference in molar heat capacity between adsorbed state and gaseous state [161]. It was also concluded that the strength of acid sites should preferably be determined by the heat curve obtained at 473 K rather than at 313 K because of more selective adsorption. Similar conclusions were derived by Stach et al. [166] for the chemisorption heats of NH_3 on H-Y zeolite (Si/Al = 2.4) measured at different temperatures. The authors suppose that at the low temperature of 303 K equilibrium was never reached but a stationary state was measured.

It can be concluded that adsorbed ammonia has sufficient mobility at 423 K to equilibrate with the catalyst surface on the time scale of microcalorimetric measurements, and these measurements provide an effective method for quantifying acid site distributions of solid acid catalysts. However, when using pyridine it is preferable to choose a higher temperature, e.g., around 573 K.

In the NH_3-TPD profiles of H-zeolite, such as H-ZSM-5 and H-mordenite, peaks are generally observed in two temperature regions, namely l-peak and *h*-peak. Signal in the high temperature region is attributed to desorption of NH_3 from strong Brönsted and Lewis sites, which are of catalytic importance. The NH_3-TPD spectra can be obtained by varying the duration and/or temperature of evacuation of NH_3 saturated samples before TPD measurements, which has significant influence on *l*-peak, as shown in Fig. 3.12 [171].

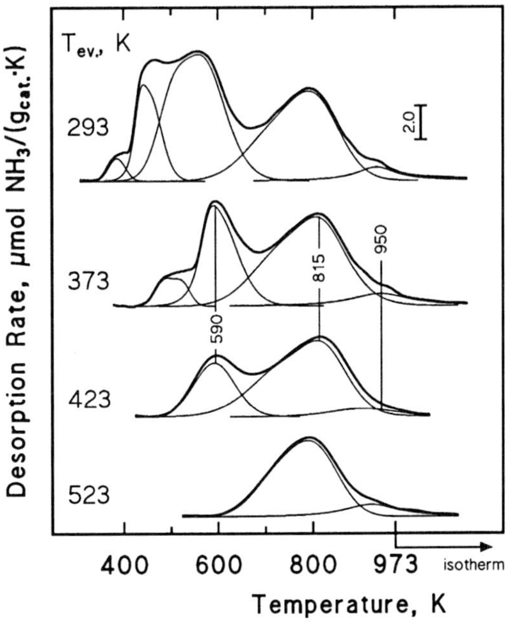

Fig. 3.12 NH_3-TPD curves obtained on H-ZSM-5 zeolite after evacuating the NH_3-loaded samples at the indicated temperature T_{ev}. The thin lines under the curves give the best-fit component peaks [171]

3.4.2 Influence of the Pretreatment of the Sample

The significance of the differential heats of adsorption for Brönsted and Lewis sites cannot be compared easily. For Brönsted sites the differential heat is the difference between the enthalpy of dissociation of the acid hydroxyl and the enthalpy of protonation of the probe molecule. For Lewis sites the differential heat of adsorption represents the energy associated with the transfer of electron density towards an electron deficient, coordinatively unsaturated site, and probably an energy term related to a relaxation of the strained surface [172]. Increasing the pretreatment temperature modifies the surface acidity of solids.

High-temperature calcination is a well-known method of reducing total acidity via dehydroxylation and dealumination. In principle, calorimetry should make it possible to determine the number of "strong" Brönsted sites, assuming that activation at 673 K before NH_3 adsorption gives rise to a maximum in H^+ concentration, i.e., that no dehydroxylation occurs at that temperature. This is a crude approximation but can be considered as valid for comparison of different samples. After calcination at increasing temperatures, dehydroxylation of the zeolite is observed: above 673 K, the number of Brönsted acid sites decreases, while that of strong Lewis acid sites increases. However, a limited dealumination occurs and the constraining character of the intracrystalline voids increases. Microcalorimetric studies of ammonia adsorption confirm the very strong acidic character of the sites and show their dependence in strength and heterogeneity upon calcination temperature [90, 173, 174], as shown in Fig. 3.13.

It has been established that calcination at 1073 K before NH_3 adsorption leads to the presence of fewer acid sites but that the remaining ones are stronger [173]. Moreover, when the samples were further rehydrated at room temperature and outgassed again at 673 K, the original curve obtained after pretreatment at 673 K was not restored. It was concluded that such a heat treatment had irreversibly modified the material [173]. These results were confirmed by Jozefowicz et al. [175], who have shown that, upon dehydroxylation at 1073 K, all the ZSM-5 samples studied had lost a considerable proportion of Brönsted sites, and there were no Brönsted sites of homogeneous energy distribution (plateau at 150 kJ mol^{-1}) left. Only the acidic sites of broad energy distribution remained unaffected by high temperature treatment, i.e., the sites with heats of adsorption between about 135 and 80 kJ mol^{-1} [175].

Figure 3.14 [175] shows ammonia TPD curves for four ZSM-5 zeolites activated at 673 K and at 1073 K, respectively. The relative mass signal intensities of $m/e = 16$ are plotted as a function of temperature. All the curves are normalized to each other and give an indication of the total number of acidic sites that desorb ammonia for each zeolite. After calcination at 1073 K the number of Brönsted sites has been reduced as evidenced by decrease in area of the 500–700 K temperature maxima; also, the distribution of the energies of these sites has been broadened. At the same time, the number of very strong Lewis sites has increased as indicated by

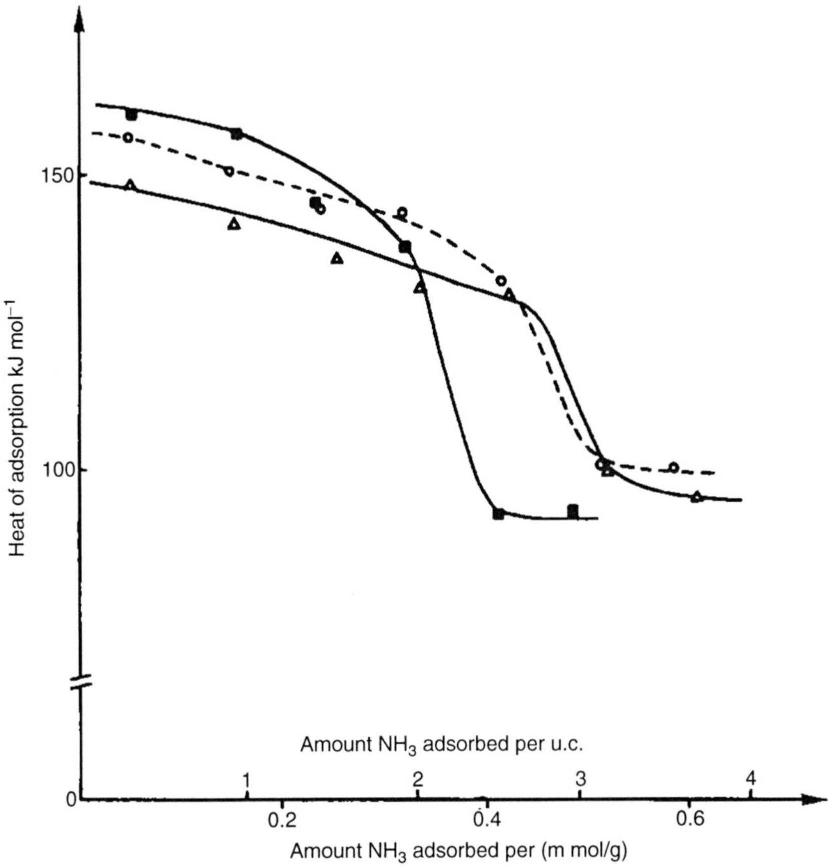

Fig. 3.13 Variation of the heats of adsorption as a function of NH_3 uptake increments. The calorimeter was maintained at 416 K. The H-ZSM-5 (Si/Al = 11) sample was initially outgased at 743 (Δ), 923 (○), and 1073 K (■) [174]

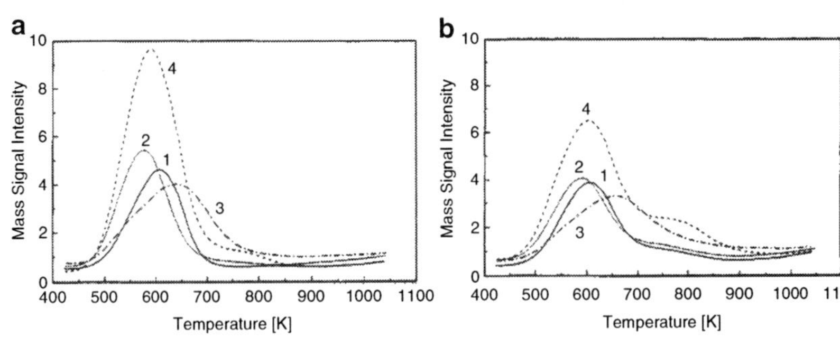

Fig. 3.14 TPD of ammonia after activation at (a) 673 K and at (b) 1073 K and adsorption/desorption at 423 K: curve 1, HZSM(1) (Si/Al = 33.4); curve 2, HZSM(2) (Si/Al = 33.7); curve 3, HZSM(3) (Si/Al = 20.9); curve 4, HZSM(4) (Si/Al = 13.9) [175]

the arisal of the broad maxima or more clearly defined sholders in the 700–850 K temperature region [175].

3.4.3 Influence of the Proton Exchange Level

The effect of proton exchange level, or sodium content, on H-Y zeolites has been the subject of numerous studies [6]. The acid form is mostly obtained by decomposing the ammonium form obtained from the Na form by cation exchange, so that the acidity varies with the exchange level. Generally, an increase in acidity with increasing proton exchange was measured; however, uncertainty exists as to the strength of the acid sites introduced at different exchange levels.

The differential heat curves for pyridine adsorption at 473 K on (Na,H)-Y zeolites with various Na contents (84, 68 and 39% levels of proton exchange, respectively) have been studied by Chen et al. [147]. Increasing the protonation level from 39% to 68% of exchange resulted in a fairly uniform generation of acid sites with strength above 110 kJ mol^{-1}. Upon decreasing the sodium content from 68% to 84% of exchange, the last sites to be proton-exchanged were slightly more energetic than those previously formed, around 130 kJ mol^{-1}. However, one might question the suitability of the probe for this study since, as said above, some hydroxyl groups are inaccessible to pyridine adsorption because of their location in the H-Y lattice.

The interaction of ammonia with zeolite Y (Si/Al = 2.4) was also compared at 100% proton exchange and 85% proton exchange by van Santen [164], who noted a significant increase (of 20 kJ mol^{-1}) in the initial heat of adsorption due to protonation of ammonia.

The effect of the exchange degree (of sodium cations by protons) on the heterogeneity of acid sites in a ZSM-5 sample has been studied [176] by adsorption of various probe molecules in order to determine the selectivity of adsorption on their acid centers of variable nature, such as Na^{+} in Na forms and acidic OHs in the corresponding H forms. Using ammonia as a basic probe at 423 K, it has been shown (see Fig. 3.15) that the numbers of medium and strong sites increase with decationization, while the number of weak sites decreases slightly. At low exchange levels most of the acid sites are rather weak. While this population of weak sites remained almost constant with the exchange level, the population of stronger sites increased progressively up to the point where, for extensively exchanged samples, the strongest sites became predominant. The population of sites presenting heats of adsorption above 150 kJ mol^{-1} illustrates the dramatic effect of removal of the very last sodium ions on the acid strength, not only for the newly created sites but also for the preexisting ones.

A study of mordenite zeolites with various exchange levels by IRMS-TPD of ammonia revealed that Brönsted acid sites consisted of two kinds of OH bands at

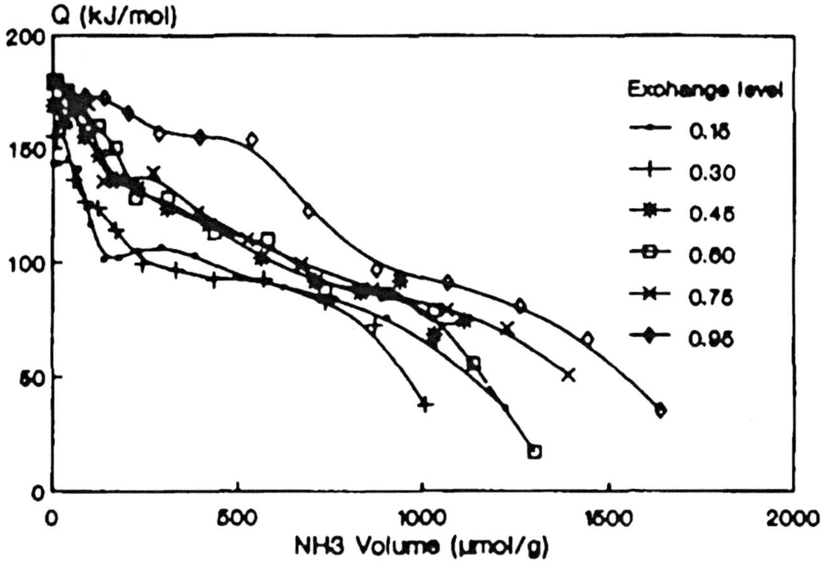

Fig. 3.15 Differential heats of adsorption of NH_3 for NaZSM-5 zeolites exchanged to 15 (•), 30 (+), 45 (*), 60 (□), 75 (x) and 95% (◇) [176]

high and low wavenumbers, ascribable to OH bands situated on 12- and 8-member rings (MR) of the mordenite structure, respectively. The amount and strength of these Brönsted hydroxyls were measured quantitatively based on a theoretical equation using a curve fitting method. Up to ca. 30% of the exchange degree, NH_{4+} was exchanged with Na^+ on the 12-MR to arrive at saturation; therefore, in this region, the Brönsted acid site was situated on the large pore of 12-MR. The NH_4^+ cation was then exchanged with Na^+ on 8-MR, in an amount that eventually exceeded that on 12-MR. In the 99% NH_4-mordenite, Brönsted acid sites were located predominantly on the 8-MR more than on the 12-MR. Irrespective of the NH_4^+ exchange degree, the strengths ΔH of Brönsted acid sites were 145 and 153 kJ mol^{-1} on the 12- and 8-MR, respectively; that is, the Brönsted acid sites on the 8-MR were stronger than those on the 12-MR [177]

Trunschke et al. [43] have studied the acidic properties of HNaY zeolites with different degrees of exchange by means of two complementary techniques involving temperature-programmed desorption (TPD) of ammonia, namely, FTIR-TPD and conventional TPD. The results obtained by TPD are shown on Fig. 3.16. For the sodium form of the zeolite, no adsorption of ammonia was observed under the same experimental conditions. In HNaY zeolites with low exchange levels only a single type of OH group was observed. The analysis of the spectra in the region of NH bending modes during NH_3-FTIR-TPD on zeolites with higher proton content (40% or higher) showed that ammonia was bound in two different adsorption states. This assessment was further confirmed by evaluation of the OH stretching region.

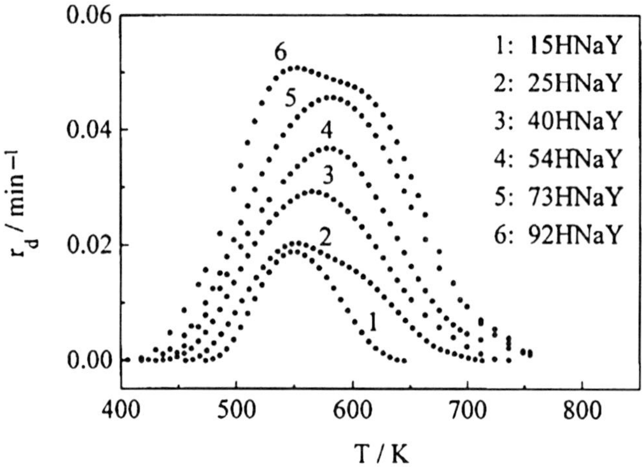

Fig. 3.16 Ammonia desorption profiles of the HNaY zeolites with different degrees of exchange [43]

3.4.4 Influence of the Si/Al Ratio and Dealumination

The Si/Al ratio plays a significant role, since the aluminum atom is directly related to the acidic site and accounts for the formation of carbenium and/or carbonium ions or possibly cation radicals inside the zeolite. Dealumination processes can promote porous structure modifications, which may improve some interesting properties of zeolites, like thermal and hydrothermal stability, acidity, catalytic activity, resistance to aging and low coking rate, and matter transfer. However, a severe dealumination may also cause a loss of crystallinity.

Different dealumination processes have been proposed, namely steaming and acid treatments, as well as reactions with $SiCl_4$ or SiF_6^{2-}. From many experimental and theoretical investigations of the acidity of zeolites with different framework aluminum contents, it has been concluded that the number of bridging hydroxyl groups increases with rising number of Al atoms in the lattice [178]. In contrast, the dependence of the acid strength on the aluminum content is more complicated, since from theoretical considerations an increasing acid strength is expected when the number of aluminum atoms decreases, whereas studies of the Al topology in the framework have often led to a curve presenting a maximum for the number of strong sites [178, 179]. The effect of steaming on the number and strength of acid sites is apparent from a comparison of the differential heat curves for dealuminated zeolites. The microcalorimetric curves also show that the strength of sites corresponding to the intermediate plateau region first increases and then progressively decreases with steaming severity. The dependence of the acid strength distribution of dealuminated faujasites on the Si/Al ratio is depicted in Fig. 3.17, in which the number of strong acid sites presents a distinct maximum. The abscissa

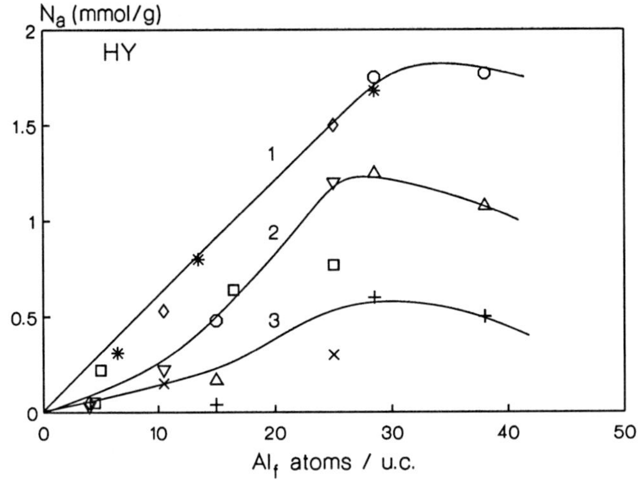

Fig. 3.17 Acid strength distribution dependence on the framework aluminum content per unit cell in dealuminated Y zeolites. *Curve 1*: total number of acid sites with Q > 80 kJ mol^{-1} (*,○,◊). Curve 2: number of acid sites with Q > 100 kJ mol^{-1} (∇, Δ). Curve 3: number of acid sites with Q > 120 kJ mol^{-1} (+, ×, □) from [178] Differential molar heats of ammonia adsorption were measured at 423 K

of the maximum corresponds to an Al content of about 29/u.c. (and Si/Al ratio around 5.5) for dealuminated Y zeolites. These values are in good agreement with the ones predicted by Barthomeuf [180] who used topological density theory to calculate a theoretical threshold for H-Y zeolites, found to lie at $m_{\mathrm{lim}} = 0.150$ (which corresponds to Si/Al = 5.8).

The removal of aluminum from Y zeolite crystals leads to products with high framework Si/Al ratios. The dealumination may be performed using one of the following procedures: thermochemical treatment of NH_4,Na-Y (steaming and acid treatments), aluminum extraction from Na-Y with EDTA in aqueous solutions, isomorphous substitution with $(NH_4)_2SiF_6$, aluminum substitution by silicon through treating Na-Y with $SiCl_4$ vapor at high temperature. Some of the aluminum atoms are released from the framework and form non-framework aluminum-containing species; thus the Si/Al ratio of the framework must be determined not by chemical analysis but by NMR or IR spectroscopy. The non-framework aluminum species can be eliminated by treatment with diluted hydrochloric acid.

The dealumination process is associated with a change in the porosity within the crystals and may sometimes cause a drastic loss of crystallinity. Microporous adsorbents of the faujasite type are so arranged that the Si/Al ratio increases as the number of cations and the average electrostatic field within the framework decrease. To assess the effect of the Si/Al ratio on the activity and acidity of Y zeolites, it is desirable to compare samples with similar extents of exchange, since the degree of exchange has a significant influence on the catalytic and acidic properties of faujasites.

Steam dealumination of H-Y zeolite is known to cause a progressive destruction of weak and intermediate sites while producing new stronger sites. Microcalorimetric measurements of ammonia and pyridine adsorption have shown the presence in samples containing extra-framework aluminum of sites with adsorption heats that are much higher than those observed for samples containing only framework Al [62, 88, 181, 182]. Moreover, steamed samples possess sites presenting a wide distribution of acid strength. The high initial heats of ammonia or pyridine adsorption that can be observed on steamed zeolites in comparison to unsteamed samples may be attributed to Lewis acid centers or to a combination of Lewis and Brönsted sites [181, 182].

The acidity of H-Y zeolites with different Si/Al ratios (2.4, 5.6 and 12.0) was characterized by calorimetric measurements of ammonia chemisorption at 423 K by Stach et al. [166]. Dealumination of Y zeolites was found to decrease the total number of acid sites and to generate very strong acid sites with $Q_{\text{diff}} > 120$ kJ mol^{-1}. The relative proportion of these very strong acid sites went through a maximum.

Microcalorimetric studies by Auroux et al. [181] of dealuminated Y zeolites prepared by steaming and subsequent acid leaching in order to remove (partially or totally) the extra-framework species generated by steaming have attributed the initial strong sites to Lewis acidity (alumina phases or non-framework aluminum). This result agrees with most IR studies which confirm that a fraction of the Lewis acid sites generated by dealumination are stronger than the Brönsted acid sites of pure Y zeolites. The samples which had similar total (chemical analysis) and framework (NMR) Si/Al ratios presented a plateau in their acid strength distribution, whereas the other samples showed a more heterogeneous distribution [181], as shown in Fig. 3.18.

An important increase of the initial heat values and of the site strength heterogeneity was observed for samples presenting many extra-framework aluminum species. Samples subjected to a moderate dealumination and nearly total removal of the extra-framework aluminum displayed a homogeneous acid strength, slightly but significantly higher than that of the parent H-Y material (150 kJ mol^{-1} instead of 140). The number of acid sites was of course reduced in accordance with the dealumination level. The residual structural acid sites remained essentially equivalent, even though alteration of their local environment inevitably resulted in a slight acid strength enhancement, well below that observed when guest aluminum species were present (in which case heats might attain 200 kJ mol^{-1}). Further dealumination generated samples characterized by heat curves presenting a narrower plateau. At high Si/Al ratios (>50) the acidity was weakened; this suggests that extreme rarefaction of the structural sites greatly depressed their strength. Thus, it appears that an optimum site density is a prerequisite for strong acidity [181].

Increasing severity of steam treatment of H-Y was associated with decreases in total acidity by all reported studies. In particular, the correlation observed in the above-mentioned study by Auroux et al. [181] between an increasing steam dealumination and a decrease in the number of sites exhibiting intermediate strength is confirmed by the results of Chen et al. [62]. Moreover, the heat of

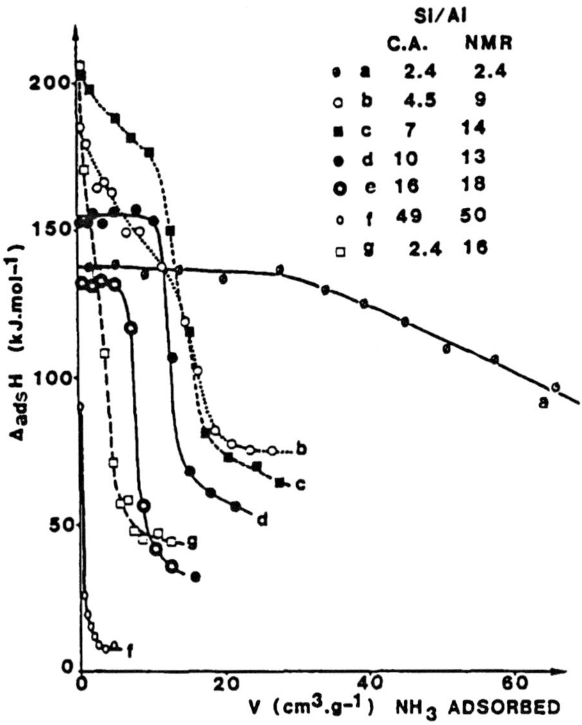

Fig. 3.18 Differential heats of ammonia adsorption over HY zeolites with various Al contents as a function of coverage [181]

ammonia adsorption at zero coverage, measured at 423 K, first increased with Si/Al ratio and then decreased upon further dealumination beyond Si/Al ratios of 10. The strength of intermediate sites, corresponding to Brönsted acid sites, also passed through a maximum at a Si/Al ratio equal to 10.

The reason for the decrease in Brönsted acid strength with severe steam treatment is not very clear. For faujasite type zeolites, Barthomeuf [180, 183], using the topological density theory, obtained Si/Al(F) (Si/Al ratio of the framework) values of about 6 at which point framework Al would have 0 NNN (next-nearest neighbor) Al atoms. For this value, the Al atoms should be far enough from each other for the interactions between them to become negligible. Y zeolites with higher Si/Al(F) ratios would, therefore, be expected to have constant Brönsted acid strength, which is not the case in most of the calorimetric studies [62, 181, 184, 185]. However this is not surprising because the topological density theory assumes ideal distribution of aluminum all over the crystal. Yet it is known from various ESCA studies that crystals are frequently far more aluminum rich at the surface than expected. Incorporation of aluminum is favored at the latest stages of crystallization. Therefore, dealumination, except at very high levels, does not yield homogeneously dispersed aluminum throughout the crystals. Since the surface is much more Al-rich, the overall Si/Al is well above the actual one at the surface.

Dealumination generally brings about a decrease in the acid site concentration. However, the extent of the indicated decrease varies with the kind of base probe, and a significant change was observed by Mitani et al. [184] in the ratio of acid site concentrations when titrated with pyridine instead of ammonia.

The acidity of commercial dealuminated H-Y zeolites (USY) has been studied by ammonia adsorption calorimetry and single ammonia TPD runs by Handy et al. [186]. All TPD profiles were represented by models consisting of two or three distinct types of sites, while the differential heat plots implied a continuous distribution of sites in the 160–80 kJ mol^{-1} range for all catalysts. A dealumination of increased severity resulted in a decrease in total site density for sites of strength less than 100 kJ mol^{-1}. A general correlation could be established between the amount adsorbed with heats of adsorption greater than 100 kJ mol^{-1} and the framework Al^{3+} site density, although the ratio of the amount of ammonia adsorbed per framework Al^{3+} site was less than one.

The effect of dealumination on NH_3-TPD profiles of HY zeolites is shown at Fig. 3.19 [187]. Evidently *l*-peak became smaller upon dealumination, while the *h*-peak increased up to a maximum for Si/Al = 15 and then decreased upon further dealumination.

Figure 3.20 shows the effect of varying the Si/Al ratio of a MFI sample on its TPD profile. As shown by the figure, both the *l*- and *h*-peaks became smaller upon dealumination. A curve-fitting analysis led to the determination of average adsorption heats that were almost constant for all investigated MFI samples, ca. 130 kJ mol^{-1} [34].

The acidity of H-ZSM-5 zeolites synthesized with different Al contents has been characterized by microcalorimetric measurements of the differential heats of adsorption of ammonia [90]. The strength of the strongest acid sites increased

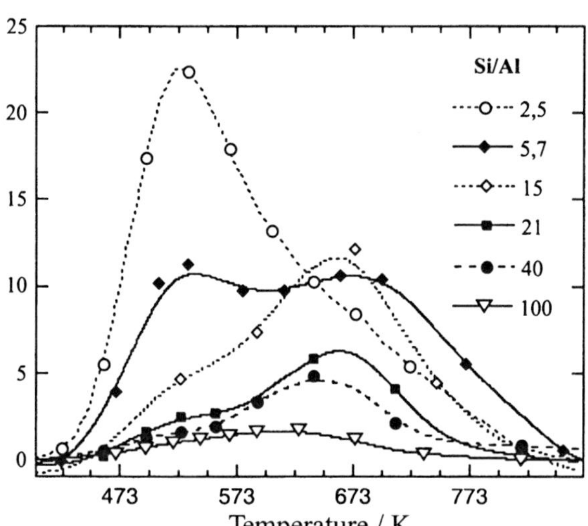

Fig. 3.19 Influence of dealumination on NH_3-TPD profiles of HY zeolites [187]

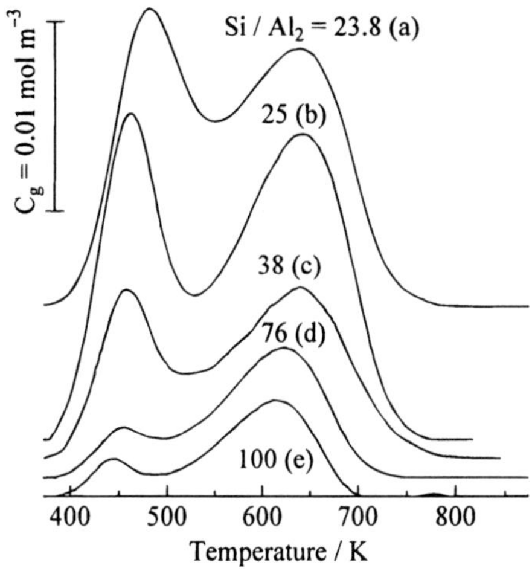

Fig. 3.20 NH_3-TPD profiles observed on H-MFI with Si/Al ratios: (**a**) 12.9; (**b**) 12.5; (**c**) 19; (**d**) 38 and (**e**) 50 [34]

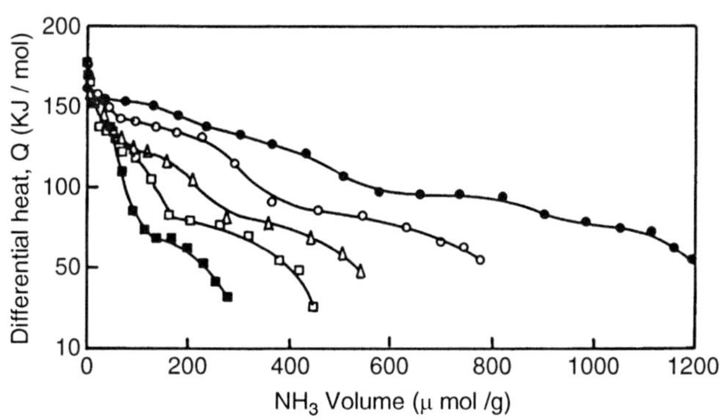

Fig. 3.21 Differential heats of ammonia adsorption at 393 K on H-ZSM-5 zeolites with Si/Al ratios: (●) 14; (○) 25; (Δ) 37.5; (□) 50 and (■) 75 [188]

with the Al content to a maximum for Si/Al = 17.5 and then decreased notably. The total acidity increased regularly with Al content. The importance of selecting appropriate Si/Al ratios for specific catalytic applications is therefore obvious.

The spatial distribution of acidity in acid-treated ZSM-5 zeolite is not uniform [162], the most acidic sites, and particularly Lewis sites, being located within the channel structure of the material where they are not always readily accessible to ammonia molecules at 416 K.

Fig. 3.21 represents the differential heats of adsorption on dealuminated H-ZSM-5 zeolite samples at 393 K versus the adsorbed amount of ammonia [188]. It appears that the adsorption heat decreases gradually with the amount of adsorbed ammonia and exhibits various steps, attributed to populations of sites of different strengths. A heterogeneous distribution of sites is clearly evidenced by the shape of the curves and can attributed to the presence of EFAL. The samples adsorb ammonia with initial differential heats that vary between 160 and 177 kJ mol^{-1}. Karge et al. [158] have reported that, in first approximation, the sites evolving Q_{diff}>150 kJ mol^{-1} can be assigned to strong Lewis sites, while the sites with Q_{diff} = 150–100 kJ mol^{-1} typically correspond to strong Brönsted acid sites. This assumption is supported by a comparison of the numbers of acidic sites obtained from IR and microcalorimetric measurements. The initial heat of adsorption for the zeolite with Si/Al = 75 is lower than that of the other zeolites except for Si/Al = 14, and the decrease in heat with coverage is steeper. The heat evolved falls abruptly from an initial value of 150 kJ mol^{-1} to 70 kJ mol^{-1} at around 50% coverage.

Microcalorimetric measurements of the differential heats of adsorption of NH_3 at 423 K were also performed by Witzel et al. [189] on H-ZSM-5 samples with various Si/Al ratios (22.5, 33.6 and 33.3) activated in high vacuum at high temperature (1075 K). The heat values reported were respectively 175, 168, and 160 kJ mol^{-1}.

3.4.5 *Influence of Substitution by Other Cations*

The nature of the exchanged cation is one of the key points that determine acidity in zeolites. It is very important to find an acidic probe able to distinguish the alkali cations from the basic sites.

In the low coverage region, the heats of ammonia adsorption by Li-X or Y zeolites observed by Khvoshev et al. [93] were substantially higher than those for adsorption by Na-X or Y zeolites. The basic features of the differential heat curves were preserved when the change was made from Na to Li zeolites. These results were explained using concepts such as the possible non-identity of the effective charges of Na^+ and Li^+ cations in zeolites, and also on the basis of an analysis of the different positions of the cations in the crystal lattice. The heat of NH_3 adsorption on Na-X zeolites was greater than that on Na-Y over the entire range of coverage [93].

The acidic properties of calcium-exchanged Na-Y zeolite were also investigated by means of *n*-butylamine desorption using differential scanning calorimetry [190]. The enthalpies of these processes were shown to be proportional to the acid strength in each specific temperature range.

A La,Na-Y sample (73% La exchange) was studied by Karge et al. [158] after activation at 675 K and at an adsorption temperature of ammonia of 425 K. The differential heat curve showed a few strong Lewis sites (Q_{diff} > 150 kJ mol^{-1}) but no obvious plateau of differential heat. The continuously decreasing part of the curve with Q_{diff} = 150–80 kJ mol^{-1} was ascribed to the interaction of NH_3 with Brönsted acid OH groups in the supercages, by analogy with the results obtained on H-Y, but these were present in considerably lower amounts than in H-Y.

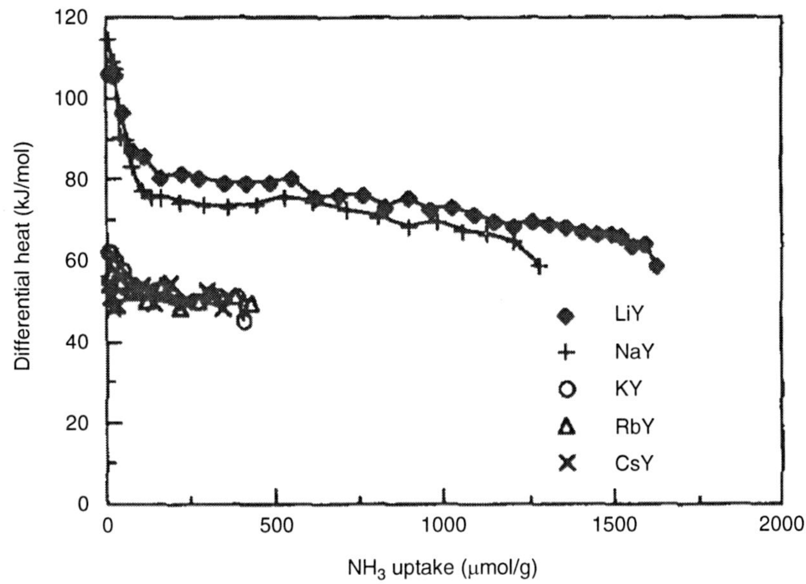

Fig. 3.22 Differential heat of adsorption versus NH_3 uptake on alkali exchanged Y zeolites. Pretreatment temperature, 673 K; adsorption temperature, 353 K [191]

The acid-base properties of alkali-metal ion exchanged X and Y zeolites were also investigated by ammonia and sulfur dioxide adsorption microcalorimetry, in parallel with the study of a catalytic reaction, namely 4-methylpentan-2-ol conversion [191]. The heats of NH_3 adsorption were found to decrease in the sequence from Li to Cs (Fig. 3.22). Li and Na zeolites presented much higher heats of NH_3 adsorption and greater coverage at the same equilibrium pressure than the other zeolites.

Differential heats of ammonia adsorption and NH_3-TPD profiles of HZSM-5 zeolite as well as FeZSM-5, Cu-ZSM-5 and MnZSM-5 zeolites are presented in Figs. 3.23 and 3.24. As can be seen on Fig. 3.23, the overall acidity of these ion-exchanged samples was not significantly modified. However, changes in the Q_{diff} vs. NH_3 uptake profiles, particularly in their middle parts (140 – 65 kJ mol^{-1}), indicate that the distribution of strength of acid sites was affected as a result of ion exchange. Ion exchange with Cu and Mn resulted in enhanced heterogeneity of the acid site strength, as confirmed by the NH_3-TPD profiles, which present poorly separated desorption peaks and even a single very broad peak for sample MnZSM-5 (Fig. 3.24) [192].

The acidity and strength distribution of acid sites of a ZSM-5 prepared by hydrothermal synthesis, and zinc and gallium incorporated ZSM-5 catalysts prepared by incipient wet impregnation, were compared to those of homologues prepared by a co-synthesis method. Ammonia adsorption microcalorimetry showed that samples prepared by co-synthesis exhibited lower acidity when compared to the others [193].

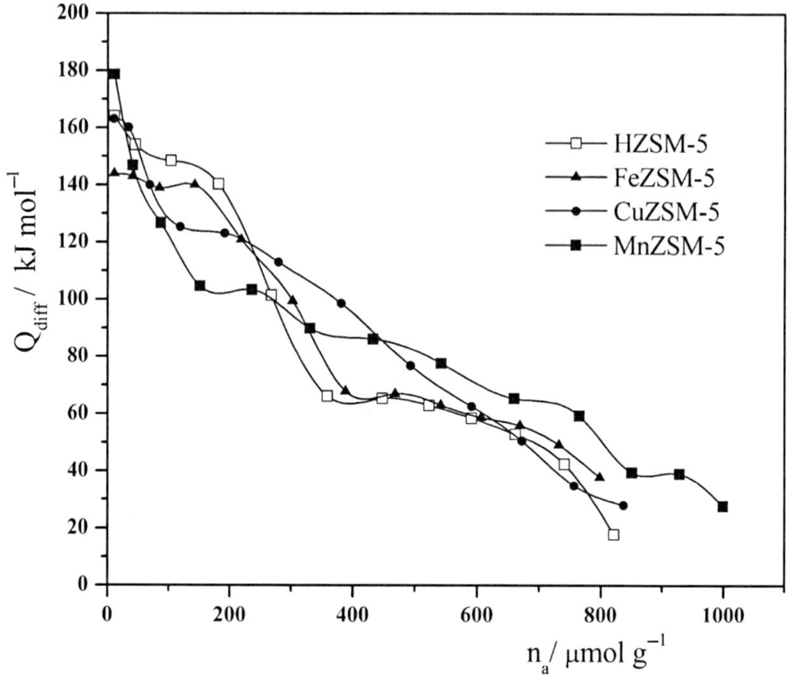

Fig. 3.23 Differential heat of adsorption versus NH_3 uptake on HZSM-5 and ion exchanged ZSM-5 zeolites [192]

A calorimetric and IR study of the adsorption of N_2O and CO at 303 K on Cu(II)-exchanged ZSM-5 zeolites with different copper loadings has been performed by Rakic et al. [194]. The active sites for both N_2O and CO are Cu (I) ions, which are present as a result of the pre-treatment in vacuum at 673 K. The measured amounts of chemisorbed species in the investigated systems and the values of differential heats of adsorption of both nitrous oxide (between 80 and 30 kJ mol^{-1}) and carbon monoxide (between 140 and 40 kJ mol^{-1}) demonstrate the dependence of the adsorption properties on the copper content. The samples were additionally characterized by ammonia adsorption microcalorimetry at 423 K [194].

A calorimetric and spectroscopic study of the coordinative unsaturation of copper (I) and silver (I) cations in ZSM-5 zeolite has been performed by Bolis et al. [195] using adsorption of NH_3 at room temperature. Adsorption microcalorimetry made it possible to characterize both quantitatively and energetically the amino-complexes formed at the cationic sites. The heats of formation of the different amino-complexes were found to be in the 130–50 kJ mol^{-1} interval for both kinds of Me(I) sites, depending on the number of ligands progressively bound, in spite of the different stoichiometries of the species formed.

CO adsorption microcalorimetry has been used in combination with the IR technique to picture the bonding nature of silver ion-exchanged ZSM-5 type zeolites (Ag-ZSM-5) [196]. CO adsorption at 301 K gave rise to a large heat

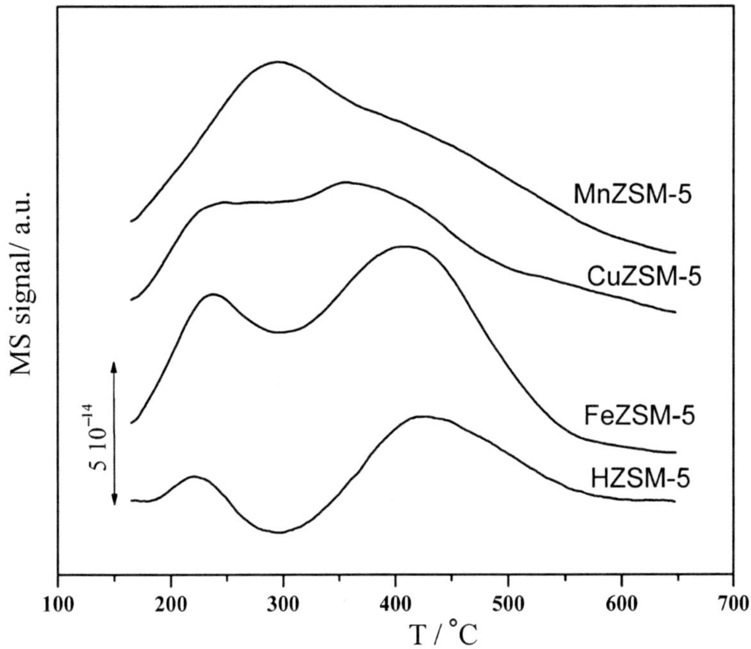

Fig. 3.24 NH_3-TPD profiles for HZSM-5 and ion exchanged ZSM-5 zeolites [192]

evolution, in the range of 100–80 kJ mol^{-1}, silver ions exchanged for protons (acting as Brönsted acid sites) being responsible for these strong adsorption sites. These adsorption energies are nonetheless lower than in the case of CO adsorption on copper ion-exchanged ZSM-5 (Cu-ZSM-5) (i.e., 140–110 kJ mol^{-1}).

3.4.6 *Influence of Coking*

The most frequent cause of catalyst deactivation in acid-catalyzed hydrocarbon reactions is coke formation, i.e., the deposition of carbonaceous residues [197]. The coke gradually formed over the catalyst may cover some of the acid sites (preferentially strong acid sites) or it may block the zeolite channels and make the acid sites inaccessible to reactants and other reaction intermediates. This, in turn, affects the activity and shape selectivity of the catalyst. Thus, the coke formed on the catalysts has been the subject of various studies dealing with the effect of acidity on both activity and coke deactivation [198].

The deposition of carbonaceous residues, leading to aging and modifications in the acidic properties of ZSM-5 zeolite, has been investigated during the reaction of methanol conversion to hydrocarbons [199]. As can be seen from Fig. 3.25, strong and medium acidic sites are poisoned by carbonaceous residues. The ammonia adsorption capacity (by medium and strong acidic sites) is decreased by about 40%

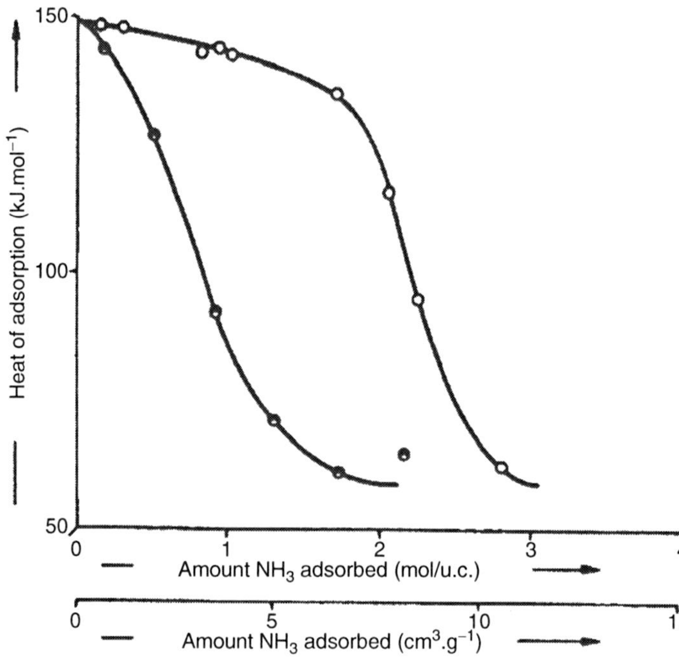

Fig. 3.25 Differential heats of ammonia adsorption on H-ZSM-5 (at 423 K) as a function of ammonia coverage. ○ Fresh material activated under vacuum at 673 K; ● catalyst "coked" by methanol at 643 K and outgassed at 673 K [199]

and the strong acidic sites have been clearly poisoned to a large extent even though the strongest sites ($Q_{ads} \approx 150$ kJ mol^{-1}) are still present.

In order to understand the effects of acidity and acid strength distribution on the nature of coke formed over ZSM-5 zeolite during the reaction of *n*-heptane aromatization, a series of ZSM-5 samples dealuminated to different degrees by steaming have been prepared and studied by microcalorimetry [198]. Microcalorimetric ammonia adsorption studies carried out at 448 K indicated the presence of very strong acid sites ($\Delta H > 140$ kJ mol^{-1}) in the catalysts steamed at 573 and 673 K. Steam treatment of the samples at various temperatures resulted in changes in initial heats of adsorption to 148, 155, 133 and 135 kJ mol^{-1}, respectively, for steaming temperatures increasing from 573 to 873 K. These variations in acidity influenced the rate and composition of the coke deposits. The aromatic nature of the coke was found to depend mainly on the density of acid sites rather than on the strength distribution.

Calorimetric measurements performed on ZSM-5 samples used for methanol, C_2H_4 and C_2H_4 + CH_3OH conversions have shown that no strong acidic sites remained accessible to NH_3 if C_2H_4 conversion was performed, while acid sites were still accessible after CH_3OH or C_2H_4 + CH_3OH reactions [173]. This was presumably due to the formation of linear polymeric residues which filled the

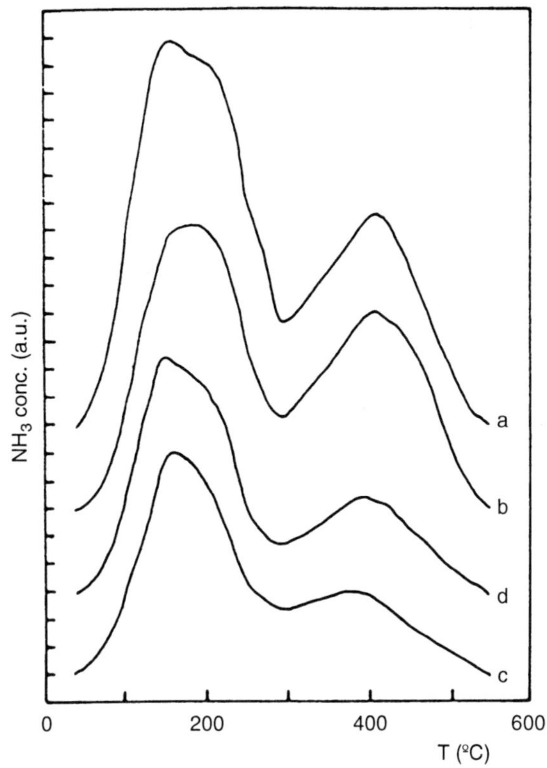

Fig. 3.26 NH_3-TPD of the fresh and coked ZSM-5 samples: (**a**) fresh, (**b**) M1, (**c**) T1, and (**d**) I1 [200]

channels when C_2H_4 was used alone. These results also explained the rapid deactivation of the ZSM-5 zeolite sample for the C_2H_4 conversion.

Uguina et al. [200] have been investigated controlled coking of ZSM-5 using mesitylene (M), toluene (T) and isobutene (I) as coke precursors. The amount, nature and location of carbonaceous residues strongly depend on the hydrocarbon used as coke precursor. NH_3-TPD has been used to measure the acidity of the fresh and coked ZSM-5 samples (Fig. 3.26). It can be observed that in the ZSM-5 samples coked with toluene (T) and isobutene (I) the number of strong acid sites accessible to NH_3 is quite lower than in the fresh ZSM-5, whereas the sample coked with mesitylene (M) presents a similar acid site population. The position of the peak maximum is the same in both the fresh ZSM-5 and the sample coked with M, but shifts to a temperature 15°C lower in the samples coked with T and I.

The shift in the position of the peak maximum shows a lower strength of the acid sites remaining in these samples: coke from T and I deactivates preferentially stronger acid sites. Then, a certain deactivation by site coverage, as well as pore blockage, must be considered. On the other hand, NH_3-TPD of samples coked with M is almost identical to that of the uncoked ZSM-5. This fact suggests that the closure of the pore openings produced by deposition of carbonaceous species is not enough to limit the entry of NH_3 into the zeolite pore structure.

3.5 Conclusion

This chapter has presented a comparison of the two most widely used thermal techniques for the study of acid/base properties of zeolites: TPD and adsorption calorimetry.

The differences in the number of acid sites determined by microcalorimetry and TPD may be ascribed to the different experimental conditions used during desorption or pre-treatment stages in each of these methods. Therefore, much attention should be paid to this aspect when comparing the results of surface acidity tests reported in the literature.

TPD is a powerful technique, but the best ways of using it to characterize acidity are still being developed; given its relative simplicity and low cost, this technique will continue to find more applications in the future.

Adsorption calorimetry may be used successfully in order to probe the strength of acid sites, to precisely determine their number, and to obtain the relative populations and strengths of Brönsted and Lewis acid sites. The combination of microcalorimetry with other techniques, preferably with IR spectroscopy, provides direct information about the molecular nature of the adsorbed species. In general, adsorption calorimetry provides a much better description of the surface acid site strength distribution than TPD.

References

1. Karge HG, Dondur V, Weitkamp J (1991) J Phys Chem 95:283
2. Farneth WE, Gorte RJ (1995) Chem Rev 95:615
3. Auroux A (2002) Top Catal 19:205
4. Auroux A (1997) Top Catal 4:71
5. Bennici S, Auroux A (2009) In: Hargreaves J, Jackson D (eds) Oxides/Oxidation catalysis, Vol 1, Chapter 9, Wiley, Weinheim, p 391
6. Auroux A (2008) Molecular Sieves- Science and technology. Acidity and basicity, vol 6. Springer, Heidelberg, p 45
7. Brunner E (1995) J Mol Struct 355:61
8. Kustov LM (1997) Top Catal 4:131
9. Stöcker M (1996) Micropor Mat 6:235
10. Guimon C, Martinez H (2003) Recent Res Devel Catal 2:81
11. Barthomeuf D (1992) Stud Surf Sci Catal 37:157
12. Paukshtis EA, Yurchenko EN (1983) Russ Chem Rev 52:242
13. Jacobs PA, Beyer HK (1979) J Phys Chem 83:1174
14. Pöppl A, Rudolf T, Michel D (1998) J Am Chem Soc 120:4879
15. Remy MJ, Stanica D, Poncelet G, Feijen EJP, Martens GPJ, JA JPA (1996) J Phys Chem 100:12440
16. Shannon RD, Gardner KH, Staley RH, Bergeret G, Gallezot P, Auroux A (1985) J Phys Chem 89:4778
17. Zecchina A, Spoto G, Bordiga S (2005) Phys Chem Chem Phys 7:1627
18. Shen J, Auroux A (2004) In: Occelli M (ed) Fluid catalytic cracking VI, preparation and characterization of catalysts, Proceedings of the international symposium, New York, 7–11 September 2003, Elsevier, Amsterdam, p 35; Stud Surf Sci Catal 149:35

19. Corma A (1997) Chem Rev 97:2373
20. Falconer JL, Schwarz JA (1983) Catal Rev-Sci Eng 25:141
21. Hunger B, Hoffmann J (1988) J Therm Anal 33:933
22. Dondur V, Karge HG (1987) Surf Sci 189/190:873
23. Karge HG, Dondur V (1990) J Phys Chem 94:765
24. Gorte RJ (1996) Catal Today 28:405
25. Sharma SB, Meyers BL, Chen DT, Miller J, Dumesic JA (1993) Appl Catal A 102:253
26. Cvetanović RJ, Amenomiya Y (1967) Adv Catal 17:103
27. Cvetanović RJ, Amenomiya Y (1972) Catal Rev 6:21
28. Katada N, Niwa M (2004) Catal Surv Asia 8:161
29. Rac V, Rakić V, Gajinov S, Dondur V, Auroux A (2006) J Therm Anal Cal 84:239
30. Bhatia S, Beltramini J, Do DD (1990) Catal Today 7:309
31. Niwa M, Iwamoto M, Segawa K (1986) Bull Chem Soc Jpn 59:3735
32. Niwa M, Katada N, Sawa M, Murakami Y (1995) J Phys Chem 99:8812
33. Gorte RJ (1999) Catal Lett 62:1
34. Katada N, Igi H, Kim J-H, Niwa M (1997) J Phys Chem B 101:5969
35. Joly J-P, Perrard A (2001) Langmuir 17:1538
36. Andersen PJ, Kung HH (1995) Catalysis. Royal Soc Chem 11:441
37. Woolery GL, Kuehl GH, Timken HC, Chester AW (1997) Zeolites 19:288
38. Bagnasco G (1996) J Catal 159:249
39. Igi H, Katada N, Niwa M (1999) In: Tracy MMJ, Marcus BK, Bisher ME, Higgins JB (eds) Proceeding of the 12th International Zeolite Conference, Materials Research Society, Warrendale, p 2643
40. Miyamoto T, Katada N, Kim J-H, Niwa M (1998) J Phys Chem B 102:6738
41. Gaillard F, Abdat M, Joly J-P, Perrard A (2004) Appl Surf Sci 238:91
42. Miessner H, Kosslick H, Lohse U, Parlitz B, Tuan V-A (1993) J Phys Chem 97:9741
43. Trunschke A, Hunger B (2002) Top Catal 19:215
44. Niwa M, Nishikawa S, Katada N (2005) Micropor Mesopor Mater 82:105
45. Derouane EG (1987) Chem Phys Lett 142:200
46. Derouane EG (1998) J Mol Catal A-Chem 134:29
47. Derouane EG, Chang CD (2000) Micropor Mesopor Mater 35–36:425
48. Hunger B, Heuchel M, Clark LA, Snurr RQ (2002) J Phys Chem B 106:3882
49. Biaglow AI, Gittleman C, Gorte RJ, Madon GJ (1991) J Catal 129:88
50. Auroux A, Jin YS, Vedrine JC, Benoist L (1988) Appl Catal 36:323
51. Aboul-Gheit AK, Aboul-Fotouh SM, Abdel-Hamid SM, Aboul-Gheit NAK (2006) Appl Catal A-General 297:102
52. Martin A, Wolf U, Berndt H, Lücke B (1993) Zeolites 13:309
53. Narayanan S, Deshpande K (1996) Appl Catal A-General 135:125
54. Robb GM, Zhang W, Smirniotis PG (1998) Micropor Mesopor Mater 20:307
55. Zhang W, Burckle EC, Smirniotis PG (1999) Micropor Mesopor Mater 33:173
56. Makowski W (2007) Thermochim Acta 454:26
57. Stach H, Lohse U, Thamm H, Schirmer W (1986) Zeolites 6:74
58. Klyachko AL, Brueva TR, Mishin IV, Kapustin GI, Rubinshtein AM (1978) In: Fejes P (ed) Proceedings of the International Symposium of Zeolites, Szeged, Hungary, 11–14 Sept 1978; Acta Phys et Chem Nova Series 24:183
59. Siril PF, Davison AD, Randhawa JK, Brown DR (2007) J Mol Catal A - Chem 267:72
60. Parillo DJ, Gorte RJ (1998) Thermochim Acta 312:125
61. Damjanović Lj, Auroux A (2008) In: Brown M, Gallagher P (eds) The handbook of thermal analysis & calorimetry, Vol 5, Further advances, techniques and applications, Chapter 11, Elsevier, Amsterdam, p 387
62. Chen D, Sharma S, Cardona-Martinez N, Dumesic JA, Bell VA, Hodge GD, Madon RJ (1992) J Catal 136:392

63. Klyachko AL, Brueva TR, Mishin IV, Kapustin GI, Rubinshtein AM (1978) Acta Phys Chem 24:183
64. Auroux A, Huang M, Kaliaguine S (1996) Langmuir 12:4803
65. Ferino I, Monaci R, Rombi E, Solinas V (1998) J Chem Soc Faraday Trans 94:2647
66. Kapustin GI, Brueva TR, Klyachko AL, Rubinshtein AM (1981) Kinet Katal 22:1247
67. Brueva TR, Kapustin GI, Mishin IV (2001) Thermochim Acta 379:35
68. Auroux A (1994) In: Imelik B, Vedrine JC (eds) Catalyst characterization: physical techniques for solid materials. Plenum Press, New York, p 611
69. Chen DT, Zhang L, Yi C, Dumesic JA (1994) J Catal 146:257
70. Siperstein F, Gorte RJ, Myers AL (1999) Langmuir 15:1570
71. Dunne JA, Rao M, Sircar S, Gorte RJ, Myers AL (1997) Langmuir 13:4333
72. Brown DR, Groszek AJ (2000) Langmuir 16:4207
73. Groszek AJ (1998) Thermochim Acta 312:133
74. Aguayo AT, Gayubo AG, Erena J, Olazar M, Arandes JM, Bilbao J (1994) J Chem Tech Biotechnol 60:141
75. MYu S, VYu B (1999) Kinet Catal 40:819
76. Groszek AJ (1999) In: Dabrowski A (ed) Adsorption and its application in industry and environmental protection, Elsevier, Amsterdam, 1999, p 143; Stud Surf Sci Catal 120A:143
77. Zajac J, Dutartre R, Jones DJ, Rozière J (2001) Thermochim Acta 379:123
78. Drago RS, Dias SC, Torrealba M, de Lima L (1997) J Am Chem Soc 119:4444
79. Drago RS, Dias SC, Mc Gilvray JM, Mateus ALML (1998) J Phys Chem B 102:1508
80. Savitz S, Myers AL, Gorte RJ, White D (1998) J Am Chem Soc 120:5701
81. Savitz S, Myers AL, Gorte RJ, White D (1999) Micropor Mesopor Mater 31:211
82. De Macedo JL, Dias SCL, Dias JA (2004) Micropor Mesopor Mater 72:119
83. Silvestre-Albero J, Sepulveda-Escribano A, Rodriguez-Reinoso F (2002) In: Rodriguez-Reinoso F, McEnaney B, Rouquerol J, Unger KK (eds), Proceedings of the 6th International Symposium on the Characterization of Porous Solids (COPS-VI), Alicante, Spain, 8–11 May 2002, Elsevier, Amsterdam, 2002, p 107; Stud Surf Sci Catal 144:107
84. Coker EN, Karge HG (1997) Rev Sci Instrum 68:4521
85. Klyachko AL, Brueva TR, Rubinshtein AM (1979) Kinet Catal 20:1256
86. Gravelle PC (1972) Adv Catal 22:191
87. Mitani Y, Tsutsumi K, Takahashi H (1986) Colloid Polymer Sci 264:445
88. Biaglow AI, Parrillo DJ, Kokotailo GT, Gorte RJ (1994) J Catal 148:213
89. Parrillo DJ, Lee C, Gorte RJ (1994) Appl Catal A: Gen 110:67
90. Auroux A, Gravelle PC, Védrine JC, Rekas M (1980) In: Rees LVC (ed) Proceedings of the 5th International Zeolite Conference, Naples, Italy, 2–6 June 1980 Heyden, London, p 433
91. Babitz SM, Williams BA, Kuehne MA, Kung HH, Miller JT (1998) Thermochim Acta 312:17
92. Tsutsumi K, Kawai T, Yanagihara T (1994) In: Hattori T, Yashima T (eds) Proceedings of the International Symposium on Zeolites and microporous crystals, Nagoya, Japan, 22–25 August 1993, Elsevier, Amsterdam, p 217; Stud Surf Sci Catal 83:217
93. Khvoshchev SS, Vasil'eva EA (1982) Izvest Akad Nauk SSSR Ser Khim 5:973
94. Guil JM, Guil-López R, Perdigón-Melón JA, Corma A (1998) Micropor Mesopor Mater 22:269
95. Corma A, Chica A, Guil JM, Llopis FJ, Mabilon G, Perdigón- Melón JA, Valencia S (2000) J Catal 189:382
96. Kunkler PJ, Downing RS, van Bekkum H (2001) In: van Bekkum H, Flanigen EM, Jacobs PA, Jansen JC (eds) Introduction to zeolite science and practice, Elsevier, Amsterdam, p 987; Stud Surf Sci Catal 137:987
97. Guil JM, Guil-López R, Perdgón-Melón JA (2000) Stud Surf Sci Catal 130:2927
98. Gorte RJ, White D (2000) Micropor Mesopor Mater 35–36:447
99. Yang L, Trafford K, Kresnawahjuesa O, Sepa J, Gorte RJ, White D (2001) J Phys Chem B 105:1935

100. Yoda E, Kondo JN, Domen K (2005) J Phys Chem B 109:1464
101. Ruthven DM (2001) In: Robson H (ed) Verified syntheses of zeolitic materials. Elsevier, Amsterdam, p 61
102. Mittelmeijer-Hazeleger MC, van der Linden B, Bliek A (1995) J Porous Mater 2:25
103. Makowski W, Kuśtrowski P (2007) Micropor Mesopor Mater 102:283
104. Corma A, Corell C, Pérez-Pariente J, Guil JM, Guil-López R, Nicolopoulos S, Gonzales Calbert J, Vallet-Regi M (1996) Zeolites 16:7
105. Eder F, Lercher JA (1997) J Phys Chem B 101:1273
106. Eder F, Lercher JA (1997) Zeolites 18:75
107. He YJ, Nivarthy GS, Eder F, Seshan K, Lercher JA (1998) Micropor Mesopor Mater 25:207
108. Stach H, Thamm H, Fiedler K, Grauert B, Wieker W, Jahn E, Öhlmann G (1986) In: Murakami Y, Iijima A, Ward JW (eds) New developments in zeolite science and technology; Proceedings of the 7th International Zeolite Conference, Tokyo, Japan, 17–22 August 1986, Elsevier, Amsterdam, Kodansha, Tokyo, p 539; Stud Surf Sci Catal 28:539
109. Pires J, Brotas de Carvalho M, Ramoa Ribeiro F, Derouane E (1989) Appl Catal 53:273
110. Thamm H, Jerschkewitz HG, Stach H (1988) Zeolites 8:151
111. Nieβen W, Karge HG, Jozefowicz L (1992) In: Suzuki M (ed) Proceedings of the 4th international Conference on fundamentals of adsorption, Kyoto, Japan, 17–22 May 1992, Kodansha, Tokyo, 1993, p 475
112. Weber G, Bouvier F, Simonot-Grange MH (1996) In: Proceedings of AFCAT and GEFTA meeting, 24–26 Sept 1996, Freiburg, Germany, p 85, 177
113. Jänchen J, Stach H, Busio M, van Wolput JHMC (1998) Thermochim Acta 312:33
114. Parrillo DJ, Gorte RJ (1993) J Phys Chem 97:8786
115. Parrillo DJ, Lee C, Gorte RJ, White D, Farneth WE (1995) J Phys Chem. 99:8745
116. Lee C, Parrillo DJ, Gorte RJ, Farneth WE (1998) J Am Chem Soc 118:3262
117. Gorte RJ, White D (1997) Top Catal 4:57
118. Knözinger H (1997) In: Ert G, Knözinger H, Weitkamp J (eds) Handbook of heterogenous catalysis, vol 2. VCH, Weinheim, p 707
119. Knözinger H, Huber S (1998) J Chem Soc Faraday Trans 94:2047
120. Webster CE, Drago RS, Zerner MC (1998) J Am Chem Soc 120:5509
121. Boreave A, Auroux A, Guimon C (1997) Micropor Mater 11:275
122. Shete BS, Kamble VS, Gupta NM, Kartha VH (1998) J Phys Chem 102B:5581
123. Kuroda Y, Mori T, Yoshikawa Y, Kittaka S, Kumashiro R, Nagao M (1999) Phys Chem Chem Phys 1:3807
124. Fanson PT, Stradt MW, Lanterback J, Nickolas Delgass W (2002) Appl Catal B: Environ 38:331
125. Shimokanwabe M, Hirano K, Takezawa N (1998) Catal Today 45:117
126. Zhanpeisov NU, Ju WS, Matsouka M, Anpo M (2003) Struct Chem 14:247
127. Solans-Monfort X, Sodupe M, Branchadell V (2003) Chem Phys Lett 368:242
128. Su BL, Norberg V (1998) Langmuir 14:2352
129. Turnes-Palomino G, Fisicaro P, Bordiga S, Zecchina A, Giamello E, Lamberti C (2000) J Phys Chem B 104:4064
130. Sobalik Z, Dedecek J, Ihonnikov I, Wichterlova B (1998) Micropor Mesopor Mater 21:525
131. Wichterlova B, Dedecek J, Sobalik Z, Vondrova A, Klier K (1997) J Catal 169:194
132. Dedecek J, Sobalik Z, Tvarnzkova Z, Kancky D, Wichterlova B (1995) J Phys Chem 99:16327
133. Kuroda Y, Yoshikawa Y, Kumashiro R, Nagao M (1997) J Phys Chem B 101:6497
134. Gervasini A, Picciau C, Auroux A (2000) Micropor Mesopor Mater 35–36:457
135. Olson D, Reischman P (1996) Zeolites 17:434
136. Eder F, Stockenhuber M, Lercher JA (1997) J Phys Chem B 101:5414
137. Millot B, Methivier A, Jobic H (1998) J Phys Chem B 103:3210
138. Makowski W, Majda D (2004) Thermochim Acta 412:131
139. Savitz S, Myers AL, Gorte RJ (2000) Micropor Mesopor Mater 37:33

140. Serykh A, Kazansky V (2004) Phys Chem Chem Phys 6:5250
141. Sepa J, Lee CC, Gorte RJ, White D (1996) In: Hightower JW, Delgass WN, Iglesia E, Bell AT (eds) Proceedings of the 11th congress on catalysis, Baltimore, Maryland, USA, June 30–July 5 1996, PO 144
142. Kotrla J, Kubelková L, Lee CC, Gorte RJ (1998) J Phys Chem B 102:1437
143. Lee CC, Gorte RJ, Farneth WE (1997) J Phys Chem B 101:3811
144. Sepa J, Lee CC, Gorte RJ, White D, Kassab E, Evleth EM, Jessri H, Allavena M (1996) J Phys Chem 100:18515
145. Stach H, Jänchen J (1991) Zeolites 12:152
146. Li J, Davis RJ (2003) Appl Catal A-General 239:59
147. Chen DT, Sharma SB, Filimonov I, Dumesic JA (1992) Catal Lett 12:201
148. Parrillo DJ, Gorte RJ, Farneth WE (1993) J Am Chem Soc 115:12441
149. Parrillo DJ, Gorte RJ (1992) Catal Lett 16:17
150. Auroux A (1988) In: Grobet PJ, Mortier WJ, Vansant EF, Schulz-Ekloff G (eds) Innovation in zeolite materials science, Proceedings of the International Symposium, Nieuwpoort, Belgium, 13–17 September 1987, Elsevier, Amsterdam, 1988, p 385; Stud Surf Sci Catal 37:385
151. Auroux A, Shi ZC, Echoufi N, Ben Taarit Y (1989) In: Karge HG, Weitkamp J (eds) Zeolites as catalysts, sorbents and detergent builders; Proceedings of the International Symposium, Würzburg, Germany, 4–8 September 1988, Elsevier, Amsterdam, 1989, p 377; Stud Surf Sci Catal 46:377
152. Yang S, Navrotsky A (2000) Micropor Mesopor Mater 37:175
153. Moïse JC, Bellat JP, Méthivier A (2001) Micropor Mesopor Mater 43:91
154. Busco C, Barbaglia A, Broyer M, Bolis V, Foddanu GM, Ugliengo P (2004) Thermochim Acta 418:3
155. Gonzalez MR, Sharma SB, Chen DT, Dumesic JA (1993) Catal Lett 18:183
156. Bolis V, Broyer M, Barbaglia A, Busco C, Foddanu GM, Ugliengo P (2003) J Mol Catal A: Chem 204–205:561
157. Takaishi T, Tsutsumi K, Chuubachi K, Matsumoto A (1998) J Chem Soc. Faraday Trans 94:601
158. Karge HG, Jozefowicz LC (1994) In: Weitkamp J, Karge HG, Pfeifer H, Hölderich W (eds) Zeolites and related microporous materials: State of the art 1994; Proceedings of the 10th International Zeolite Conference, Garmisch-Partenkirchen, 17–22 July 1994, Elsevier, Amsterdam, 1994, p 685; Stud Surf Sci Catal 84:685
159. Spiewak BE, Handy BE, Sharma SB, Dumesic JA (1994) Catal Lett 23:207
160. Kapustin GI, Kustov LM, Glonti GO, Brueva TR, VYu B, Klyachko AL, Rubinshtein AM, Kazanskii VB (1984) Kinet Katal 25:959
161. Tsutsumi K, Mitani Y, Takahashi H (1983) Bull Chem Soc Japan 56:1912
162. Auroux A, Bolis V, Wierzchowski P, Gravelle PC, Védrine JC (1979) JCS Faraday Trans II 75:2544
163. Huang M, Auroux A, Kaliaguine S (1995) Micropor Mat 5:17
164. van Santen RA (1994) In: Jansen JC, Stöcker M, Karge HG, Weitkamp J (eds) Advanced zeolite science and applications, Elsevier, Amsterdam, p 273; Stud Surf Sci Catal 85:273
165. Lohse U, Parlitz B (1989) J Phys Chem 93:3677
166. Stach H, Wendt R, Lohse U, Jänchen J, Spindler H (1988) Catal Today 3:431
167. Mellot CF, Cheetham AK, Harms S, Savitz S, Gorte RJ, Myers AL (1998) Langmuir 14:6728
168. Matsumoto A, Tsutsumi K (1995) J Chem Soc Faraday Trans 91:1707
169. Simonot-Grange M–H, Bertrand O, Pilverdier E, Bellat J–P, Paulin C (1997) J Thermal Anal 48:741
170. Cardona-Martinez N, Dumesic JA (1992) Adv Catal 38:149
171. Lónyi F, Valyon J (2001) Micropor Mesopor Mater 47:293
172. Auroux A, Muscas M, Coster DJ, Fripiat JJ (1994) Catal Lett 28:179

173. Védrine JC, Auroux A, Coudurier G (1984) In: White TE et al. (eds) Catalytic materials, relationship between structure and reactivity. A.C.S. Symposium Series 248, 13:254
174. Védrine JC, Auroux A, Bolis V, Dejaifve P, Naccache C, Wierzchowski P, Derouane EG, van Hoff JCH (1979) J Catal 59:248
175. Jozefowicz LC, Karge HG, Coker EN (1994) J Phys Chem 98:8053
176. Muscas M, Dutel JF, Solinas V, Auroux A, Ben Taarit Y (1996) J Mol Catal A-Chem 106:169
177. Niwa M, Suzuki K, Katada N, Kanougi T, Atoguchi T (2005) J Phys Chem B 109:18749
178. Stach H, Jänchen J, Lohse U (1992) Catal Lett 13:389
179. Stach H, Jänchen J, Jerschkewitz HG, Lohse U, Parlitz B, Hunger M (1992) J Phys Chem 96:8480
180. Barthomeuf D (1988) In: Ward JW (ed) Catalysis 1987, Proceedings of the 10th north american meeting of the catalysis society, San Diego, CA, USA, 17–22 May 1987, Elsevier, Amsterdam, 1988, p 177; Stud Surf Sci Catal 38:177
181. Auroux A, Ben Taarit Y (1987) Thermochim Acta 122:63
182. Williams BA, Babitz SM, Miller JT, Snurr RQ, Kung HH (1999) Appl Catal A: Gen 177:161
183. Barthomeuf D (1990) In: Barthomeuf D, Derouane EG, Hoelderich W (eds) Guidelines for mastering the properties of molecular sieves: relationship between the physicochemical properties of zeolitic systems and their low dimensionality, vol 11. Plenum, New York, p 426
184. Mitani Y, Tsutsumi K, Takahashi H (1983) Bull Chem Soc Jpn 56:1921
185. Mishin IV, Klyachko AL, Brueva TR, Tkachenko OP, Beyer HK (1993) Kinet Catal 34:502
186. Handy BE, Jacobo A, Cardenas Galindo MG, Gonzalez M, Llanos ME, Guzman M, Hernandez F (2002) Top Catal 19:249
187. Delahay G (2004) In: Thibault-Starzyk F (ed) Les Matériaux Micro et Mésoporeux, Caractérisation, EDP sciences. Les Ulis, France, p 234
188. Narayanan S, Sultana A, Le QT, Auroux A (1998) Appl Catal A-Gen 168:373
189. Witzel F, Karge HG, Gutsze A (1992) In: von Ballmoos R, Higgins JB, Treacy MMJ (eds), Proceedings of the 9th international zeolite conference, Montreal, Canada, 5–10 July 1992, Butterworth-Heinemann, Boston, 1993, Vol 2, p 283
190. de Araujo AS, Fernandes VJ Jr, Fernandes GJT (1997) J Therm Anal 49:567
191. Auroux A, Artizzu P, Ferino I, Monaci R, Rombi E, Solinas V (1997) Micropor Mater 11:117
192. Rac V, Rakić V, Gajinov S, Dondur V, Auroux A (2006) J Therm Anal Cal 84:239
193. Viswanadham N, Muralidhar G, Prasada Rao TSR (2004) J Mol Catal A-Chem 223:269
194. Rakic V, Dondur V, Gajinov S, Auroux A (2004) Thermochim Acta 420:51
195. Bolis V, Bordiga S, Turnes Palomino G, Zecchina A, Lamberti C (2001) Thermochim Acta 379:131
196. Kuroda Y, Onishi H, Mori T, Yoshikawa Y, Kumashiro R, Nagao M, Kobayashi H (2002) J Phys Chem B 106:8976
197. Karge HG (2001) In: van Bekkum H, Flanigen EM, Jacobs PA, Jansen JC (eds) Introduction to zeolite science and practice, 2nd edn. Elsevier, Amsterdam, p 707; Stud Surf Sci Catal 137:707
198. Sahoo SK, Viswanadham N, Ray N, Gupta JK, Singh ID (2001) Appl Catal A-Gen 205:1
199. Dejaifve P, Auroux A, Gravelle PC, Védrine JC, Gabelica Z, Derouane EG (1981) J Catal 79:123
200. Uguina MA, Serrano DP, Van Griieken R, Vènes (1993) Appl Catal A 99:85

Chapter 4
Electron Microscopy and Imaging

Christine E. Kliewer

Abstract This chapter gives a concise overview of various zeolite morphologies and continues with a general discussion regarding the historical development of electron microscopes and the integration of associated elemental analysis instrumentation. Various specialized forms of electron microscopy (e.g., high resolution imaging, electron tomography, etc.) are reviewed with respect to zeolites. Zeolite-specific sample preparation techniques for scanning electron microscope (SEM), transmission electron microscopy (TEM), and scanning transmission electron microscopy (STEM) are discussed in detail. General SEM, TEM and high angle annular dark field STEM imaging protocols for zeolite examination are reviewed with an emphasis on obtaining general microstructural information, pore/channel morphology, and small metal particle dispersion information. Elemental analysis of zeolites using energy dispersive spectrometry and electron energy loss spectrometry is examined.

4.1 Zeolite Morphologies

Zeolites are well-ordered aluminosilicates built from corner-shared SiO_4 or AlO_4 tetrahedra [1–3]. These cage-like materials may have uni- or multi-dimensional channels [2, 3] and may exist in a variety of morphologies (e.g., plate-like, cube-shaped, etc.) [2, 4–6]. Primary zeolite particles can be relatively large, ~1 μm [1, 7]. Thus, secondary electron imaging (SEI) in the SEM provides very useful information about the general microstructure [1]. The zeolite unit cell, however, is significantly smaller, typically ~1–3 nm, [3] and thus is better characterized by TEM [1–5, 7–13].

C.E. Kliewer
ExxonMobil Research & Engineering, 1545 Route 22 East, Annandale, NJ 08801
e-mail: chris.e.kliewer@exxonmobil.com

A.W. Chester and E.G. Derouane (eds.), *Zeolite Characterization and Catalysis*,
DOI 10.1007/978-1-4020-9678-5_4, © Springer Science+Business Media B.V 2009

4.2 Electron Microscopy

4.2.1 General Historical Background

Over the years, a wide variety of electron microscopy (EM)-based acronyms have evolved. A listing of some of the most common EM abbreviations is presented in Table 4.1. A number of these acronyms are used extensively throughout this text and all appear quite frequently throughout the literature. Thus, the abbreviations are presented here for the reader's general convenience.

Characterization studies typically attempt to relate specific aspects of a material's microstructure (e.g., particle shape, pore size, etc.) to its physical properties (e.g., activity, selectivity, etc.). Hence, imaging techniques have been a fundamental component of analytical efforts for many years. Because the earliest studies were conducted using optical microscopy, minimal structural information was obtained due to limitations set by the wavelength of light, ~400 nm [8]. However, by the mid-1920s de Broglie hypothesized that electrons exist as waves and that their wavelengths are shorter than visible light (Table 4.2) [8, 9]. The idea of using electrons as a microscope illumination source developed around that general time, and by the early 1930s Knoll and Ruska had constructed the first TEM [8, 14]. These events put scientists on the path towards improved materials' characterization via high resolution imaging in the electron microscope (Figs. 4.1 and 4.2).

Because the early TEM's were generally designed after the basic concept of a transmitted light microscope (Fig. 4.3) [12, 13], both instruments have an illumination source, a series of lenses and apertures situated within a column, a specimen stage, a viewing port, and, typically, an image collection system [12, 13, 15, 16]. These two instruments, however, vary in a several key aspects. First, unlike the optical microscope which uses glass lenses to focus light traveling through the microscope column, the TEM uses electromagnetic lenses to guide electrons through its column [8–13, 15]. Second, unlike the optical microscope column which operates at atmospheric pressure, the TEM column is held under high vacuum to prevent the electron beam from being scattered by gas molecules [5–13, 15].

Improvements in the TEM arena continued and by the 1940s, scientists became interested in examining materials under more realistic (e.g., thermal fluxuation, environmental control, etc.) conditions [17]. While the TEM column's high vacuum constraints posed an issue, self-enclosed stages for gas and liquid phase studies [17] as well as sample holders for designed for various heating, cooling, and straining experiments were developed [8, 17]. Significant bodies of work describing these in situ TEM experiments were published [17], and over time, both in situ [18–20] and a variety of ex situ [21–24] techniques progressed. By the 1990s, sufficient interest in environmental TEM (ETEM) prompted advances leading to the differentially pumped TEM column configuration [25, 26]. This drive ultimately resulted in the development of the commercially available, high resolution ETEMs on the market today.

Table 4.1 Common acronyms used in electron microscopy

TEM	
ADF	annular dark field
BF TEM	bright field TEM
ETEM	environmental TEM
HAADF	high angle annular dark field
IVEM	intermediate voltage TEM
HREM	high resolution electron microscopy
HVEM	high voltage TEM
STEM	scanning TEM
TEM	transmission electron microscopy
TEMT	TEM tomography
SEM	
BEI	backscattered electron imaging
ESEM	environmental SEM
SEI	secondary electron imaging
SEM	scanning electron microscopy
FESEM	field emission SEM
VPSEM	variable pressure SEM
ELEMENTAL ANALYSIS	
EDS	energy dispersive spectrometry
EELS	electron energy loss spectrometry

Table 4.2 Relationship between electron wavelength and accelerating voltage

Accelerating voltage (kV)	Electron wavelength, *l* (nm)
100	0.0037
200	0.00251
500	0.00142
1,000	0.00087

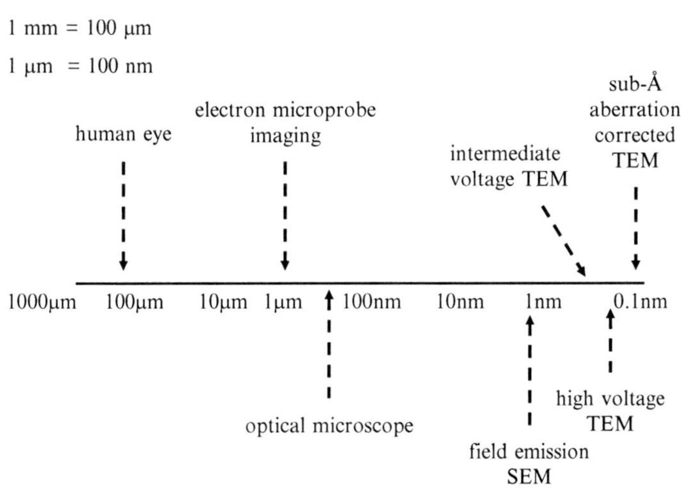

Fig. 4.1 Schematic presents the increase in resolution from ~0.1 mm using the unaided human eye to sub-Ångstrom using aberration corrected TEM

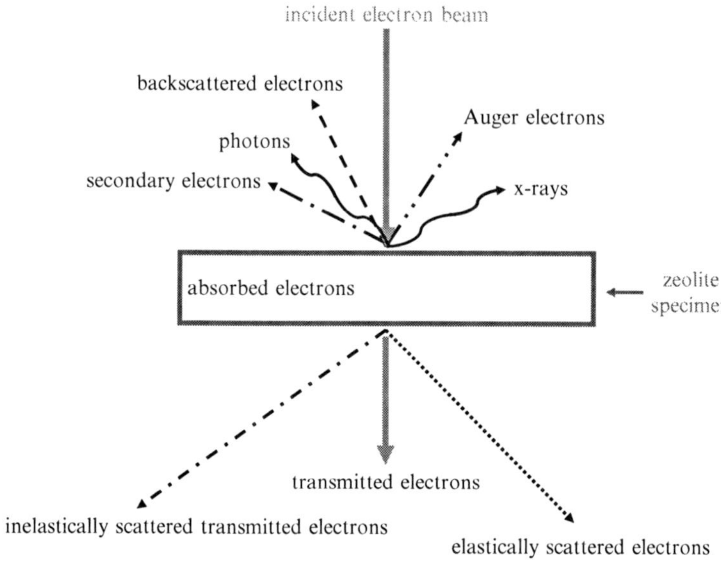

Fig. 4.2 Schematic view presents signals generated when electron interacts with a material

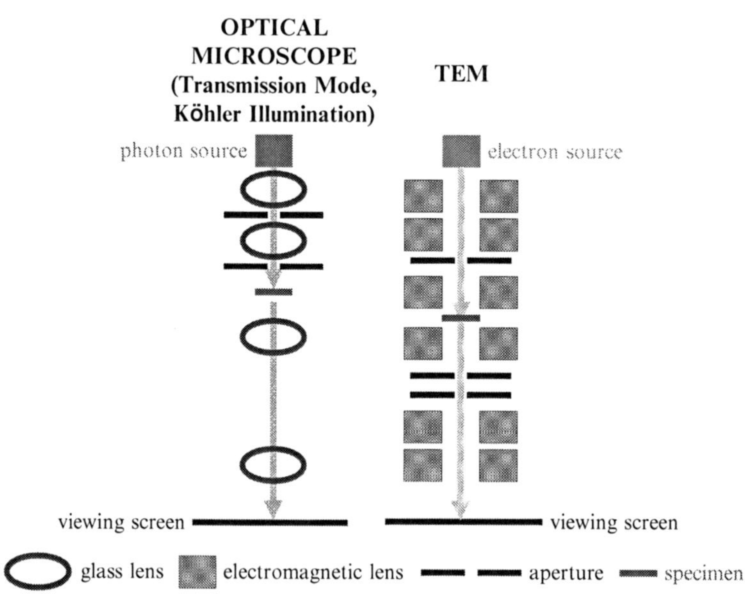

Fig. 4.3 Schematic views compare the optical microscope operated in the transmitted light mode with Köhler illumination to a transmission electron microscope (TEM)

As improvements in TEM's continued the goal of obtaining higher resolution images followed. While it was long known that sub-Ångstrom resolution was theoretically achievable even within the early TEMs [8, 27], ultra-high resolution was limited by aberrations inherent in the electron microscope's optical system [8, 27]. A serious drive towards eliminating these imperfections began in the 1960s with the development of an aberration correction system [8, 28, 29]. However, sub-Ångstrom imaging efforts met limited success due to the severe inherent instabilities common to electron microscopes during that period. [29] By the 1970s, however, atomic resolution imaging was possible using high voltage ($\geq$1000 kV) electron microscopes (HVEM's) [8]. Continued advances in instrumentation eventually resulted in intermediate voltage (300–400 kV) electron microscopes (IVEMs) and even 200 kV TEMs in which atomic level resolution imaging was routinely possible [8].

Beginning in approximately the mid-1980s personal computers (PCs) were used to control slow speed CCD cameras that were being interfaced to TEMs [30]. These digital imaging systems were optimized through the early 1990s [30]. Increases in PC power along with improvements in stage stability continued throughout the 1990s and into the early 2000s. These advances (e.g., digital imaging, computing power, computer-controlled TEM stages with high stability, etc.) drove the materials science community to explore TEM tomography (TEMT) making these studies more "routine" [31, 32]. Further, by the turn of this century, continued improvements in TEM instrumentation and renewed interest in aberration correction capabilities combined to make sub-Ångstrom resolution imaging possible [33–39].

As TEMs were being built and improved, the SEM evolved [13, 40]. By the mid-1930s, Knoll designed a microscope in which an electron probe could be simultaneously scanned over a material's surface and synchronized with a cathode ray tube (CRT) to produce an image [15, 40]. Similar to the TEM, the SEM operates under high vacuum and uses an electron-based illumination source, electromagnetic lenses, and apertures [13, 15, 40]. Unlike the TEM, however, the SEM can house bulk specimens ranging from a few millimeters to centimeters in diameter [13].

In 1960, Everhart and Thornley developed a more efficient secondary electron detector for the SEM, and by the mid 1960s, SEMs became commercially available as seen in Fig. 4.4 [40]. The development of temperature-controlled stages in combination with variable pressure SEMs (VPSEM) in the early 1990s enabled electron beam sensitive materials to be examined more easily, making environmentally-based SEM (ESEM) studies more accessible to the average electron microscopist [40]. As in the case of the TEM, the SEM vacuum system continued to improve with time and eventually allowed the routine use of higher brightness filaments (lanthanum hexaboride and field emission) in both academic and industrial communities [40]. Finally, the development of a commercially available field emission SEM (FESEM) has made higher resolution, low voltage ($<\sim$5 kV) imaging of electron beam sensitive specimens commonplace [40].

By the late 1930s, the scanning probe technology was integrated into the TEM design [40]. This resulted in the development of the dedicated scanning transmission electron microscope (STEM) and the hybridized TEM/STEM (Fig. 4.5) [40]. Various types of specialized STEM detectors (bright field (BF), annular dark field (ADF), high angle annular dark field (HAADF)) were developed [8]. While unscattered and weakly scattered transmitted electrons are collected by the BF STEM

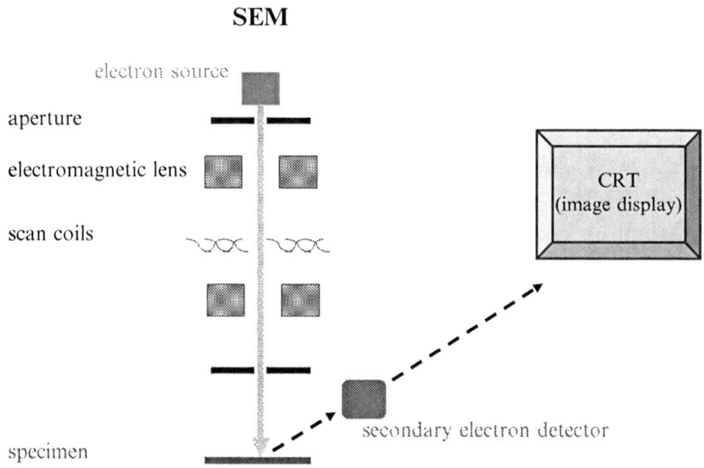

Fig. 4.4 Schematic view presents a scanning electron microscope (SEM)

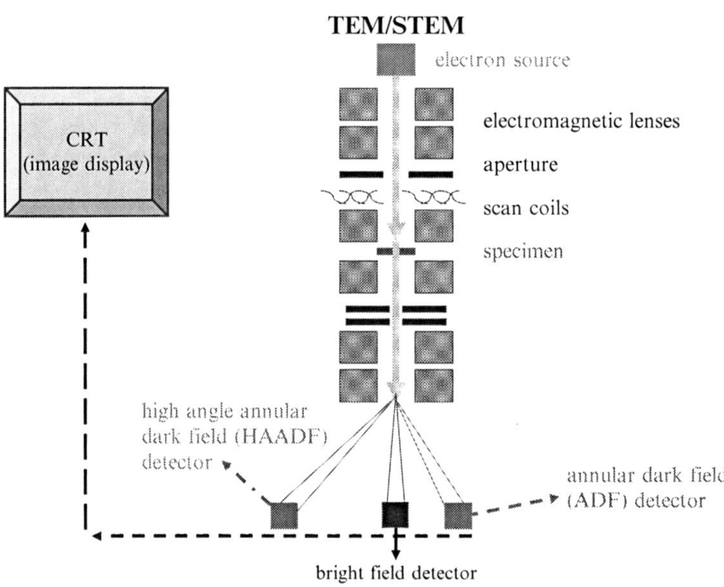

Fig. 4.5 Schematic view presents a hybridized transmission electron microscope (TEM)/scanning transmission electron microscope (STEM)

detector, more highly scattered electrons are collected by either the ADF or HAADF STEM detectors [8]. Thus, BF STEM images generally have relatively poor contrast, while ADF and particularly HAADF STEM images have strong atomic number-dependent contrast [8]. As in the case of the TEM, both electron tomography [41–46] and sub-Ångstrom resolution via aberration correction have

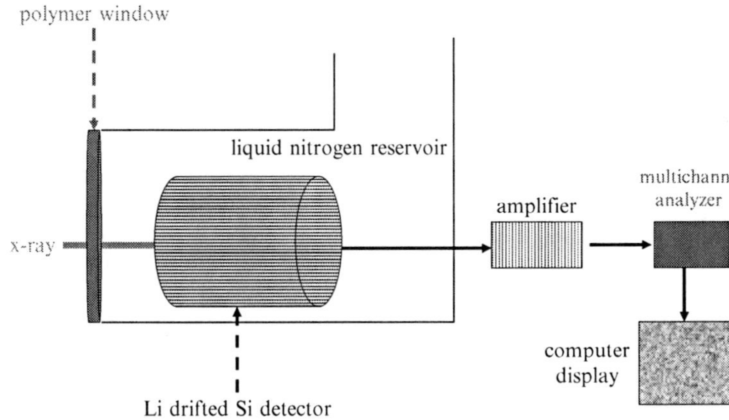

Fig. 4.6 Schematic view presents an energy dispersive spectrometry (EDS) system

been extended to both the dedicated STEM and the hybridized TEM/STEM [33, 47–52].

In addition to imaging, high spatial resolution elemental analysis is often desirable during characterization. As early as the late 1800s, Starke's showed that an electron source could generally distinguish materials of different compositions [15]. However, it was not until the early 1900s that Kaye demonstrated the use of electrons to produce X-rays that were characteristic of specific elements [15]. While curved X-ray spectrometers were designed in the 1920s [13], it was not until the late 1940s that Castaing and Guinier constructed an electron-based instrument equipped with an optical imaging system that provided elemental analyses via characteristic X-rays [13, 15, 53]. These efforts rightly earned Castaing recognition as a leader in the field of microanalysis [13, 15, 40, 54–58], even though an earlier instrument using electron energy loss for elemental identification was developed by Hillier [13, 15, 40, 54–58]. Since then, the addition of an elemental analysis system (e.g., energy dispersive spectrometry (EDS), electron energy loss spectrometry (EELS), etc.) to an electron microscope has become commonplace.

Both EDS (Fig. 4.6) and EELS (Fig. 4.7) detectors provide unique elemental information about the specimen. However, only EDS analyses, which can be conducted equally well on both thick and thin samples, can be conducted in both the SEM and the TEM. Thus, EDS systems are the most frequently purchased electron microscope accessory for elemental analysis. Because EELS requires very thin specimens [8, 56] and the analysis is significantly more complicated than EDS [8, 56], and is less common. Nevertheless, the information provided by EELS can be highly valuable. [8] Thus, in recent years both techniques have been utilized in conjunction with advances in TEM, dedicated STEM, and hybridized TEM/STEM instrumentation. The potential for combining EELS data with electron tomographic studies has been evaluated [58], and improvements in EELS [59, 60] and X-ray [61] data in aberration corrected electron microscopes have been made.

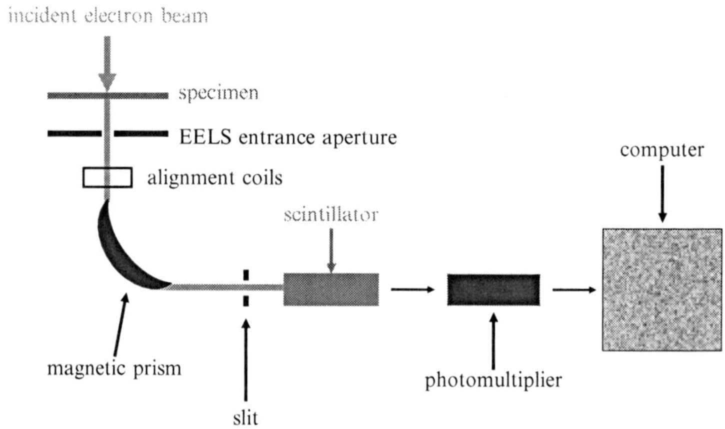

Fig. 4.7 Schematic view presents an electron energy loss spectrometry (EELS) system

4.2.2 Scanning Electron Microscopy of Zeolites

4.2.2.1 Zeolite Sample Preparation Methods for Scanning Electron Microscopy

Because bulk materials can be analyzed in the SEM, highly extensive sample preparation is generally not required [40, 53, 62–64]. Zeolites typically crystallize as micron-sized aggregates in powder form. Because these materials are not electrically conductive in the SEM, it is advantageous to disperse large aggregates into smaller fragments that can have better contact with a conductive surface (e.g., SEM stub, carbon paint, carbon tape, etc.). Dispersion is generally accomplished by using an agate mortar and pestle to gently crush a small amount (~0.01 g) of the zeolite powder into fine pieces. In some cases, it is sufficient to dust a thin coating of the crushed material directly either onto a piece of double stick carbon tape that has been affixed to a standard SEM mount or to brush a small amount of the crushed fines onto the tacky surface of a carbon-painted SEM stub (Fig. 4.8) [40, 64]. In other instances, the crushed zeolite particles may need to be transferred into a vial of a non-interacting solvent (e.g., isopropyl alcohol, ethanol, etc.) [40, 53, 64]. A sufficient concentration of the crushed powder should be placed into the solvent to create a dilute (relatively clear) solution, and the resulting mixture should be sonicated to disperse the zeolite particles [52, 64]. A few drops of solution can be pipetted onto a flat, highly polished SEM mount and allowed to air dry (Fig. 4.9) [53, 64].

In some instances, the aforementioned preparations may be directly transferred into the SEM for examination [64]. However, in other cases, a thin metal or carbon coating may still be required to dissipate charging effects [40, 53, 63, 64]. Thus, evaporation of the thinnest possible metal or carbon coating under the highest achievable vacuum is desirable [40, 63, 64]. For optimal results, gold/palladium

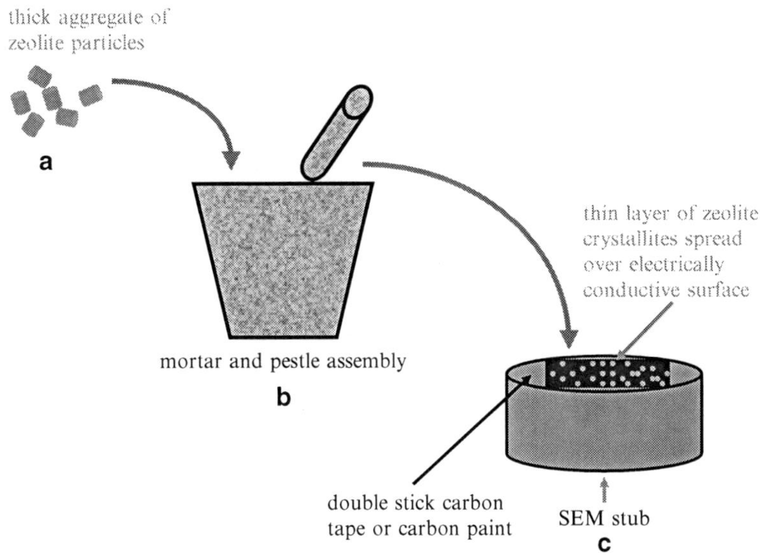

Fig. 4.8 Schematic presents the dispersion sample preparation technique for SEI of a zeolite in the scanning electron microscope. (**a**) Zeolite aggregates are (**b**) gently broken apart into small crystallites using a mortar and pestle. (**c**) A thin layer of crystallites is transferred onto either a piece of double stick carbon tape affixed to a standard SEM stub or a somewhat dry film of carbon paint coated onto a standard SEM stub

sputter coating may be used for intermediate magnification imaging studies while high vacuum chromium deposition should be used for higher resolution imaging [40, 63]. Depending upon the requirements on the X-ray microanalysis study, either chromium or carbon may be selected [40, 63]. Because the metal coatings consist of small islands that can agglomerate over time and thereby reduce the ability of the conductive layer to dissipate charge, the specimen should be imaged in the electron microscope as soon as possible after being prepared [63].

4.2.2.2 Secondary Electron Imaging of Zeolites in the Scanning Electron Microscope

While many different signals are generated in the SEM, it is generally the secondary electrons (SEs) that are used to image zeolites (Fig. 4.10) [1, 4–6, 13]. Because secondary electrons are, by definition, low energy electrons, they are generated from the top few nanometers of the material's surface [12, 13, 15, 40, 62–64], making contrast in the secondary electron image (SEI) highly dependent upon the crystallite topography (e.g., overall crystal size/shape, general surface roughness, etc.) [1, 12, 13, 40, 63, 64]. Thus, SEs produce images that appear to have a three-dimensional nature and thereby lend themselves to a relatively straightforward, intuitive interpretation [1, 12, 13, 40, 63, 64].

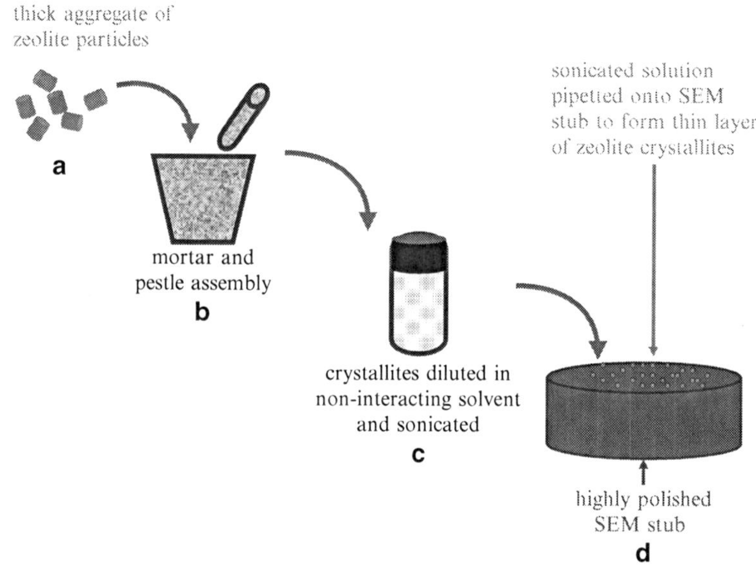

Fig. 4.9 Schematic presents the dispersion and dilution sample preparation technique for SEI of a zeolite in the scanning electron microscope. (**a**) Zeolite aggregates are (**b**) gently broken apart into small crystallites using a mortar and pestle. (**c**) A sufficient concentration of the gently crushed zeolite is transferred into a vial containing a non-interacting solvent to create a dilute (slightly cloudy) solution. The solution is sonicated. (**d**) Droplets of sonicated solution are pipetted onto a highly polished SEM stub and allowed to air dry

In the case of zeolites, the general crystal morphology is typically characterized in the SEM using the SEI [1, 4, 6, 40, 65–67]. Cube-like morphologies have been observed in faujasite crystals [1, 66], while plate-like structures have been noted in SSZ-25 [4], MCM-47 [65], and levyne [66]. An examination of SAPO-11, SAPO-31, and SAPO-41 crystallites indicated different microstructures in each case [6]. Higher resolution SEI data has even revealed channel openings in SBA-15 [68]. The reader is strongly encouraged to review 66, 68 to obtain a clearer understanding of the potential of SE imaging with respect to zeolite morphology. The SEI macro-scale information (e.g., general crystallite size, overall shape, etc.) can be highly useful in selecting the optimal TEM sample preparation technique. However, it is important to note that the SEI typically reflects the growth rate of specific crystal planes, and thus this data may or may not be related to the actual crystal structure [69].

To characterize the general features of zeolite crystals (e.g., size, shape, etc.), SEM low (<~5 kV) operating voltages are typically desirable [1, 40, 64]. These lower accelerating voltages minimize the creation of backscattered electrons (BEs), which are generated deeper within the crystallite, and which, when formed, can mask surface details in the SEI [13, 40, 62–64]. To further improve the SEI integrity, it is necessary to work with an uncoated zeolite [1, 40, 64]. This insures that the SEI does not reflect features of the coating material and that the zeolite's fine surface structures are not hidden by the metal or carbon deposition layer [40].

Fig. 4.10 Secondary electron image presenting general morphology of faujasite crystallites. SEI courtesy Tom Bruno, ExxonMobil Research and Engineering

To visually enhance specific surface morphologies, such as small steps, it can be useful to tilt the zeolite in the SEM [1, 13].

4.2.3 Transmission Electron Microscopy of Zeolites

4.2.3.1 Zeolite Sample Preparation Methods for Transmission Electron Microscopy

Because TEM requires specimens that are ~100 nm or less in thickness, extensive sample preparation can be required [1, 8–12, 62, 64, 70–128]. Zeolite crystals typically present as powders which are too agglomerated to directly image in the TEM without some form of preparation. As in the case of the SEM sample preparation previously described (Sect. 4.2.2.1), a small amount (~0.01 g) of powder can be gently crushed into fragments at ambient temperature using an agate mortar and pestle [10, 70, 72, 74], although crushing under liquid nitrogen conditions has also been explored [75]. In some cases, it is sufficient to simply dust the crushed material directly onto a standard holey-carbon-coated TEM grid by sliding the grid into a pile of the crushed fines [64, 65, 74]. This action allows small

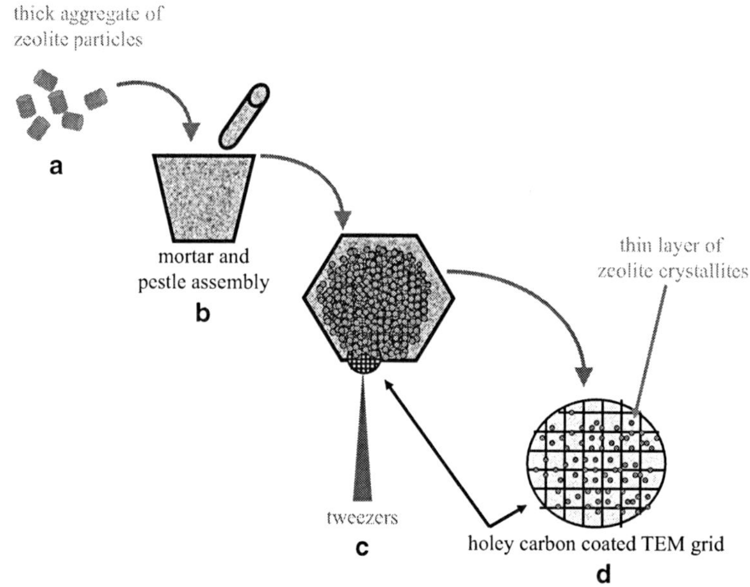

Fig. 4.11 Schematic presents the dispersion sample preparation technique for transmission electron microscopy of a zeolite. (**a**) Zeolite aggregates are (**b**) gently broken apart into small crystallites using a mortar and pestle. (**c**) A holey carbon coated TEM grid is carefully slid underneath the pile of crushed zeolite fines, (**d**) allowing small zeolite crystallites to adhere to the carbon film by electrostatic attraction. The TEM grid is removed and reinserted into the pile of crystallites several additional times. The grid is then ready for transfer into the TEM

zeolite crystallites to adhere to the carbon film by electrostatic attraction. The TEM grid, however, may need to be inserted into the pile of crushed files several times to insure sufficient specimen pickup (Fig. 4.11). While in most cases, researchers directly examine the crushed material in the TEM, some investigators have advocated coating the crushed material on the TEM grid with a protective carbon layer prior to TEM characterization [75]. In a few other instances, the crushed zeolite particles may need to be transferred into a non-interacting solvent (e.g., isopropyl alcohol, ethanol, etc.) to create a dilute suspension similar to that previously described (Sect. 4.2.2.1) for SEM sample preparation [8, 60, 64, 70, 72, 74]. Subsequent to sonication, either a droplet of the suspension may be pipetted onto a standard holey carbon-coated TEM grid [69, 72] or the grid may be directly dipped into the solution [60, 74]. After air drying, the grid can be transferred directly into the TEM [8, 60, 64, 72, 74].

In still other instances, it may be desirable to view elongated zeolite structures in all possible orientations, to make relatively thick (>100 nm) zeolite crystals thinner than can be achieved by simple crushing techniques, or to maintain the spatial relationship between zeolite particles within an extrudate. Thus, in these cases, the zeolite may be prepared for TEM examination using ultramicrotomy [8, 64, 70–73, 75]. This procedure typically involves gently crushing a small amount (~0.01 g) of zeolite into fines using an agate mortar and pestle as previously described (Sect. 4.2.2.1) [72]. The fines are transferred into a standard, embedding mold (typically a BEEM® capsule) and encased in a hard mixture of an appropriate TEM embedding resin (typically LR White®) [64, 70–74]. After curing, the zeolite/resin composite block is released from the BEEM® capsule. The end of the block housing the zeolite particles is trimmed to form a small (~0.1 to ~0.2mm), trapezoidal face [71–74]. The resin block encased zeolite is then ready for sectioning. The specimen-containing block is clamped into the microtome and aligned such that the zeolite-containing face is parallel to the microtome knife edge. The boat is filled with deionized water until the liquid surface appears to shimmer when examined under reflected light. A 55° materials science diamond knife is used to cut electron transparent sections from the resin/zeolite block [71–74]. Thin resin-encased zeolite sections are floated onto the water's surface. Once a continuous ribbon of electron transparent sections has formed, the slices are ready for collection onto a standard holey carbon-coated TEM grid [72]. Section pickup can be accomplished by carefully sliding the TEM grid into the water-filled boat at the point furthest from the knife edge. The submerged grid can then be carefully guided towards the ribbon of sections. Once the grid is located directly beneath the ribbon, the grid can be lifted upwards towards the sections, and the thin slices positioned in the center of the grid (Fig. 4.12). In some instances, the ribbon of sections is repelled by the TEM grid. This phenomenon can generally be overcome by removing the TEM grid from the water-filled boat and dipping the grid into a beaker of acetone for a few seconds. The grid can then again be submerged into the boat and re-positioned under the thin sections for collection as previously described (Sect. 4.2.3.1).

Once the thin sections have been successfully positioned onto the TEM grid, the grid is placed on a clean piece of filter paper. Subsequent to air drying, the grid is transferred directly into the TEM [72–74]. In most cases, the holey carbon support film provides sufficient electrical conductivity to prevent the specimen from building up electrical charge during the TEM characterization. However, in some instances, an additional conductive coating is required. This can be accomplished by high vacuum evaporation of a thin carbon layer over the specimen [70].

4.2.3.2 Imaging Zeolites in the Transmission Electron Microscope

BF TEM imaging is important with regard to many aspects of zeolite characterization (Fig. 4.13) [76–79]. First, TEM reveals defect structures (e.g., dislocations, stacking faults, twins, etc.) and allows detailed characterization of these features [7, 80–93]. Second, TEM clearly shows the effects of dealumination treatment on zeolite microstructures [93–101]. Third, TEMT imaging reveals zeolite porosity in three dimensions [96–99, 101, 102] and metal particle positions within pores [102]. However, it must be noted that TEMT studies on fine structures require ~141 images to produce a quality reconstruction [31, 32, 96–99, 101, 102]. Thus, TEMT analyses still have very limited usefulness with respect to many zeolites, which

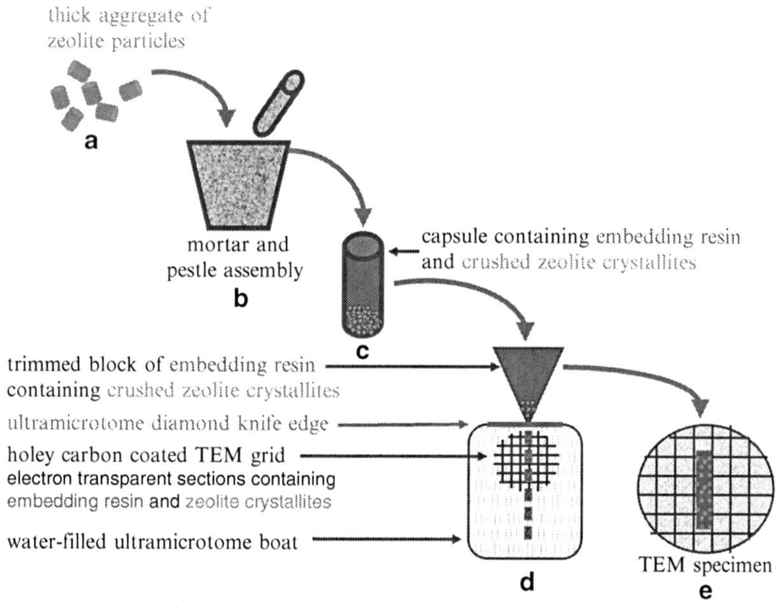

Fig. 4.12 Schematic presents the ultramicrotomy sample preparation technique for transmission electron microscopy of a zeolite. (**a**) Zeolite aggregates are (**b**) gently broken apart into small crystallites using a mortar and pestle. (**c**) Crushed zeolite fines are embedded in a resin block. The resin block face containing the zeolite crystallites is trimmed to form a trapezoid. The block is mounted and in the ultramicrotome and the trapezoidal face is aligned with the diamond knife edge. (**d**) A ribbon of electron transparent sections forms on the water's surface behind the diamond knife edge as sectioning progresses. (**e**) The TEM grid is submerged into the water and the thin sections are collected

easily damage under the electron beam [31, 32]. Lastly, high resolution TEM (HRTEM) imaging shows critical pore structures, lattice planes, and atomic arrangements within materials [103–128]. The reader is strongly encouraged to review any of the cited references for examples of the aforementioned structures.

High resolution TEM (HRTEM) imaging, while a highly powerful analytical technique [1–5, 7, 68, 78, 103–128], must be conducted with extreme caution with zeolites since electron beam damage can severely affect the material's microstructure (Fig. 4.14) [1–3, 7, 8, 129–138]. Depending upon the accelerating voltage of the TEM, different electron damage mechanisms become increasingly important [2, 8, 136]. Knock-on damage, the displacement of atoms from the crystal lattice, becomes dominant at higher accelerating voltages, while radiolysis and bond breaking, occur (more important) at lower accelerating voltages [2, 8, 136]. In general, electron beam damage occurs more rapidly in zeolites having a higher water concentration [2, 129, 130, 132, 136], decreased silica to alumina ratio [2, 132, 134], and lower levels of rare earth additions [2]. Further, it has even been shown that zeolite-supported, small Pt particles can sinter during the course of a TEM study [112, 134].

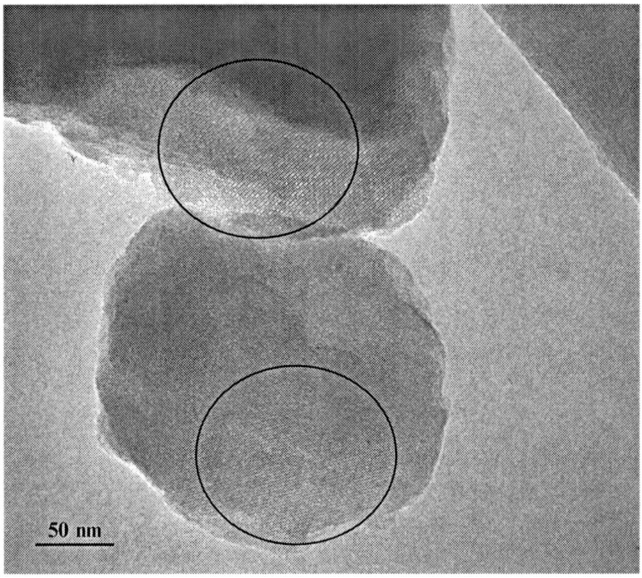

Fig. 4.13 Bright field TEM image presenting lattice image (circled areas) of faujasite crystallite

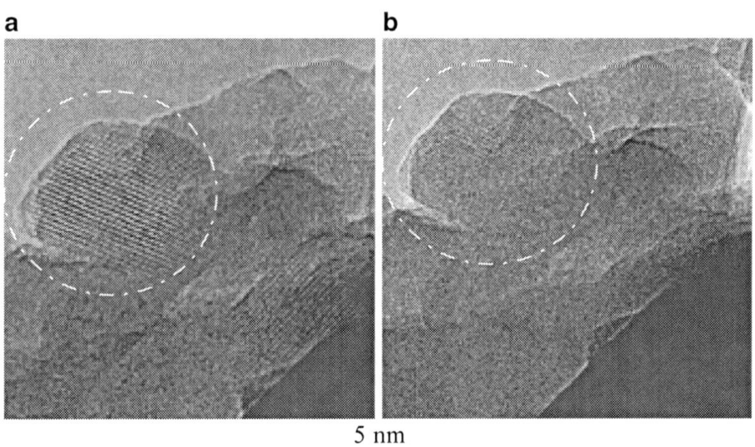

Fig. 4.14 Bright field TEM images presenting faujasite crystallite (**a**) prior to and (**b**) subsequent to electron beam damage. Note the loss of structural information due to electron beam damage effects in the circled area

Low-dose imaging techniques, developed in the 1970s, can significantly reduce electron beam damage effects [8, 71, 135, 139–142]. The low-dose protocol typically involves randomly examining material in the TEM's diffraction mode until a crystallite oriented in the desired direction (positioned with regard to a specific zone axis) is located [139]. The TEM is switched from diffraction mode

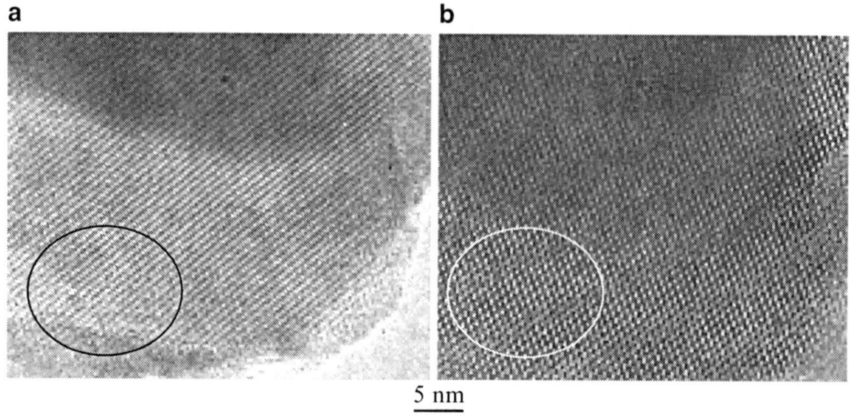

Fig. 4.15 Bright field TEM images presenting faujasite crystallite (**a**) prior to and (**b**) subsequent to applying a median filter. Note the enhancement of image details in the circled area subsequent to applying the filter

into the BF mode, and an image is collected at the minimum magnification required to achieve the required resolution using as little illumination as possible [139]. Because the resulting image generally has a low signal-to-noise ratio, image processing algorithms, such as real space averaging, may be required to enhance the features of interest (Fig. 4.15) [8, 142–144].

Further, HRTEM image interpretation must be approached with caution. The kinematic approximation utilizes the weak phase object approach and thus suggests that single scattering events result in image intensities that directly relate to electron wave amplitudes [8, 10]. However, this approximation is only accurate in limited cases where specimens are very thin [3, 8, 10]. More commonly, the dynamical approach is required, and image simulations (e.g., multislice, reciprocal space, Fourier transform, etc.) are necessary for accurate high resolution data interpretation [8].

Relative to TEM, STEM imaging can, in certain cases, have significant benefits with regard to zeolite characterization [8, 41–46] and zeolite-supported small metal clusters [145, 146]. First, the rastered STEM probe can reduce electron beam damage relative to the continuous, static illumination used during HRTEM imaging [8]. Second, HAADF STEM images contain minimal diffraction contrast, and thus they may be particularly useful for mitigating diffraction contrast effects in TEMT images of highly crystalline, porous materials [41–46]. Third, HAADF STEM images are particularly sensitive to variation in average atomic number and thus give rise to an exceptionally high signal-to-noise images, which reveal metal particle size and location [145, 146]. The reader is strongly encouraged to directly examine 41, 46 and 145, 146 for state-of-the art TEMT of zeolites and HAADF STEM images of zeolite-supported metal clusters, respectively. However, it remains important to note that even in HAADF STEM TEMT studies issues surrounding electron beam damage may remain simply due to prolonged data acquisition times required to collect the ~141 images required to obtain a quality reconstruction of small structures. Third, HAADF STEM image contrast is highly atomic number dependent, making these data particularly valuable in examining zeolite-supported small metal particles [8, 41–44, 136]. HAADF STEM is best conducted in a well-aligned [147] field emission instrument using the smallest probe size necessary to produce high signal-to-noise ratio images [8].

4.2.4 Elemental Analyses of Zeolites

4.2.4.1 Energy Dispersive Spectrometry (EDS) of Zeolites

Elemental analysis is frequently useful in both SEM- and TEM-based studies. EDS is based on the collection of X-rays generated as incident electrons impinge upon specimens in both the SEM and TEM [8, 13, 15, 40, 53, 62–64, 137]. When the interaction energy between the incident electron and an atom in the specimen is sufficiently high, an electron in a specimen atom can be ejected, creating a vacancy and ionizing the atom [8, 13, 15, 40, 53, 137]. To return the atom to its ground, an outer shell electron can transfer into the vacancy created by the ejected electron [8, 13, 15, 40, 53, 137]. As the electron transfers from the outer shell to the lower energy shell, a characteristic X-ray is emitted [8, 13, 15, 40, 53, 137]. Collection of these X-rays is most commonly done using a Li-drifted Si detector system connected to an analyzer [8, 13, 15, 40, 53, 62, 64, 137]. The X-ray information is ultimately displayed as a series of spectral peaks located at specific energies (Fig. 4.16) [8, 13, 15, 40, 53, 62, 64, 137].

While the spectral data can provide some very general level of information regarding the elements present within a material, it alone is not sufficient for an accurate elemental analysis [8, 13, 15, 40, 62, 64, 137]. In the SEM, meaningful elemental analyses require standards with a well-known composition that is close to that of the unknown [8, 13, 15, 40, 53, 62, 64, 137]. Additionally, both the unknown

Fig. 4.16 Typical spectrum from energy dispersive spectrometry (EDS) analysis presenting Gaussian peaks at characteristic energies

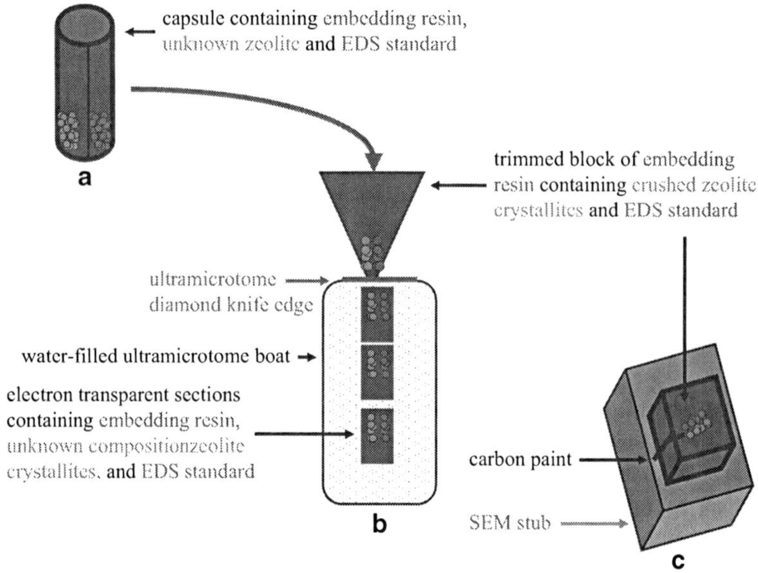

Fig. 4.17 Schematic presents the ultramicrotomy sample preparation technique for elemental analysis by energy dispersive spectrometry of a zeolite of unknown composition. Zeolite and a standard material with well known composition similar to that of the zeolite are prepared for ultramicrotomy as outlined in Fig. 4.12. (**a**) Crushed fines of both the zeolite and the standard are embedded in a resin block. The resin block face containing crystallites of the zeolite and the standard is trimmed to form a trapezoid. The block is mounted in the ultramicrotome and the trapezoidal face is aligned with the diamond knife edge. (**b**) A ribbon of electron transparent sections containing crystallites of both the zeolite and the standard float on the water's surface behind the diamond knife edge. (**c**) In the case of an SEM analysis, the resin block is removed from the microtome and affixed to a standard SEM stub such that its cut surface is facing upwards. The block is coated with a thin layer of either carbon or metal, and a conductive layer of carbon paint is drawn from the cut resin block face to the SEM stub. In the case of a TEM analysis, a holey carbon coated grid is submerged into the water and the thin sections are collected as described in Fig. 4.12

and the standards must have flat, smooth surfaces to minimize the formation of spurious X-ray peaks in the spectral data [8, 13, 15, 40, 53, 62, 64, 137]. If the zeolite crystal surface is not smooth, an appropriate surface can easily be achieved using the ultramicrotomy protocol previously described (Sect. 4.2.3.1) or by following standard metallographic procedure for mounting and polishing a hard material. The embedded and microtomed/polished zeolite must be lightly metal or carbon coated to prevent charging. The coated zeolite/resin block can be mounted into a standard SEM holder and an electrically conductive line extending from the zeolite/resin face to the metal stub can be drawn with carbon paint. The entire block can then be transferred into the SEM. If possible, both the standard and the unknown should be mounted side-by-side on the SEM stub or in a resin block (Fig. 4.17).

To obtain an accurate quantitative analysis in the zeolite, locate a flat area of the standard and check the EDS calibration at the accelerating voltage that will be used in the quantitative analysis, typically ~2 times the value of the highest energy peak of the elements being analyzed [40, 64]. If the calibration is not correct, re-calibrate the EDS system. Locate a smooth, homogenous region of the standard. Set spectral data acquisition parameters such that a dead time of ~20–30% is achieved and continue to collect data until all peaks contain a statistically significant number of counts [40, 64]. Once the data collection has been completed the standard spectrum can be stored as a standard [64]. Locate a uniformly flat area, homogeneous regions of the unknown specimen and repeat the spectral collection procedure described above using the same accelerating voltage and beam current [64]. Save the spectral data from the unknown. As a best practice, locate a flat, homogeneous region of the standard and re-acquire another set of spectral data at the same accelerating voltage and beam current used in the previous spectral collections.

The unknown spectra and the final spectral data collected from the standard are analyzed using original spectral data from the standard in combination with the full quantitative analysis algorithms within the EDS system [40, 62, 64]. Peak identification, background subtraction, and atomic number, absorption, and fluorescence corrections are all necessary for a quantitative analysis [40, 53, 62, 64, 137]. Analysis of the final spectral data collected on the standard against the original spectral data collected from the same specimen serves as a "quantitative analysis quality check."

The same general concepts of X-ray microanalysis in the SEM apply to the TEM and the STEM. However, qualitative and quantitative analyses in the TEM and the STEM require an understanding of spectral artifacts that may arise from the electron microscope column itself [8, 137]. Quantitative analyses mandate standards having both nanoscale homogeniety and thicknesses similar to those of the unknown [8, 62, 137]. As in the SEM, if possible, it is ideal to have both the standards and the unknown mounted on the same TEM grid (Fig. 4.17).

Thus, to obtain a quantitative EDS microanalysis in the TEM or the STEM, regions within the standard having homogeneous composition must be selected, and data collection times must be set to insure that all the peaks in the spectra contain a statistically significant level of counts [8, 62]. Subsequent to storing the spectral data from the standards, regions of uniform thickness (similar to those analyzed in the standards) and homogeneous composition in the unknown must be located. Spectral acquisition conditions (e.g., accelerating voltage, beam current, etc.) in the unknown must be identical to those of the standard [8, 62]. The spectral data from the unknown should be stored. As a best practice, regions within the standards having nanoscale homogeniety should be re-analyzed using spectral the same acquisition parameters as the previous collections, and the data stored.

As in the case of quantitative X-ray microanalysis in the SEM, the TEM and STEM analyses require the use of analysis algorithms within the EDS system [8, 62, 64]. Peak identification, background subtraction, atomic number, absorption, and fluorescence corrections are all necessary [8, 53, 62, 64, 137]. Analysis of the final spectral data collected on the standard against the original spectral data collected from the same specimen again serves as a quantitative analysis quality check. As a note of caution, even after following the appropriate microanalysis protocols, the

X-ray analysis of zeolites in the TEM and STEM can still be in error due to elemental migration during the analysis [132]. As Si migrates out of the analysis region, the Al to Si ratio increases over the course of the data acquisition leading to incorrect calculations in the zeolite composition [132]. While this effect is not generally noticeable in an SEM analysis due to the relatively large microanalysis volume, it is important and must be carefully considered in both qualitative and quantitative TEM and STEM analyses. However, general X-ray mapping in conjunction with STEM imaging can be useful in identifying small metal clusters supported on zeolites [145, 146] and in nanostructured materials [148]. The reader is strongly encouraged to examine 145, 146, 148 for state-of-the art STEM/EDS elemental mapping.

4.2.4.2 Electron Energy Loss Spectrometry (EELS) of Zeolites

In the TEM, elemental information can also be obtained through EELS [8, 57, 62]. EELS generally has higher spatial resolution and better detectability than EDS making it particularly useful for analyses in which low concentrations of a particular element are present [8, 62]. Examination of the EELS spectrum reveals that it comprises several distinct regions including the zero-loss peak, low-loss energy peaks, higher loss energy peaks, and fine structure (Fig. 4.18) [8, 57, 62, 63]. The zero-loss peak is composed of elastically scattered as well as unscattered electrons and contains little useful information [8, 57, 62, 63]. The low-loss energy peaks (<~50 eV) arise from interactions with weakly bound outer shell electrons and can provide information about conduction and valence bands [8, 57, 62, 63]. The higher loss energy peaks (>~50 eV) are generated by interactions with inner shell electrons and generally provide the most useful elemental information [8, 62, 63]. Lastly, the fine structure, known as the extended loss near edge structure (ELNES) provides

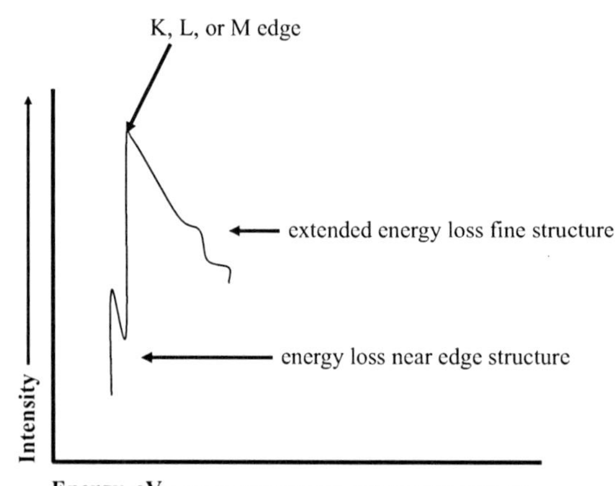

Fig. 4.18 Schematic presents electron energy loss spectrum and indicates distinct features (energy loss near edge structure, edge, and extended energy loss fine structure) within the spectrum

bonding information [8, 62, 63]. EELS quantification is very difficult, requiring a very thin specimen (<~50 nm) as well as standards having thicknesses and bonding similar to the unknown [8, 57, 62].

While quantitative EELS analysis is possible, it is generally extremely complex and time consuming [8, 57, 62, 63, 150, 151]. Thus, a semi-qualitative analysis is discussed here. Before beginning any EELS study, it is necessary to set the parameters required for the specific analysis (e.g., slit width, energy range, etc.) and then check the system calibration under these conditions [57, 63, 149–151]. An efficient calibration check protocol involves using the zero loss and carbon peaks associated with the carbon film supporting the TEM specimen [57, 63]. For any EELS analysis, it is important to locate a thin (~<10 nm), homogeneous area of the material that is positioned over a hole in the carbon support film. Collect and save spectral data from the region analyzed [62, 63].

In all cases, after collecting data, identify the spectral edges using the EELS Atlas [151]. A background fit and subtraction must be completed next [57, 62, 63]. Due to the complex nature of the EELS background, it is generally advantageous to select regions of the spectra preceding the edge of interest by ~50 to ~100eV and use curve fitting algorithms associated with the EELS software to fit the background beneath the region of interest [62]. Use the EELS software algorithm to compare edges in the unknown. In all EELS analyses of beam sensitive materials such as zeolites, the possibility of beam damage must be considered and may be a significant limitation to the EELS analysis of a zeolite [62, 152]. However, EELS can be successfully used in conjunction with STEM imaging to identify elements [62]. The reader is strongly encouraged to examine reference 149 for start-of-the art EELS imaging of nanostructured materials.

References

1. Terasaki O, Ohuna T (1995) What can we observe in zeolite related materials by HRTEM? Catal Today 23:201
2. Terasaki O, Ohsuna T, Alfredsson V, Bovin JO, Watanabe D, Tsuno K (1991) The study of zeolites by HVHREM. Ultramicroscopy 39:238
3. Pan M, Crozier PA (1993) Quantitative imaging and diffraction of zeolites using a slow-scan CCD camera. Ultramicroscopy 52:487
4. Chan IY, Labun PA, Pan M, Zones SI (1995) High-resolution electron microscopy characterization of SSZ-25 zeolite. Micropor Materials 3:409
5. Sakamoto Y, Kaneda M, Terasaki O, Zhao DY, Kim JM, Stucky G, Shin HJ, Ryoo R (2000) Direct imaging of the pores and cages of three-dimensional mesoporous materials. Nature 408:499
6. Meriaudeau P, Tuan VA, Nghiem VT, Lai SY, Hung LN, Naccache C (1997) SAPO-11, SAPO-31, and SAPO-41 molecular sieves: synthesis, characterization, and catalytic properties in n-octane hydroisomertization. J Catal 169:55
7. Terasaki O (1994) Fine structures of zeolites. J Electron Microsc 43:337
8. Williams DB, Carter CB (1996) Transmission electron microscopy – A textbook for materials science. Plenum, New York

9. Edington JW (1976) Practical electron microscopy in materials science. Van Nostrand Reinhold, New York
10. Hirsch P, Howie A, Nicholson RB, Pashley DW, Whelan MJ (1977) Electron microscopy of thin crystals. Krieger, New York
11. Thomas G, Goringe MJ (1979) Transmission electron microscopy of materials. Wiley, New York
12. Sawyer LC, Grubb DT (1987) Polymer microscopy. Chapman and Hall, London
13. Heinrich KFJ (1981) Electron beam X-Ray microanalysis. Van Nostrand Reinhold, New York
14. Hull D (1975) Introduction to Dislocations. In: Owen WS, Eng D (eds) International Series on Materials Science and Technology, vol 16. Pergamon Press, Oxford
15. Scott VD, Love G (1983) Quantitative electron-probe microanalysis. Wiley, New York
16. Bradbury S (1989) An introduction to the optical microscope (revised edition). In: Christopher Hammond Microscopy handbooks 1, Oxford University Press, Oxford
17. Butler EP, Hale KF (1981) Dynamic Experiments in the Electron Microscope. In: Glauert AM (ed) Practical Methods in Electron Microscopy, vol 9. North Holland, Amsterdam
18. Gai PL (1981) Dynamic studies of metal oxide catalysts: MoO_3. Philos Mag A 43:841
19. Gai PL, Boyes ED, Bart JCJ (1982) Electron microscopy of industrial catalysts. Philos Mag A 45:531
20. Gai-Boyes PL (1992) Defects in oxide catalysts: fundamental studies of catalysts in action. Catal Rev Sci Eng 34:1
21. Chan I (1985) Controlled atmosphere specimen holder for transmission electron microscopy. J Electron Microsc Tech 525–532
22. Parkinson GM, White D (1986) The application of a controlled atmosphere reaction cell for studying electroactive polymers by TEM. In: Proceedings of the XI international congress on electron microscopy, Koyoto, Japan, 331
23. Allard LF, Ailey KS, Dayte AK, Bigelow WC (1979) An ex-situ reactor with anaerobic specimen transfer capabilities for TEM study of reactive (catalyst) systems. In: Microsc Microanal Proc 5(2):595
24. Kliewer CE, Kiss G, DeMartin GJ (2006) Ex-situ transmission electron microscopy: a fixed bed reactor approach. J Microsc Microanal 12:135
25. Boyes ED, Gai PL (1997) Environmental high resolution electron microscopy and applications to chemical science. Ultramicroscopy 67:219
26. Gai PL (1998) Direct probing of gas molecule-solid catalyst interactions on the atomic scale. Adv Mater 10:1259
27. Scherzer O (1949) The theoretical resolution limit of the electron microscope. J Appl Phys 20:20
28. Krivanek OL, Dellby N, Lupini AR (1999) Towards sub-Å electron beams. Ultramicroscopy 78:1
29. Rose H (2002) Correction of aberrations - past, present and future. Microsc Microanal Proc 8(2):6
30. Krivanek OL, Mooney PE (1993) Applications of slow-scan CCD cameras in transmission electron microscopy. Ultramicroscopy 49:95
31. Koster AJ, Ziese U, Verkleij AJ, Janssen AH, de Jong KP (2000) Three-dimensional transmission electron microscopy: a novel imaging and characterization technique with nanometer scale resolution for materials science. J Phys Chem B 104:9368
32. de Jong KP, Koster AJ (2002) Three-dimensional electron microscopy of mesoporous materials – recent strides towards spatial imaging at the nanometer scale. Chem Phys Chem 3:776
33. Zach J, Haider M (1995) Correction of spherical and chromatic aberration in a low voltage SEM. Opik 98:(3)112–118
34. O'Keefe MA, Nelson EC, Wang YC (2001) Sub-Ångstrom resolution of atomistic structures below 0.8Å. Phil Mag B 81:1861–1878

35. Hutchison JL, Titchmarsh JM, Cockayne DJH, Mobus G, Hetherington CJ, Doole RC, Hosokawa F, Hartel P, Haider M (2002) Application of a C_s corrected HRTEM in materials science. In: Microsc Microanal Proc 8(2)
36. Nellist PD, Chisholm MF, Dellby N, Krivanek OL, Murfitt MF, Szilagyi ZS, Lupini AR, Borisevich A, Sides WH, Jr PSJ (2004) Direct sub-Ångstrom imaging of a crystal lattice. Science 305:1741
37. O'Keefe MA, Allard LF, Blom DA (2005) HRTEM imaging of atoms at sub-Ångstrom resolution. J Electron Microsc 54(3):169–180
38. Lehmann M, Lichte H (2005) Electron holographic material analysis at atomic dimensions. Cryst Res Technol 40 Numbers 1 & 2:149
39. Kirkland AI, Meyer RR, Shery Chang LY (2006) Local measurements and computational refinement of aberrations for HRTEM. J Microsc Microanal 12:461
40. Goldstein JI, Newbury DE, Joy DC, Lyman CE, Echlin P, Lifshin E, Sawyer L, Michael JR (2003) Scanning electron microscopy and X-ray microanalysis. Plenum, New York
41. Midgley PA, Weyland M, Thomas JM, Johnson BFG (2001) Z-contrast tomography: a technique in three-dimensional nanostructural analysis based on Rutherford scattering. Chem Comm 10:907–908
42. Weyland M, Midgley PA, Thomas JM (2001) Electron tomography of nanoparticle catalysts on porous supports: a new technique based on Rutherford scattering. J Phys Chem B 105:7882
43. Midgley PA, Thomas JM, Laffont L, Weyland M, Raja R, Johnson BFG, Khimyak T (2004) High-resolution scanning transmission electron tomography and elemental analysis of zeptogram quantities of heterogeneous catalyst. J Phys Chem B 108:4590
44. Thomas JM, Midgley PA, Yates TJV, Barnard JS, Raja R, Arslan I, Weyland M (2004) The chemical application of high-resolution electron tomography: bright field or dark field? Angew Chem Int Ed 43:6745
45. Yates TJV, Thomas J, Fernandez JJ, Terasaki O, Ryoo R, Midgley PA (2006) Three-dimensional real-space crystallography of MCM-48 mesoporous silica revealed by scanning transmission electron tomography. Chem Phys Lett 418:540
46. Weyland M (2002) Electron tomography of catalysts. Topics Catal 21(4):175–183
47. Batson PE, Dellby N, Krivanek OL (2002) Sub-ångström resolution using aberration corrected electron optics. Nature 418:617
48. Pennycook SJ, Lupini AR, Varela M, Borisevich A, Peng Y, Chisholm MF, Dellby N, Krivanek, OL, Nellist PD, Szilagy SZ, Duscher G (2003) Sub-Angström resolution through aberration-corrected STEM. Microsc Microanal Proc 9(2):926
49. Muller H, Uhlemann S, Haider ML (2006) Advancing the hexapole C_s-corrector for the scanning transmission electron microscope. J Microsc Microanal 12:421
50. Mitsuishi K, Takegucki M, Kondo Y, Hosokawa F, Okamoto K, Sannomiya T, Hori M, Iwama T, Kawazoe M, Furuya K (2006) Ultrahigh-vacuum third-order spherical Aberration (C_s) corrector for a scanning transmission electron microscope. J Microsc Microanal 12:456
51. Blom DA, Allard LF, Mishina S, O'Keefe MA (2006) Early results from an aberration-corrected JEOL 2200FS STEM/TEM at Oak Ridge National Laboratory. Microsc Microanal Proc 12:483
52. Muller DA, Mkhoyan KA, Cha JJ, Intaraprasonk V, Xin H (2007) Image contrast in sub-Angström ADF-STEM. Microsc Microanal Proc 13(2):124
53. Chandler JA (1981) X-Ray Microanalysis in the Electron Microscope. In: Glauert AM (ed) Practical methods in electron microscopy Series 5 Part 11 North Holland, Amsterdam
54. Hillier J (1943) On microanalysis by electrons. Phys Rev 64:318
55. Hillier J, Baker RF (1944) Microanalysis by means of electrons. J Appl Phys 15:663
56. Hillier J (1947) Electron probe analysis employing X-ray spectrography. US Patent 2,418,029, 25 March 1947

57. Budd PM, Goodhew PJ (1988) Light element analysis in the transmission electron microscope: WEDX and EELS. In: Christopher Hammond Microscopy handbooks 16, Oxford University Press, Oxford
58. Weyland M, Midgley PA (2003) Examining the possibilities and pitfalls of three dimensional energy filtered transmission electron microscopy (3D-EFTEM). Microsc Microanal Proc 9 (Suppl 2):148
59. Dellby N, Krivanek, OL, Murfitt M, Nellist PD, Szilagyi ZS (2003) Aberration-corrected STEM for elemental mapping. Microsc Microanal Proc 9(2):924
60. Findlay S, Oxley MP, Allen LJ, Lupini AR, Pennycook SJ (2003) Atomic-resolution EELS in aberration-corrected STEM. Microsc Microanal Proc 9(2):852
61. Watanabe M, Ackland DW, Burrows A, Kiely CJ, Williams DB, Krivanek OL, Dellby N, Murfitt MF, Szilagyi Z (2006) Improvements in the X-ray analytical capabilities of a scanning transmission electron microscope by spherical-aberration correction. J Microsc Microanal 12:515
62. Williams DB (1984) Practical analytical electron microscopy in materials science. Electron Optics Publishing Group, New Jersey
63. Newbury DE, Joy DC, Echlin P, Fiori CE, Goldstein JI (1986) Advanced scanning electron microscopy and X-ray microanalysis. Plenum, New York
64. Lyman CE, Newbury DE, Goldstein JI, Williams DB, Romig AD Jr, Armstrong JT, Echlin P, Fiori CE, Joy DC, Lifshin E, Peters KR (1990) Scanning electron microscopy, X-ray microanalysis, and analytical electron microscopy: A laboratory workbook. Plenum, New York
65. Burton A, Accardi RJ, Lobo RF (2000) MCM-47: a highly crystalline silicate composed of hydrogen-bonded ferrierite layers. Chem Mater 12:2936
66. Ghobarkar H, Schaf O, Knauth P (2001) Zeolite synthesis by the high pressure hydrothermal method: synthesis of natural 6-ring zeolites with different void systems. Agnew Chem Int Ed 40(20):3831
67. Ghobarkar H, Schaf O, Paz B, Knauth P (2003) Zeolite synthesis by simulation of their natural formation conditions: from macroscopic to nanosized crystals. J Solid State Chem 173:27
68. Terasaki O, Ohsuna T (2003) TEM study on zeolite fine structures: homework from Cambridge days. Topics Catal 24:13–18
69. Carpenter GB (1969) Principles of crystal structure determination. Benjamin, New York
70. Goodhew PJ (1972) Specimen Preparation in Materials Science. In: Glauert AM (ed) Practical Methods in Electron Microscopy Series 1 Part I North Holland, Amsterdam
71. Csencsits R, Schooley C, Gronsky R (1985) An improved method for thin sectioning of particulate catalysts. J Electron Microsc Techn 2:643
72. Csencsits R, Gronsky R (1988) Preparation of zeolites for TEM using microtomy. In: Bravman JC, Anderson RM, McDonald ML (eds) Specimen Preparation for transmission electron microscopy of materials. Proceedings of materials research society, vol 115, Boston, MA, 1987, p 103
73. Ekmejian EM, Bulko JB (1988) Preparation techniques for structural characterization of powdered and composite materials. In: Bravman JC, Anderson RM, McDonald ML (eds.) Specimen preparation for transmission electron microscopy of materials. Proceedings of materials research society, vol 115, Boston, MA, 1987, p 87
74. Rice SB, Treacy MMJ (1988) The art of the possible; an overview of catalyst specimen preparation techniques for TEM studies. In: Bravman JC, Anderson RM, McDonald ML (eds) Specimen preparation for transmission electron microscopy of materials. Proceedings of Materials Research Society, vol 115, Boston MA, 1987, p.15
75. Tesche B, Jentoft FC, Schlog R, Bare SR, Nemeth LT, Valencia S, Corma A (2007) Novel preparation steps and imaging procedures suitable for electronmicroscopical characterization of zeolite structures. Microsc Microanal Proc 13(2):560CD

76. Dayte AK, Smith DJ (1992) The study of heterogeneous catalysts by high-resolution transmission electron microscopy. Catal Rev Sci Eng 34:129–178
77. Thomas JM (1999) Design, synthesis, and in-situ characterization of new solid catalysts. Angew Chem Int Ed 38:3589
78. Thomas JM, Terasaki O, Gai P, Zhou W, Gonzalez-Calbet J (2001) Structural elucidation of microporous and mesoporous catalysts and molecular sieves by high resolution electron microscopy. Acc Chem Res 34:583
79. van Donk S, Jansses AH, Bitter JH, de Jong KP (2003) Generation, characterization, and impact of mesopores in zeolite catalysts. Catal Rev 45(2):297–319
80. Thomas JM, Millward GR (1982) Direct, real-space determination of intergrowths in ZSM-5/ZSM-11 catalysts. J Chem Soc Chem Comm 1380–1383
81. Dominguez JM, Acosta DR, Schifter I (1983) On the crystallography of twinned ZSM-5 type zeolites. J Catal 83:480
82. Gramlich-Meier R, Meier WM, Smith BK (1984) On faults in the framework structure of the zeolite ferrierite. Zeitschrift fur Kristallographic 169:201
83. Sanders JV (1985) Crystallographic faulting in the mordenite group zeolites. Zeolites 5:81
84. Millward GR, Ramdas S, Thomas JM (1985) On the direct imaging of offretite, cancrinite, chabazite, and other related ABC-6 zeolites and their intergrowths. Proc Roy Soc London A399:57
85. Zandbergen HW, Engelen CWR, van Hooff JHC (1986) Electron microscopic study of defects and metal particles in zeolites. Appl Catal 25:231
86. Treacy MMJ, Newsam JM, Beyerlein RA, Leonowicz ME, Vaughan DEW (1986) The structure of zeolite CSZ-1 interpreted as a rhombohedrally distorted variant of the faujasite framework. J Chem Soc Chem Comm 1211–1214
87. Treacy MMJ, Newsam JM (1988) Two new three-dimensional twelve-ring zeolite frameworks of which zeolite beta is a disordered intergrowth. Nature 332:249
88. Terasaki O, Thomas JM, Millward GR, Watanabe D (1989) Role of high-resolution electron microscopy in the identification and characterization of new crystalline microporous materials: "reading off" the structure and symmetry elements of pentasil molecular sieves. Chem Mater 1(1):158–162
89. Lobo RF, Pan M, Chan I, Li HX, Medrud RC, Zones SI, Crozier PA, Davis ME (1993) SSZ-26 and SSZ-33: two molecular sieves with intersecting 10- and 12- ring pores. Science 262:1543–1546
90. Terasaki O, Ohsuna T, Alfredsson V, Bovin JO, Watanabe D, Carr SW, Anderson MW (1993) Observation of spatially correlated intergrowths of faujastic polytypes and the pure end members by high resolution electron microscopy. Chem Mater 5:452
91. Szostak R, Pan M, Lillerud KP (1995) High-resolution TEM imaging of extreme faulting in natural zeolite tschernichite. J Phys Chem 99:2104
92. Wright PA, Zhou W, Perez-Pariente J, Arranz M (2005) Direct observation of growth defects in zeolite beta. J Am Chem Soc 127:494
93. Choi-Feng C, Hall JB, Huggins BJ, Beyerlein RA (1993) Electron microscope investigation of mesopore formation and aluminum migration in USY catalysts. J Catal 140:395
94. Sasaki Y, Suzuki T, Takamura Y, Saji A, Saka H (1998) Structure analysis of the mesopore in dealuminated zeolite Y by high resolution TEM observation with slow scan CCD camera. J Catal 178:94
95. Tromp M, van Bokhoven JA, Garriga Oostenbrink MT, Bitter JH, de Jong KP, Koningsberger DC (2000) Influence of the generation of mesopores on the hydroisomertization activity and selectivity of n-hexane over Pt/mordenite. J Catal 190:209
96. Koster AJ, Ziese U, Verklij AJ, Janssen AH, de Jong KP (2000) Three-dimensional transmission electron microscopy: a novel imaging and characterization technique with nanometer scale resolution for materials science. J Phys Chem B 104:9368

97. Janssen AH, Koster AJ, de Jong KP (2001) Three-dimensional transmission electron microscopic observations of mesopores in dealuminated zeolite. Y. Angew Chem 113 (6):1136
98. Janssen AH, Koster AJ, de Jong KP (2001) Imaging the mesopores in zeolite Y with three-dimensional transmission electron microscopy. Stud Surf Sci Catal 135:14
99. Janssen AH, Koster AJ, de Jong KP (2002) On the shape of mesopores in zeolite Y: a three-dimensional transmission electron microscopy study combined with texture analysis. J Phys Chem 106:11905
100. Sato K, Nishimura Y, Matsubayshi N, Imamura M, Shimada H (2003) Structural changes of Y zeolites during ion exchange treatment: effects of Si/Al ratio of the starting NaY. Micropor Mesopor Mater 59:113
101. Janssen AH, Van Der Voort P, Koster AJ, de Jong KP (2002) A 3D-TEM study of the shape of mesopores in SBA-15 and modified SBA-15 materials. Chem Comm 15:1632–1633
102. Janssen AH, Yang CM, Wang Y, Schuth F, Koster AJ, de Jong KP (2003) Localization of small metal (oxide) particles in SBA-15 using bright-field electron tomography. J Phys Chem B 107:10552
103. Terasaki O, Thomas JM, Millward GR (1984) Imaging the structures of zeolite L and synthetic mazzite. Proc Roy Soc London A395:153
104. Newsam JM, Treacy MMJ, Koetsier WT, DeGruyter CB (1988) Structural characterization of zeolite beta. Proc Roy Soc London A420:375
105. Alfredsson A, Terasaki O, Bovin JO (1990) On crystal structure imaging of silicate by HREM. J Solid State Chem 84:171
106. Kresge CT, Leonowicz ME, Roth W, Vartuli JC, Beck JS (1992) Ordered mesoporous molecular sieves synthesized by a liquid-crystal template mechanism. Nature 359:710
107. Terasaki O (1993) HREM study on the fine structures of zeolites and materials confined in their spaces: are zeolites good enough as containers for confined spaces? J Solid State Chem 106:190
108. Leonowicz ME, Lawton JA, Lawton SL, Rubin M (1994) MCM-22: a molecular sieve with two independent multidimensional channel systems. Science 264:1910
109. Lobo RF, Tsapatsis M, Freyhardt CC, Khodabandeh S, Wagner P, Chen CY, Balkus KJ, Jr ZSI, Davis ME (1997) Characterization of the extra-large pore zeolite UTD-1. J Am Chem Soc 119:8474
110. Nicolopoulos S, Gonzalez-Calbet JM, Vallet-Regi M, Camblor MA, Corell C, Corma A (1997) Use of electron microscopy and microdiffraction for zeolite framework comparisons. J Am Chem Soc 119:11000
111. Yang P, Zhao D, Margolese DI, Chmelka BF, Stucky GD (1998) Generalized syntheses of large-pore mesoporous metal oxides with semicrystalline frameworks. Nature 396:152
112. Jentoft RE, Tsapatsis M, Davis ME, Gates BC (1998) Platinum clusters supported in zeolite LTL: influence of catalyst morphology on performance in n-hexane reforming. J Catal 179:565
113. Wagner P, Terasaki O, Ritsch S, Nery JG, Zones SI, Davis ME, Hiraga K (1999) Electron diffraction structure solution of a nanocrystalline zeolite at atomic resolution. J Phys Chem B 103:8245
114. Thomas JM, Terasaki O, Gai PL, Zhou W, Gonzalez-Calbet J (2001) Structural elucidation of microporous and mesoporous catalysts and molecular sieves by high-resolution electron microscopy. Acc Chem Res 34:583
115. Liu Z, Ohsuna T, Terasaki O, Camblor MA, Diaz-Cabanas MJ, Hiraga K (2001) The first zeolite with three-dimensional intersecting straight-channel system of 12-membered rings. J Am Chem Soc 123:5370
116. Ohsuna T, Liu Z, Terasaki O, Hirag K, Camblor MA (2002) Framework determination of a polytype of zeolite beta by using electron crystallography. J Phys Chem B 106:5673
117. Kaneda M, Tsubakiyama T, Carlsson A, Sakamoto Y, Ohsuna T, Terasaki O (2002) Structural study of mesoporous MCM-48 and carbon networks synthesized in the spaces of MCM-48 by electron crystallography. J Phys Chem B 106:1256

118. Ohsuna T, Liu Z, Terasaki O, Hiraga K, Camblor M (2002) Framework determination of a polytype of zeolite Betz by using electron crystallography. J Phys Chem B 106:5673
119. Terasaki O, Ohsuna T, Liu Z, Kaneda M, Kamiya S, Carlsson A, Tsubakiyama T, Sakamoto Y, Inagaki S, Che S, Tatsumi T, Camblor MA, Ryoo R, Zhao D, Stucky G, Shindo D, Hiraga K (2002) Porous materials: looking through the electron microscope. Stud Surf Sci Catal 141:27
120. Slater B, Catlow CRA, Liu Z, Ohsuna T, Terasaki O, Camblor MA (2002) Surface structure and crystal growth of zeolite beta C. Angew Chem Int Ed 41(7):1235–1237
121. Kaneda M, Tsubakiyana T, Carlsson A, Sakamoto Y, Ohsuna T, Terasaki O (2002) Structural study of mesoporous MCM-48 and carbon networks synthesized in the spaces of MCM-48 by electron crystallography. J Phys Chem B 106:1256
122. Sakamoto Y, Diaz I, Terasaki O, Zhao D, Perez-Pariente J, Kim JM, Stucky GD (2002) Three-dimensional cubic mesoporous structures of SBA-12 and related materials by electron crystallography. J Phys Chem B 106:3118
123. Ohsuna T, Liu Z, Terasaki O, Hiraga K, Camblor MA (2002) Framework determination of a polytype of zeolite Beta by using electron crystallography. J Phys Chem B 106:5673
124. Liu Z, Fujita N, Terasaki O, Ohsuna T, Hiraga K, Camblor MA, Diaz-Cabanas MJ, Cheetham AK (2002) Incommensurate modulation in the microporous silica SSZ-24. Chem Eur J 8(19):4549–4556
125. Diaz I, Kokkoli E, Terasaki O, Tsapatsis M (2004) Surface structure of zeolite (MFI) crystals. Chem Mater 16:5226
126. Corma A, Diaz-Cabanas MJ, Rey F, Nicolopoulus S, Boulahya K (2004) ITQ-15: the first ultralarge pore zeolite with a bi-directional pore system formed by intersecting 14- and 12-ring channels, and its catalytic implications. Chem Comm 12:1356–1357
127. Gramm F, Baerlocher C, McCusker LB, Warrender SJ, Wright PA, Han B, Hong SB, Li Z, Ohsuna T, Terasaki O (2006) Complex zeolite structure soled by combining powder diffraction and electron microscopy. Nature 444:79
128. Baerlocher C, Gramm F, Massuger L, McCusker LB, He Z, Hovmoller S, Zou X (2007) Structure of the polycrystalline zeolite catalyst IM-5 solved by enhanced charge flipping. Science 315:1113
129. Hren JJ, Goldstein JI, Joy DC (1979) Introduction to analytical electron microscopy. Plenum, New York
130. Bursill LA, Lodge EA, Thomas JM (1980) Zeolite structures as revealed by high-resolution electron microscopy. J Phys Chem 286:111
131. Bursill LA, Thomas JM, Rao KJ (1981) Stability of zeolites under electron irradiation and imaging of heavy cations in silicates. Nature 289:157
132. Csencsits R, Gronsky R (1987) Damage of zeolite Y in the TEM and its effects on TEM images. Ultramicroscopy 23:421
133. Treacy MMJ, Newsam JM (1987) Electron beam sensitivity of zeolite L. Ultramicroscopy 23:411
134. Pan M, Cowley JM, Chan IY (1990) HREM imaging of small Pt clusters dispersed in Y-zeolites. Catal Lett 5:1
135. Rice SB, Koo JY, Disko MM, Treacy MMJ (1990) On the imaging of Pt atoms in zeolite frameworks. Ultramicroscopy 34:108
136. Sasaki Y, Suzuki T, Ikuhara Y (1995) Direct observation of channel structures in zeolite Y and A with a slow-scan, charge-coupled-device camera. J Am Ceram Soc 78(5):1411–1413
137. Blanford CF, Carter CB (2003) Electron radiation damage of MCM-41 and related materials. Microsc Microanal Proc 9:245
138. Van Tenddeloo G, Lebedev OI, Kremer SPB, Kirschhock CEA, Aerts A, Villani K, Martens JA (2003) Tilting silicate nanoslabs into 3D mosaics. In: Microsc Microanal Proc 9(2):2166
139. Williams RC, Fisher HW (1970) Electron microscopy of tobacco virus under conditions of minimal beam exposure. J Mol Bio 52:121

140. Pan M, Crozier PA (1993) Low-dose high-resolution electron microscopy of zeolite materials with a slow-scan CCD camera. Ultramicroscopy 48:332
141. Pan M, Crozier PA (1993) Quantitative imaging and diffraction of zeolites using a slow-scan CCD camera. Ultramicroscopy 52:487
142. Ohnishi N, Hiraga K (1996) Slow-scan CCD camera analysis of electron diffraction and high-resolution micrographs of zeolite TPA/ZSM-5. J Electron Microsc 45:85
143. Chan YI, Csencsits R, O'Keefe MA, Gronsky R (1987) Computer-simulated images of platinum clusters in the channels of Y zeolites: zone-axis results. J Catal 103:466
144. Russ JC (1995) The image processing handbook. CRC Press, Boca Raton
145. Borisevich AY, Wang S, Rashkeev SN, Pantelides ST, Sohlberg K, Pennycook SJ (2004) Single-atom sensitivity for solving catalysis problems. Microsc Microanal Proc 10(2):460
146. Borisevich AY, Lupini AR, Rashkeev SN, Wang S, Veith GM, Dudney NJ, Pantelides K, Pennycook SJ (2005) Aberration-corrected STEM for understanding of the catalytic mechanisms and development of new catalysts. Microsc Microanal Proc 11(2):218
147. Kotaka Y, Yamazaki T, Watanabe K, Kikuchi Y, Nakanishi N, Hashimoto I (2005) Measurement of 3rd order spherical aberration coefficient for scanning transmission electron microscopy. In: Microsc Microanal Proc 11(2):2166
148. Bentley J, Al-Sharab JF, Cosandey F (2006) EDS and EELS spectrum imaging of nanostructured Al_2O_3-MgO-$MgAl_2O_4$. Microsc Microanal Proc 12(2):144
149. Egerton RF, Chen SC (1994) Characterization of an analytical electron microscope with a NiO test specimen. Ultramicroscopy 55:43
150. Disko MM (1986) Practical methods for quantitative and fine structure analysis of energy loss spectra. In: Romig AD Jr, Chambers WF (eds.) Microbeam analysis. San Francisco Press, San Francisco, CA, USA 429
151. Ahn CC, Krivanek OL (1983) EELS atlas. Gatan Inc., Warrendale, PA
152. Egerton RF (2002) The future of EELS. In: Microsc Microanal Proc 8(2):464

Chapter 5
Infrared and Raman Spectroscopy

Xinsheng Liu

Abstract A brief introduction of applications of infrared and Raman spectroscopic techniques to zeolites, zeolite chemistry and the chemistry that occurs in zeolite pores and channels is given in this chapter. The basic information, obtained by using these techniques, of zeolite frameworks including isomorphous substitution, perturbation of zeolite frameworks by foreign cations, and cation locations, zeolite surfaces including OH groups, acid and basic sites and adsorbed species on their surfaces, and reaction mechanisms occurring in zeolites, is highlighted with examples to show the usefulness of these techniques.

5.1 Introduction

In this chapter, we will briefly introduce various applications of infrared and Raman spectroscopy in zeolites, zeolite chemistry and the chemistry that occurs in zeolite pores and channels. We have assumed that the readers have knowledge of the basic principles of infrared and Raman spectroscopy. Due to limited space, we do not intend to give a comprehensive review of the field but rather very briefly introduce the most widely used infrared and Raman sampling techniques for zeolites and then discuss the various applications of infrared and Raman spectroscopy to zeolite chemistry, including framework vibrations, surface properties, adsorption and catalysis.

X. Liu
BASF Catalysis Research, R&D Center, 25 Middlesex-Essex Turnpike, Iselin, NJ 08830, USA
e-mail: xinsheng.liu@basf.com

A.W. Chester and E.G. Derouane (eds.), *Zeolite Characterization and Catalysis*,
DOI 10.1007/978-1-4020-9678-5_5,

5.2 Infrared Spectroscopy

Infrared spectroscopy (IR) is one of the most widely used techniques in zeolite chemistry and catalysis [1, 2]. Due to its ready availability of instrumentation, IR has been used to almost all aspects of zeolite chemistry, for information on zeolite formation, framework vibrations, surface property, adsorption and catalysis. IR spectroscopy is arbitrarily divided into three categories based on the instrumental designs: near-IR (>3,000 cm^{-1}), mid-IR (4,000–400 cm^{-1}) and far-IR (<300 cm^{-1}). In general, near-IR spectroscopy deals with overtones of fundamental vibrations and provides information on adsorbed species such as water, organic molecules and small gas molecules, etc. in zeolite cavities or channels. Mid-IR spectroscopy is fruitful for zeolite chemistry and provides information on surface OH groups, adsorbed molecules and framework vibrations. Far-IR spectroscopy has been used to study framework oxygen and charge-balancing cations in zeolite structures.

Although the IR technique is versatile for studying zeolite structures and surfaces, the information provided by this technique is not always straightforward. In many cases, probe molecules with specific vibrational spectroscopic properties have to be used.

There are several sampling techniques in IR spectroscopy: KBr wafer technique, attenuation total reflection (ATR), diffuse reflectance and in certain cases, photoacoustic FT-IR, and time-resolved FT-IR. The KBr technique is very popular and commonly used for zeolite framework vibrations. To study surface OH groups, a pure zeolite wafer is employed. With the zeolite wafer technique, in most cases, framework vibrations cannot be obtained due to their strong absorption. The ATR technique significantly simplifies the sample preparation in contrast to the KBr technique. The zeolite powder can be examined directly without being diluted with KBr. However, due to the variation of the light penetration depth in the sample with wavelength, spectral correction is needed in order to get a comparable IR spectrum to the KBr technique. The diffuse reflectance technique is gaining in popularity in recent years due to its ease of operation. With a high-temperature and high-pressure diffuse reflectance chamber, the sample environment can be controlled, and adsorption, desorption as well as reactions can be easily studied. However, the diffuse reflectance follows the Kubelka-Monk theory [3] expressed by the equation $k/s = (1 - R_\infty)^2/2R_\infty$, where k is the absorption coefficient, and s is the scattering factor which depends on sample particle size. R is reflectance with an infinite sample thickness. From the equation, it is clear that in order to make a comparison between samples, constant particle size for samples has to be achieved. Another factor in diffuse reflectance is the IR light penetration. No study so far for IR has been reported. These complications with diffuse reflectance make quantitative measurement rather difficult. The photoacoustic technique deals with thermal waves and is suitable for samples with dark color. It is easy to operate but difficult to be used for adsorption/desorption and reaction studies due to the presence of thermal conducting gas (Hellium) in the sample cavity used for conducting thermal waves. Time-resolved FT-IR within a millisecond time regime can be achieved using the fast scan mode on a research FT-IR instrument while with the step scan technique, microsecond and nanosecond time-resolved FT-IR spectroscopy can be achieved.

5.3 Raman Spectroscopy

Raman spectroscopy has become popular in recent years due to the development in instrumentation. The replacement of the mercury lamp by lasers and the use of charge couple device (CCD) detectors and notch filters increase the Raman sensitivity and spectral range dramatically. Similar to IR, Raman spectroscopy gives vibrational information but, due to its different selection rule from that of IR, dipole moment vs. polarizability, the information obtained is complementary to IR. Raman is preferred for studying host-guest interactions in zeolites via monitoring the guest molecules due to the very weak Raman scattering nature of the "normal" zeolite frameworks.

Two types of Raman instruments, dispersive Raman (conventional) and Fourier transform Raman (FT-Raman), are available. The use of FT-Raman is to overcome the intense background fluorescence problem of zeolites under laser excitation. The conventional dispersive Raman is used for utmost sensitivity and Raman bands below 300 cm^{-1}.

Different wavelengths of lasers ranging from near IR, visible to UV are available for dispersive Raman spectroscopic studies of zeolite systems. The spectral sensitivity increases as the wavelength of the light shortens. The choice of a laser depends on the zeolite materials and also on the spectral region to be examined. For the former, the main obstacle for obtaining a Raman spectrum is the fluorescence, which, possibly, originates from inorganic impurities, residual organics after removal of templates, defects, etc., and for the latter, different lasers require different optics, and the detectors may have different sensitivities to different spectral regions. To detect low concentrations of surface species and species with inherently lower Raman intensities, a short wavelength laser is preferred. In the case where the excitation wavelength is close to the absorption band of an excited electronic state, resonance Raman effect can remarkably enhance the sensitivity.

Each laser line has its advantages and disadvantages. The UV-Raman has minimum fluorescence interference, high Raman signal and no limit on the temperature of the sample, but sample degradation during spectral collection may be serious and the resonance Raman effect may complicate the spectral interpretation. The visible Raman has no specific requirement for the optics, sample degradation is less important than the UV-Raman, and no detection problem with sample temperature lower than 900°C, but fluorescence may, indeed, be a serious problem. The near-IR Raman may avoid the fluorescence problem encountered in the visible Raman in some cases and may provide details of framework vibrations, but the sample temperature and sample heat-up may be problems.

Raman spectroscopy is a suitable technique for in situ study.

5.4 Framework Vibrations

5.4.1 General Features

The framework vibrations of zeolites are observed in the spectral region 1,400–300 cm^{-1}. Flanigen summarized the observations for most of the early-synthesized zeolites and proposed the general assignments for the IR bands [1]. These assignments are applicable to newly synthesized zeolites. Figure 5.1 gives an IR spectrum of zeolite Y used in Flanigen's review [1] to illustrate the infrared assignments.

In Fig. 5.1, two types of bands are observed for a zeolite framework: internal tetrahedra bands which are structure insensitive (1, solid bands) and external linkages, which are structure sensitive (2, dashed bands). The former is observed for all aluminosilicates and silicates including amorphous zeolite precursors while the latter is observed only when the zeolite structure is constructed. The number and position of the structure sensitive bands depend on the zeolite structure and are summarized below [1, 2].

Internal tetrahedra (cm^{-1})

Asymmetric stretching	1,250–950
Symmetric stretching	720–650
T–O bending	500–420

External linkages

Asymmetric stretching	1,050–1,150
Symmetric stretching	820–720
Double ring	650–500
Pore opening	420–300

In more recent years, theoretical calculations have been performed by several research groups [4–6] to detail the IR band assignments. Based on the calculations, it is concluded [5] that zeolite framework vibrations are strongly coupled and cannot be strictly separated as internal tetrahedral and external linkage vibrations.

5.4.2 Zeolite Formation

The zeolite framework is built up of SiO_4 and AlO_4^- tetrahedra. The AlO_4^- tetrahedra can be replaced by other three and four valence state ions, such as B, Ga, Fe, Ge, Ti, etc. These tetrahedra form primary building units such as rings and cages, and these rings and cages are further connected to form the zeolite frameworks. In all cases, the SiO_4 and AlO_4 tetrahedra combine according to Lowenstein's rule that states that two neighboring AlO_4 tetrahedra cannot connect via an oxygen bridge in the zeolite structure. Formation of zeolite frameworks can

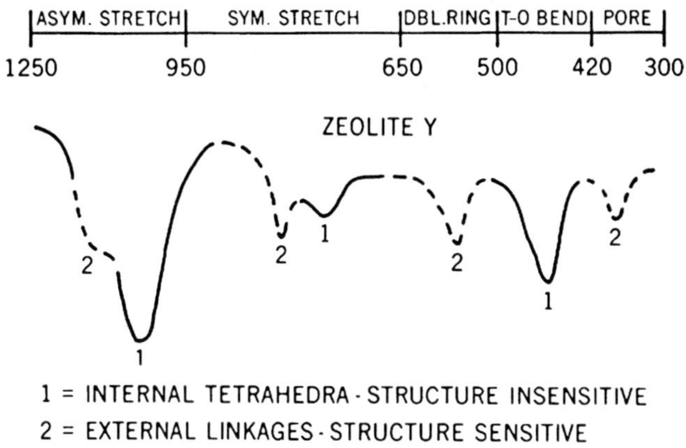

Fig. 5.1 Infrared assignments illustrated with the spectrum of zeolite Y, Si/Al=2.5 [1]. Reprinted from Zeolite chemistry and catalysis, ACS Monograph 171, with permission from American Chemical Society, Washington, DC

be monitored using IR and Raman spectroscopy due to the facts that the structure sensitive bands develop in IR spectra upon zeolite crystallization and that the chemical environmental change of the template molecule used in the crystallization leads to development of new Raman bands.

However, due to the strong adsorption of water, IR has to be an ex situ tool used to examine the zeolite structure formation after the solid phase is separated from the liquid phase at different times during the zeolite crystallization. Due to the weak Raman scattering nature of water and quartz tubes (if used as a vessel for zeolite synthesis), Raman spectroscopy can be used for in situ studies of zeolite formation as long as the synthesis setup allows. As mentioned above, silica is not a strong Raman scatter, and therefore, Raman spectroscopic study of zeolite formation is the most suitable to zeolite synthesis systems containing organic templates.

5.4.3 Framework Si/Al and Isomorphous Substitution

Each zeolite has its characteristic framework Si/Al range. Changes in framework Si/Al of a specific zeolite by direct hydrothermal synthesis are limited but the Si/Al range can be greatly expanded via post syntheses such as dealumination and realumination. When Al content in the framework increases, the bands are shifted towards lower wavenumbers due to longer Al–O bond (~1.75 Å) compared to Si–O bond (~1.62 Å). Different vibrations may show different extents of band shift. Plots of band shifts vs. mole fraction Al in zeolite frameworks give linear relationships, which are often used as an analytical method to determine zeolite framework Si/Al ratio [7–9]. However, when this method is used for Si/Al ratios, care must be taken to make sure that the zeolites are in the same cation form and have experienced similar pretreatment conditions.

Isomorphous substitution of zeolite frameworks by other elements can also lead to band position shifts. The shifts to higher or lower wavenumbers depend on the atomic weight of the element incorporated as well as on its bond length with the framework oxygen. In some cases such as Ti in TS-1 (MFI structure), a separate band at 960 cm^{-1} in their IR spectrum can even be observed and the new band intensity increases as the Ti content in the zeolite framework increases [10, 11]. A linear plot of the band intensity ratio, $I_{960\ cm^{-1}}/I_{800\ cm^{-1}}$ or $I_{960\ cm^{-1}}/I_{550\ cm^{-1}}$ vs. the mole fraction of Ti in the framework is observed, which can also be used for quantitative analysis of the framework Si/Ti ratios of TS-1 and TS-2 [10, 11].

5.4.4 Perturbation of Zeolite Frameworks

IR spectroscopy has recently been used to study location of exchanged ions in zeolites Y and siliceous zeolites such as ferrierite, Beta, modenite and ZSM-5 [12–16]. These studies are based on local framework perturbation of the zeolite frameworks by exchanged ions after dehydration. The perturbation is associated with T–O bond length, T–O–T angles, and to a lesser extent O–T–O angles (T=Si, Al) [17, 18]. The perturbation creates new bands in the spectral region of 1,000–850 cm^{-1} in their IR spectra. Table 5.1 summarizes the IR results reported in the literature [18, 19].

The band position depends on the nature of the cation and is sensitive to its coordination: either fully dehydrated and only interacting with framework oxygen ligands (bare cation) or multiligand complexes coordinating with guest molecules in addition to the framework oxygen ligands. The electron density on the cation determines the strength of interaction of the cation with the zeolite framework. If the coordination of the guest molecule to the cation increases the electron density on the cation, interaction of the cation with the zeolite framework oxygen is weakened. The band is shifted (relaxation effect) to higher wavenumbers [15, 17]. Fully hydrated zeolites in general do not have IR bands in this spectral region. For cations with redox properties such as Cu and Fe, spectral shift towards

Table 5.1 Summary of the $\nu_{T\text{–}O\text{–}T}$ values for bare metal ions in zeolites

	MFI [19] wavenumber (cm^{-1})	BEA [19]	FER [19]	MOR [19]	Y [18]
Co(II)	931–935	915–917	918–919	905	931
Ni(II)	930	915–916	914–915		931
Cu(II)	927	915	912–915		
Cu(I)	960	950	950		962
Mg(II)	935–939	925	922–923		
Mn(II)	940	924–926	927		
Fe(II)	927	913–914	913–914		
Fe(III)	890				
Zn(II)	940(?)	916	915		
Na(I)	>980	>980	>980		

higher wavenumbers may be observed upon reduction (increase electron density on the cations) [18, 19].

Different cation sites, denoted as α, β, and γ in the literature [19], in siliceous zeolites such as ferrierite, Beta, modenite and ZSM-5, are differentiated via deconvolution of the observed band. Quantitative analysis is also possible due to the existence of a linear plot of the cation loading vs. band intensity [19].

5.4.4.1 Cation Locations

The information about cation location in zeolites is normally obtained from crystallographic measurements. But in many cases where by crystallographic methods researchers have difficulty or find it is not possible to get such information, IR and Raman methods are very valuable and can be considered to be complementary to diffraction methods.

The frustrated translational motions of cations with respect to the zeolite framework fall in the far-IR spectral region and are detectable by far-IR and Raman spectroscopy [20–23]. The band position, ν, of the cation vibration was found to be related with the mass and radius of the cation, $\nu = C_B m^{-1/2} r^{-3/2}$, where C_B is the so-called Brodskii constant [20]. Studies of cation distribution among different crystallographic positions, their migrations, and sequence of replacement of cations at different sites during ion exchange have been carried out on various types of zeolites such as A, X, Y, E and ZSM-5. Different cation sites can be differentiated. Band position shift towards lower wavenumbers can occur for a particular site as the framework Si/Al ratio increases. Stepwise cation dehydration and rehydration can also be monitored. However, for highly siliceous zeolites, the far-IR technique becomes difficult due to low concentration of cations.

5.5 OH Groups

Surface OH groups in zeolites originate from several sources: acidic OH groups due to framework Al and other trivalent ions such as B, Ga, Fe, etc., silanol groups on the external surface and in the "nests" (defect sites), OH groups attached to di- or tri-valent cations due to hydrolysis, OH groups from "dangling Al," OH groups associated with non-framework Al, and OH groups of adsorbed water. All these OH groups are observable directly with IR spectroscopy. Figure 5.2 gives a schematic drawing of OH groups and Lewis acid sites of a steamed FCC catalyst that contains USY and an alumina/silica matrix together with its FT-IR spectra showing the OH groups. The assignment for the IR bands is the following:

- 1, dangling Al–OH: This assignment is still under debate and assignment to non-framework alumina has been reported [24]
- 2 and 2′, surface silanol Si–OH groups on zeolite and matrix

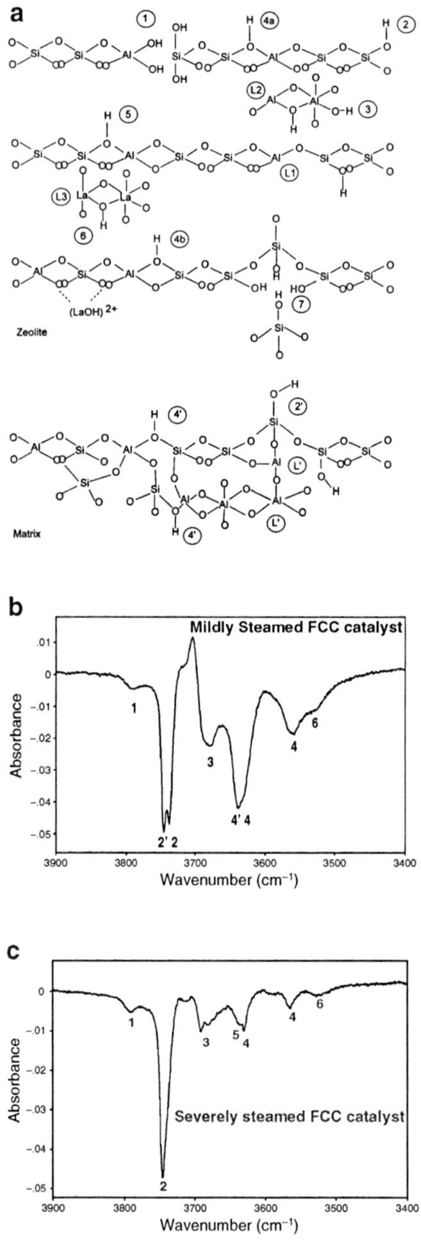

Fig. 5.2 (**a**) A schematic drawing of OH groups in a FCC catalyst. (**b**) and (**c**) FT-IR spectra of OH groups perturbed by pyridine probe of mildly and severely steamed FCC catalyst. See text for the assignment

- 3, OH groups associated with non-framework aluminum species
- 4 and 4′, acidic OH groups associated with framework Al in zeolite and in matrix
- 5, acidic OH groups associated with framework Al in zeolite but being stabilized by rare-earth ions
- 6, OH groups associated with rare-earth oxide
- 7, OH groups of the framework defect sites, or "nests" formed by steam induced framework dealumination or high temperature calcination

In Figs. 5.2b and c, examples are given to show the direct observation of some of the OH groups.

Depending on the samples and their treatment history, some bands may be observed in one case and the other bands may be observed in other cases. The OH groups attached on the charge balancing ions formed via hydrolysis can also be observed and the dehydration condition of the sample is normally critical for such observation. The "nest" is easily observed from IR spectra of highly siliceous zeolites and generally it gives a very broad band centered ~3,000 cm^{-1}, indicating hydrogen bonding of the OH groups in the "nest."

The band position of acidic OH groups depends on many factors such as the type of zeolite, the framework Si/Al ratio, the nature of the trivalent ion, the location of OH group in the zeolite structure, and the co-existence of other cations. For some zeolites such as faujasite, erionite, ZSM-20, SAPO-37, more than one type of acidic OH groups is observed due to the existence of distinct crystallographic framework oxygen in their structures [25–28]. The complication increases when a zeolite experiences dealumination or introduction of foreign ions such as rare earth (see Fig. 5.2).

Increasing framework Si/Al ratio, in general, shifts the acidic OH band position towards lower wavenumbers [29]. This seems to suggest that the higher the Si/Al ratio, the more acidic the OH group. However, this general conclusion does not always hold good. Some zeolites with similar acid strengths have a variety of OH groups [25–28, 30] and chemical composition changes and imhomogeneous distribution of Al in the zeolites may shift the band position [31]. The nature of trivalent ions in the same framework has an effect on the band position. For instance, in a ZSM-5 structure, Al–OH has its OH vibration at ~3,610 cm^{-1}, while Ga–OH at ~3,620 cm^{-1}, Fe–OH at ~3,630 cm^{-1} and B–OH at ~3,725 cm^{-1} [32]. Due to strong interactions of OH groups with the framework, the band position can shift substantially to lower wavenumbers, as seen in the case of faujasite zeolites (the lower frequency band). Such a shift does not correspond to more acidic OH groups in the zeolites and furthermore, these OH groups cannot participate in reactions with reactant molecules as readily as those with higher frequency bands due to structure constraints. The band position can also be affected by the species either created during the sample treatments such as steaming or decomposition of NH_4^+ or ion exchange [33].

It has been found that there is a general trend with respect to acidic strength: Si–OH–Al>P–OH>Al–OH>Si–OH [34].

Table 5.2 Characteristics of the probe molecules used for acidity measurements

Probe	Advantage and disadvantage
Ammonia (pK_a=9.2)	Used for B (~1,450 cm^{-1}) and L (~1,630 cm^{-1}), do not differentiate acid strength from IR spectra. Dissociative adsorption forming NH_2 and NH at high temperature, multiple interactions, water interference for L acidity
Pyridine (pK_a=5.2)	Used for B (~1,540 cm^{-1}) and L (~1,450 cm^{-1}); can differentiate different Brönsted and Lewis acid sites with different strengths (1,560–1,680 cm^{-1}); thermally stable. Size too big for high-density acidity measurements and for small pore zeolites
Acetone (pK_a=−7.2)	Carbonyl interactions with B and L shows –CO (~1,700 cm^{-1}) and OH band position shift. Highly reactive
Acetonitrile (pK_a=−10.4)	OH and CN (~2,300 cm^{-1}) band position shift; different acid strengths. Relatively high reactivity
Benzene	π-bonds to B and L give OH and CH (~670 cm^{-1}) band position shift; differentiate different B acid strengths; reveal heterogeneity of acid sites. Multiple interactions may exist; Large size
CO	CO (~2,200 cm^{-1}) and OH band position shift for L and B; differentiate different Lewis acid sites. Low temperature, CO–CO coupling

5.6 Brönsted vs. Lewis Acid Sites

Brönsted acid sites can be investigated with and without probe molecules using IR spectroscopy while Lewis acid sites can only be studied with the help of probe molecules. Raman spectroscopy can also be used for Brönsted acid sites when probe molecules are employed [35]. For Brönsted and Lewis acidity measurements, ideally, it is best if reactant molecules can be used as probe molecules. However, in most of the cases, it is difficult. The choice of a probe molecule depends on the system under investigation, but for correlating acidity data with catalyst performance, one has to keep in mind differences between the probe and reactant molecules. Lercher et al. [36] have summarized the criteria for choosing a proper probe molecule for solid acids. The probe molecule should have dominating base and rather weak acidic properties and is able to distinguish Brönsted and Lewis acid sites and differentiate acid sites of the same type but of different strengths. The size of the probe should be comparable to that of the reactant in order to minimize the effects of structure constraints.

Many probe molecules such as ammonia, pyridine, n-butylamine, acetonitrile, benzene, carbon monoxide, acetone, aldehyde, ether, alkanes and alkenes have been reported for zeolite acidity measurements in the literature [36]. They have different basicities and are used under different experimental conditions. Each one has its own characteristics and disadvantages. Table 5.2 summarizes the main characteristics of some of the probe molecules.

Very recently, small molecules such as N_2, H_2, and CH_4 have also been used as probe molecules for acidity [37]. To measure the acidity on the external surface of zeolites, large bulky probe molecules are employed which cannot penetrate the small zeolite pores [38].

Pyridine is an excellent probe [39, 40] because (1) pyridine interacts with Brönsted acid sites, forming pyridinium ion, PyH^+, and pyridine adsorbs onto Lewis acid sites to give characteristic absorption bands in its IR spectrum [40] and (2) pyridine as a probe differentiates Lewis acid sites with different strengths; and (3) pyridinium is thermally stable. Figure 5.3 gives a typical IR spectrum of pyridine adsorbed on acidic ZSM-5 zeolite. The band at ~1,540 cm^{-1} is due to pyridine on Brönsted acid sites while that at ~1,450 cm^{-1} is associated with pyridine adsorbed on Lewis acid sites. Both bands are due to the ν_{19b} ring deformation mode of pyridine [41]. The multi-bands in the spectral region 1,580–1,660 cm^{-1}, due to ν_{8a} and ν_{8b} ring vibrations [30], can also be used to differentiate pyridine adsorbed on different Brönsted and Lewis acid sites.

The Brönsted acid sites mainly originate from framework Al while the Lewis acid sites can be due to three-coordinated framework Al (never observed directly) formed upon dehydroxylation, non-framework alumina formed during dealumination, and charge balancing cations (see Fig. 5.2). The Brönsted acid strength has been correlated to charge density of the framework [30] and to the Sanderson electronegativity of zeolite frameworks [42–44], which suggests a long-range interaction and collective phenomenon for Brönsted acid strength. However, recent studies [45–48]

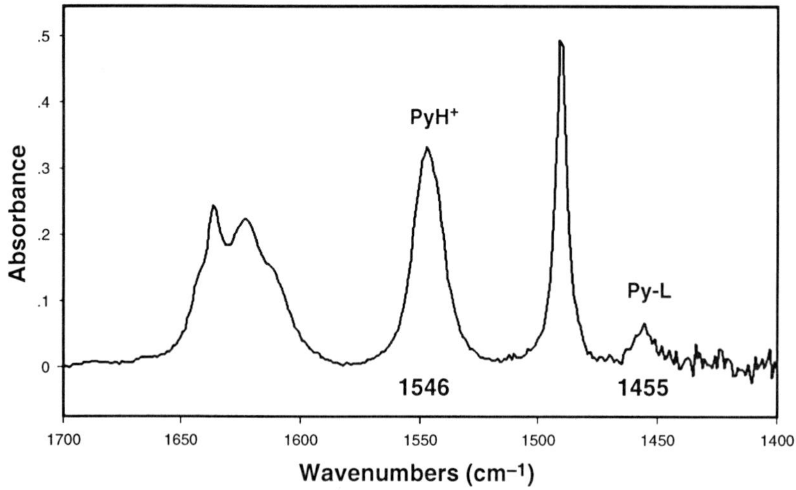

Fig. 5.3 DRIFT spectrum of pyridine adsorbed on H^+-ZSM-5 (Si/Al=35, calcined NH_4^+-ZSM-5 at 500°C for 4 h). Background subtracted. The sample was pretreated at 450°C for 1 h under dry N_2 flow before introducing pyridine vapor at RT. The spectrum was collected on a Perkin-Elmer 1,000pc instrument equipped with a SpectraTec high-temperature and high-pressure chamber and a MCD detector. 256 scans were collected for this spectrum

show that local configuration is also important. Using probe molecule such as benzene, Datka [49] has found that Brönsted acidity in zeolites is quite heterogeneous.

Using probe molecules and IR spectroscopy, the relative acid strengths of the Brönsted and Lewis acid sites can be revealed via monitoring the band position shifts of the OH groups upon adsorption and desorption of probe molecules [50]. In general, the larger the shift of the OH bands to lower wavenumber, the stronger the acid strength. The higher the desorption temperature for adsorbed probe, the stronger the acid strength. Compared to other Lewis acid sites, the charge balancing cations are weak for Lewis acids and show easy probe molecule removal upon desorption.

However, complications may be encountered when performing acid strength studies using the methods given above: (1) different steric hindrance of different zeolite structures to the probe molecule and (2) simultaneous coordination of the lone pair electron on the N of the pyridine to a Lewis acid site and hydrogen bonding of the pyridine ring to its neighboring OH group [51]. In the first case, the interaction of the probe molecules with the acid site becomes weak while in the second case, high desorption temperature is needed to remove the probe molecule.

5.7 Basic Sites

Basic sites in zeolites are basic hydroxyls (Brönsted basic sites), basic oxygens (Lewis basic sites), basic alkali metal oxides in zeolite cages [52–58], and metal clusters such as $Na_n^{(m-1)+}$ and $K_n^{(m-1)+}$, where n, $m \leq 4$ and $m<n$ [59, 60].

The most important basic sites are framework oxygens and may be qualitatively described by the intermediate Sanderson electronegativity. Barthomeuf [61, 62], and recently, Davis [63] reviewed various aspects of zeolite basicity and its applications in catalysis.

The basicity of zeolites arises from the framework negative charge due to the presence of framework Al. Thus, in a homologous series of zeolites, the zeolites with lower framework Si/Al ratio may be more basic than those with higher framework Si/Al ratio. Introducing alkali metal cations enhances the basicity of zeolites, and the more basic cations, such as Cs, more profound [61–63].

Similar to the acidity measurements, detection of the basic sites using FT-IR and Raman spectroscopic techniques also needs the help of probe molecules. However, there are not many probe molecules suitable for the basic site measurement. Commonly used probe molecules are CO_2 [54, 64–66], pyrrole [67, 68] and chloroform [69–71], but they have their drawbacks, such as decomposition and reaction with the basic sites. Pyrrole, as a probe molecule, has been used to show a trend in basicity for various zeolites [61, 72].

5.8 Adsorbed Species

Zeolite cavities and channels provide a unique environment for adsorbing molecules. The negative charged zeolite framework formed by isomorphous substitution of Al for Si requires positive charged cations to balance. The surface of the framework is essentially oxygen atoms that are polarizable and provide van der Waals interactions with adsorbed molecules. The cations can be located at different sites, depending on the particular zeolite framework, cation size, dehydration state, and nature of adsorbed molecules. The negative charge centers in the framework and the positive-charged cations are not always located close together, which creates strong electric fields and field gradients within the zeolite channels and cavities. Therefore, adsorbed species can experience a van der Waals field from the framework oxygens together with specific interactions with the charge balancing cations.

FT-IR and Raman spectroscopies are among the most widely used techniques for studying adsorption in zeolites. Except for the probe molecules used for specific properties such as acidity and basicity, FT-IR and Raman spectroscopic studies on adsorption of molecules in the zeolite cavities and channels are of paramount importance in providing fundamental understanding for separation and catalytic reactions.

5.8.1 Adsorbed Water

Zeolites always adsorb water when they are exposed to the atmosphere due to their hydrophilic properties. The adsorbed water in zeolites can be present in the forms of free water, loosely coordinated water, tightly coordinated water and even water clusters [73]. Due to the strong absorption by water in the IR spectral region, water can be studied using FT-IR technique. The water-bending vibration at ~1,640 cm^{-1} indicates the presence of adsorbed water in zeolites. Due to the hydrogen bonding of water with OH groups, the bands in the hydroxyl spectral region (3,000–4,000 cm^{-1}) are generally structureless and broad. The water bands in the near-IR spectral region are characteristic, and located at ~5,250 and ~7,090 cm^{-1}. Dehydration and rehydration of zeolites and water removal upon adsorption of organic molecules into zeolites can be monitored in situ via these vibration bands using FT-IR techniques.

The FT-IR studies [74–77] reveal that (1) dehydration and rehydration of zeolites can be a reversible process. Most of the zeolites with high Si/Al ratios fall into this category; (2) dehydration and rehydration can still be reversible but the zeolite frameworks may distort and the unit volume may decrease significantly; and (3) dehydration and rehydration may become partially irreversible and are accompanied by the breakage of T–OT bonds. When this happens, a new band around 4,520 cm^{-1} representing Si–OH bond appears in their near-IR spectra [77–79].

After dehydration, the channels and cavities of zeolites occupied by water molecules are empty and cations normally screened by water molecules are uncovered. The strong field around the cations polarizes adsorbed molecules. This is demonstrated by the appearance of IR bands for normally IR-silent molecules such as H_2, N_2, and O_2 upon adsorption in zeolites [80]. The polarization strength depends on the nature of cations (charge density), hydration state of the cations (water screening), and also on the zeolite frameworks (accessibility).

5.8.2 *Molecules with π-Complexation*

π-Complexation of olefins, alkynes and aromatics with cations in zeolites has been well studied with the FT-IR technique. Due to the loss of symmetry of the ethylene molecule, the IR silent C=C bond becomes IR active and shows band shift towards lower wavenumbers. Ethylene interactions with acidic OH groups in zeolite HY, via π-complexation, shift the C=C band from 1,625 cm^{-1} (the Raman gas-phase band) to 1,612 cm^{-1} [81, 82]. A similar phenomenon is also observed for acetylene and other alkyne molecules adsorbed in H-form zeolites [83]. This interaction can also be considered as hydrogen bonding between the adsorbed molecule and the OH group [84]. The spectral features of the OH groups perturbed by the adsorbed molecules are a band shift towards lower wavenumber and a broadening of the band. Both features are correlated with each other and depend on the proton affinity of the adsorbed (base) molecules.

If the proton affinities of the adsorbed molecules such as olefins, alkynes and aromatics are large enough, protonation of the molecules takes place. The formation of protonated olefin and alkyne molecules in H-form zeolites leads to oligomerization of the adsorbed molecules. The process can be monitored by FT-IR spectroscopic technique [83]. With increasing length of oligomeric chains, the characteristic IR bands are shifted towards lower wavenumbers. The reaction rate depends on temperature, the zeolite structure and the structure of the reaction products [84].

π-Complexation of adsorbed olefin, alkyne and aromatic molecules with other cations in zeolites also results in band position shifts of C=C, C≡C and phenyl ring towards lower wavenumbers [83, 85–87]. The more significant band position shifts are observed for transition metal cations compared to non-transition metal cations due to the existence of a back electron donation from the transition metal cations to the anti-bonding orbital (d-π* backdonation) of the molecules. Among all the metal cations studied, Ag+ is unique and shows the largest effect on the ethylene molecule [88]. This is because Ag+ has the strongest interaction with ethylene, which has been confirmed by heat of adsorption measurements [89]. The uniqueness of Ag^+ in zeolites is also shown by its "hydrophobic" properties [90]. With the hydrated form of Ag^+-zeolites, aromatic and olefin molecules can still easily be adsorbed whereas for all other cations, adsorption is inhibited due to the screening of charge by the water molecules [90].

5.8.3 *Molecules with Different Conformations*

More than one type of adsorbed molecule may exist in the zeolite channels and cavities. FT-IR spectroscopic studies clearly show this is the case [90–95]. The existence of more than one type of adsorbed species is due to (1) partial ion exchange and (2) different cation sites.

Adsorption of small molecules such as H_2 and N_2 onto transition metal cations such as Cu, Fe, Zn and Cd in zeolites forms σ-bonded complexes [96–98]. The FT-IR spectra of adsorbed molecules in these complexes exhibit significant band shifts towards lower wavenumbers with respect to the corresponding band of their gas phase molecule (up to a 220 cm^{-1}). Very recently, large band shifts, ~1,000 cm^{-1} for H_2 and ~400 cm^{-1} for C_2H_6, were observed for Cu^+-ZSM-5 and the phenomena were interpreted based on $[Cu^+(\eta^2\text{-}H_2)]$ and $[Cu^+(\eta^2\text{-}C_2H_6)]$ adsorption complexes [99].

Another important aspect of the adsorbed molecules in zeolite cavities and channels is the conformational behavior of the adsorbed molecules. Different conformations of molecules such as chlorocyclohexane and *trans*-1,4-dichlorocyclohexane inside the zeolite frameworks, ZSM-5, NaY and SiY have been studied using FT-Raman spectroscopy [100, 101]. The chlorocyclohexane molecule can adopt either axial or equatorial conformers that give distinct Raman bands at 685 and 808 cm^{-1} (ν_{22}) and at 730 and 817 cm^{-1} (ν_{21}), respectively. Upon adsorption of the molecule into the zeolite frameworks, equilibrium of the conformers is shifted towards the equatorial configuration (band relative intensity changes). Cooling down causes complete conversion to the equatorial configuration of the molecule (disappearance of the 685 and 808 cm^{-1} bands). The conformational enthalpy change can be obtained from a plot of equatorial and axial conformer band intensity ratio vs. reciprocal temperature (1/T). Raman band broadening (the CH_2 scissoring mode at 1,446 cm^{-1}) provides dynamics information of the adsorbed molecules in the zeolite cavities and channels. At room temperature, the adsorbed chlorocyclohexane molecules retain a substantial degree of molecular motion within the zeolite cavities and channels. The molecular motion can be restrained either by lowering temperature or by introducing stronger interaction centers such as Na^+. The adsorbed chlorocyclohexane molecules in NaY exist in a statically disordered state due to the strong interactions of the adsorbed molecules with Na^+.

5.9 In Situ Studies of Reaction Mechanisms

Among all the in situ techniques, FT-IR and Raman spectroscopy are still the most important characterization techniques [102]. An in situ study allows us to better understand the fundamental relations between the physicochemical properties of a catalyst and its catalytic activity and selectivity and also to know how the reactions occur on the surface of a catalyst. Further development of an in situ study is the so-called operando study, which, simultaneously, characterizes the catalyst surface with one or more spectroscopic techniques and monitors the changes in concentration of reactants and products. The operando studies allow us to characterize surface species and to identify reaction intermediates and spectators under catalyst working conditions, as well as to find out detailed reaction mechanisms.

To perform an in situ FT-IR and Raman spectroscopic study, designing a suitable in situ cell and finding out a reaction condition which compromises the spectroscopic and catalytic measurements are very important. Catalyst temperature, catalyst form, catalyst damage, catalyst activity and selectivity, reaction cell volume, spectroscopic window and safety, all need to be considered [103]. For a catalytic reaction system involving many reactants and species, breaking down the system into several simple ones and examining them separately is sometimes necessary. This can be a very important step in helping with in band identification and assignment and sometimes, it is almost the prerequisite for understanding a complex reaction system.

There are many reports in the literature on mechanistic studies of reactions in zeolites cavities and channels using in situ FT-IR and Raman spectroscopy [104–109]. Examples, just mention a few, include skeletal isomerization [104], NO_x reduction [105, 106], CO_2 hydrogenation [107], chloromethane to light olefins [108], methanol oxidation [109], etc. Due to the space limitation here, we are unable to discuss all these reactions but will just give a few examples to demonstrate the main features.

5.9.1 *Example 1: NO_x Reduction*

NO_x reduction by C_3H_6 and C_3H_8 over NiNa-mordenite catalyst [105] – The emission control of NO_x is important to our environment. Selective catalytic reduction (SCR) of NO_x with NH_3 has emerged as an effective technology using

oxides and zeolite catalysts. Reduction of NO_x by hydrocarbons over exchanged zeolites is a potential alternative to the NH_3-SCR for emission control in the presence of excess oxygen. A wide range of transition metals such as Cu, Co, Fe and Ni in combination with different zeolite structures has been found to be active catalysts for reduction of NO_x using hydrocarbons as reducing agents [106]. In Mosqueda-Jimenez et al.'s work [104] on the surface reactions during NO reduction with propene and propane on Ni-exchanged Na-mordenite, first a systematic study of adsorption of NO, $NO+O_2$, C_3H_6, C_3H_8, $C_3H_6+O_2$, $C_3H_8+O_2$ over the catalyst to assign IR bands was performed and then the reaction mechanism of 1,000 ppm NO+1,000 ppm C_3H_6 or C_3H_8, +5% O_2 was studied. Table 5.3 summarizes the species they observed during adsorption of $NO+C_3H_6+O_2$ at 150°C and the assignments based on earlier literature.

The presence of oxygen was found to be crucial for the reduction of NO with hydrocarbons on Ni-exchanged zeolites. The reaction mechanism involves the

Table 5.3 Assignment of infrared bands during adsorption of $NO+C_3H_6+O_2$ at 150°C

Wavenumber (cm^{-1})	Assignment	Species
3305	$\nu_s(NH_2)$	R–NH_2
3125	$\nu_{as}(=CH_2)$	
	ν(OH) of an allyl oxime	C=N–OH
2954	$\nu_{as}(CH_3)$	
2930	$\nu_{as}(CH_2)$	
2860	$\nu_s(CH_2)$	
2275	Out-of-phase ν(N=C=O)	Al^{3+}–NCO
2188	Out-of-phase ν(N=C=O)	Ni^{2+}–NCO
2146	ν(C≡N)	C≡N
1670	ν(C=O) of acrolein	
	ν(N=O) of organic nitrito compound	R–O–N=O
	ν(C=N) of an allyl oxime	C=N–OH
1638	ν(C=C)	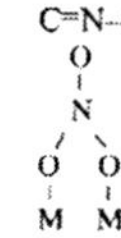
	(ν_3) Bridging bidentate nitrate	
1602	$\nu_{as}(CO_2)$ of a carboxylate	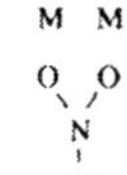
	Aromatic species or polyenes	
	$\nu_{as}(NO_2)$ of nitro compound	
1550	$\nu_{as}(NO_2)$ of organic nitro compound	R–NO_2
1520	(ν_3) Monodentate nitrate	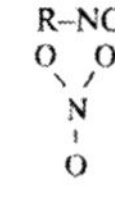
1465	ν(C–O) of an acetate group	
	CH_2, CH_3 deformation	
1436	$\nu_s(CO_2)$ of a carboxylate	
	OH deformation of an allyl oxime	C=N–OH
1386	CHO deformation of a formate group	
	CH_3 deformation	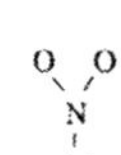
1345	$\nu_s(NO_2)$ of nitro compound	
	$\nu_s(NO_2)$ of organic nitro compound	

For the assignment references, please see [105]

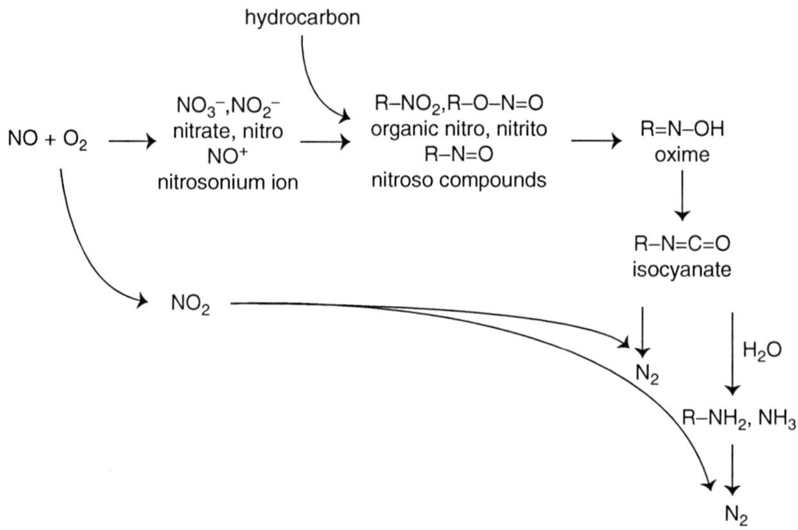

Fig. 5.4 Scheme of the reaction mechanism for the reduction of NO with hydrocarbons [105]. Reprinted from Applied Catalysis B: Environmental, with permission from Elsevier

formation of adsorbed NO_x species from reactions of NO and O_2, e.g., NO_3^- and NO_2^- (1,638, 1,602, 1,520 cm^{-1}), which react with the hydrocarbons (dehydrogenation of propane to propene may occur on transition metal sites in the first reaction step) to form organic nitrito and nitro compounds (1,670, 1,550 cm^{-1}). The presence of acid sites in the zeolite explains the formation of nitrosonium ion NO^+ (2,166 cm^{-1}) that might be also involved in the reaction mechanism as a precursor of nitroso compounds (1,897 cm^{-1}). The organo nitro, nitrito, and nitroso compounds are transformed to surface oximes (3,125, 1,670, 1,436 cm^{-1}). The surface oximes rearrange to isocyanates (2,275–2,270, 2,200–2,188 cm^{-1}). The isocyanates either react with the gas phase NO_2 through an Eley-Rideal type reaction to form N_2 or are hydrolyzed to amines and ammonia (3,305 cm^{-1}), which reduce NO_2 to N_2. Oxygenated carbon species, formate (1,386 cm^{-1}), acetate (1,465 cm^{-1}), carboxylate (1,436 cm^{-1}), and acrolein (1,670 cm^{-1}), contribute to the formation of CO_2 and might also be involved in the formation of isocyanates. A schematic drawing of the reaction mechanism is given in Fig. 5.4.

5.9.2 Example 2: CO_2 Hydrogenation

CO_2 hydrogenation in Li-promoted, zeolite-supported Rh catalysts [107] – The purpose of CO_2 hydrogenation study was to develop technologies for mitigation of CO_2 emission, which causes the global climate change [110, 111]. This study

focused on Li effect on CO_2 hydrogenation on a Rh ion-exchanged zeolite Y catalyst. It has been experimentally observed that without Li, hydrogenation of CO_2 on the Rh ion-exchanged zeolite Y catalyst gives methane as a main product. Adding Li to the catalyst, depending on the amount of Li, greatly changed the product distribution. Oxygenated products such as CO, methanol and ethanol were formed. To explain the function of Li and to understand the reaction mechanism of CO, methanol and ethanol formation, in situ FT-IR spectroscopic technique was employed. Similar to the de-NO_x study discussed above, the system was fragmented into several subsystems (or model systems) and studied separately. By using in situ FT-IR, the authors found that during the reaction, adsorbed CO species formed via dissociation of CO_2 (bands around 2,038–2,050 cm^{-1}) were the intermediate in the CO_2 hydrogenation on both RhY and Li/RhY. Li additive promotes adsorption of CO_2 in a η^2-bonded form (band at 1,618 cm^{-1}), which enhances the concentration of surface CO during the reaction (evidenced by the band intensity difference between RhY and Li/RhY), and also promotes methanol and ethanol formation. Li additive also hinders H_2 activation and prevents adsorbed CO from being dissociatively hydrogenated. The reaction mechanism of CO_2 hydrogenation on Li/RhY catalyst was proposed on the basis of the in situ FT-IR results. Figure 5.5 gives a schematic drawing of the proposed mechanisms.

The reaction mechanism in (a) explains how the CH_4 is formed on the catalyst RhY while the reaction mechanism in (b) explains how the CO and C_2H_5OH are formed. The ethanol formation proposed in (a) via a gas phase CO insertion reaction to Rh-CH_3 species was not supported by the in situ FT-IR studies but that proposed in (b) via two types of adjacent Rh sites, one is surrounded by Li and the other is not (interface), explains well the in situ FT-IR and catalyst performance results. The

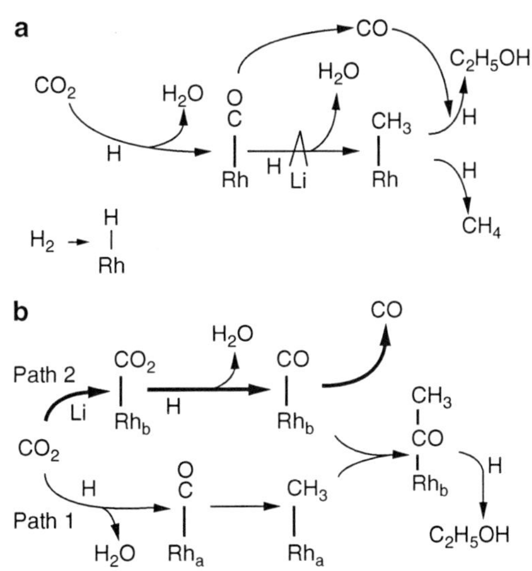

Fig. 5.5 Proposed mechanisms of CO_2 hydrogenation on Li/RhY (Rh=5 wt%, Li/Rh=10) [107]. Reprinted from Applied Catalysis A: General, with permission from Elsevier

insertion reaction between Rh_a-CH_3 and Rh_b-CO species forms acyl groups, which then desorbs as ethanol.

5.9.3 Example 3: Methanol Oxidation

Methanol oxidation in photo-activated $FeAlPO_4$-5 (AFI) [109] This is a study using time-resolved FT-IR spectroscopy. For a time-resolved spectroscopic study, initiation of reaction by different physical methods is the first step. In this example, an 8 ns 355 nm laser excitation pulse was employed. The mechanistic study was started by looking at the CH_3OH+O_2 reaction under steady-state conditions using the normal FT-IR spectroscopy with irradiation of the ligand-metal charge transfer (LMCT) absorption for 30 min or longer to find out the reaction products. Then, the time-resolved FT-IR study of the reaction was carried out with the rapid-scan mode to get about a millisecond time resolution to see the evolution of the species during the reaction. In the next step, the time-resolved FT-IR study was performed with the step-scan mode to get a microsecond time resolution to reveal the reaction intermediates. Finally, a detailed reaction mechanism was proposed.

The initiation of reaction is triggered by a laser pulse on the Fe center in the $FeAlPO_4$-5. LMCT transition leads to reduction of Fe^{3+} to Fe^{2+} and generation of a positive hole on the framework oxygen, Fe^{3+}-$O^{2-} \rightarrow Fe^{2+}$-$O^-$. The steady-state experiments reveal the presence of 4-electron transfer products HCO_2H, H_2O and HCO_2CH_3 from the photo-activated CH_3OH+O_2 reaction. On the second time scale (irradiation achieved with a train of ten consecutive laser pulses (one 8 ns pulse per 100 ms, total duration 1 s)), it was found that just after the initiation of reaction, HCO_2H (1,679 cm^{-1}) and H_2O (1,650 cm^{-1}) are dominant species together with a small amount of $CH_2{=}O$ (1,726 cm^{-1}). The formic acid then converts to formate (HCO_2^-, 1,625 cm^{-1}) within 10 s. However, the fact that the single pulse experiment does not show instantaneous appearance of HCO_2H and H_2O implies that HCO_2H and H_2O are secondary photolysis products of an intermediate with a lifetime shorter than 1 s. On a microsecond time scale (step-scan study), the intermediate was found to be HO_2CH_2OH (1,445 cm^{-1}) and was formed by direct coupling of HOO and CH_2OH radicals.

The time-resolved FT-IR study of the photo-activated methanol oxidation on $FeAlPO_4$-5 catalyst leads to the following detailed mechanism (Fig. 5.6): Fe^{3+}-$O^{2-} \rightarrow Fe^{2+}$-$O^-$ excitation results in a transiently reduced Fe center and a hole on framework oxygen. The initial electron-transfer steps are reduction of O_2 by Fe^{2+} to O^{2-} and concurrent oxidation of CH_3OH by the positive hole to methanol radical cation. The highly acidic methanol radical cation undergoes efficient proton transfer to superoxide to form HOO and CH_3O radicals. CH_3O radical is unstable and rapidly isomerizes to CH_2OH. The HOO and CH_2OH couple on the microsecond timescale to yield the 2-electron-transfer product HO_2CH_2OH, which spontaneously rearranges to HCO_2^- and H_2O within 4 s (250 K).

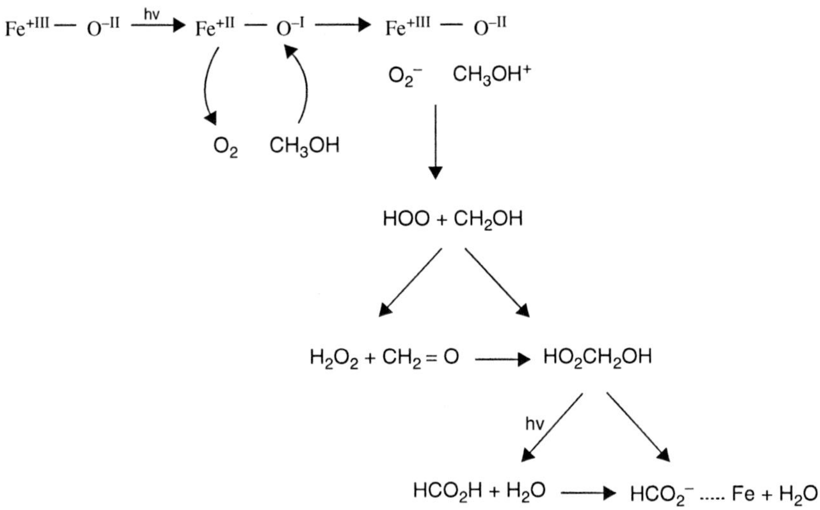

Fig. 5.6 A schematic of reaction mechanism of methanol oxidation in a photo-activated $FeAlPO_4$-5 catalyst [109]. Reprinted from Journal of Physical Chemistry A, with permission from American Chemical Society

The study also reveals that the reduced framework Fe centers in $FeAlPO_4$-5 have substantially stronger reducing power compared to the Fe_2O_3 conduction band electrons.

5.10 Concluding Remarks

FT-IR and Raman spectroscopy is useful for obtaining information on zeolites, zeolite surfaces, molecules being adsorbed in the zeolite cavities and channels, and catalytic reactions within zeolites. Knowledge of the various FT-IR and Raman techniques and their characteristics and disadvantages is essential for the proper use of these techniques. The field is still growing and new applications continue to appear. With the development of new instrumentation and new techniques, such as 2D correlation spectroscopy [112, 113], ultrafast time-resolved spectroscopy [114, 115], and operando set-ups [116, 117], FT-IR and Raman spectroscopy will continue to play an important role in zeolite field.

References

1. Flanigen EM (1976) Structural analysis by infrared spectroscopy. In: Rabo JA (ed) Zeolite chemistry and catalysis, ACS Monograph 171, American Chemical Society, Washington, DC, pp 80–117
2. Blackwell CS (1979) Investigation of zeolite frameworks by vibrational properties. J Phys Chem 83:3251–3261

3. Kubelka P, Monk F (1931) Ein Beitrag zur Optik der Farbenstriche. Z Tech Phys 12:593–601
4. Miecznikowski A, Hanuza J (1985) Application of the long chain approach to the structure and vibrational spectra of X and Y zeolites. Zeolites 5:188–193
5. Geidel E, Boelig H, Peuker Ch et al (1991) Approximate assignment of vibrational frequencies of the NaX framework. Stud Surf Sci Catal 65:511–520 (In: Öhlmann G, Pfeifer H, Fricke R (eds) Catalysis and adsorption by zeolites)
6. van Santen RA, van Beest BWH, de Man AJM (1990) On lattice dynamics, stability and acidity of zeolites. NATO ASI Ser 221:201 (In: Barthomeuf D, Derouene EG, Höelderich W (eds) Guidance for mastering the properties of molecular sieves)
7. Flanigen EM, Szymanski HA, Khatami H (1971) Infrared structural studies of zeolite frameworks. Adv Chem Ser 101:201–229 (American Chemical Society, Washington, DC)
8. Beyer HK, Belenykaja IM, Hange F et al (1985) Preparation of high-silica faujasites by treatment with silicon tetrachloride. J Chem Soc Faraday Trans 1(81):2889–2901
9. Kubelková L, Seidl V, Borbely G et al (1988) Correlations between wavenumbers of skeletal vibrations, unit-cell size and molar fraction of aluminium of Y zeolites. Removal of non-skeletal al species with H_2Na_2EDTA. J Chem Soc Faraday Trans 1 84:1447–1454
10. Vayssilov GN (1997) Structural and physicochemical features of titanium silicalites. Catal Rev Sci Eng 39:209–251, and references therein
11. Cundy CS, Forrest JO (2004) Some observations on the preparation and properties of colloidal silicalites: Part II: Preparation, characterisation and properties of colloidal silicalite-1, TS-1, silicalite-2 and TS-2. Micropor Mesopor Mater 72:67–80
12. Sárkany J, Sachtler WMH (1994) Redox chemistry of Cu/NaZSM-5: Detection of cuprous ions by FTi.r. Zeolites 14:7–11
13. Beutel T, Sárkany J, Lei GD et al (1996) Redox chemistry of Cu/ZSM-5. J Phys Chem 100:845–851
14. Chen HY, Wang X, Sachtler WMH (2000) Reduction of NO_x over zeolite MFI supported iron catalysts: Nature of active sites. Phys Chem Chem Phys 2:3083–3090
15. Sobalik Z, Sponer JA, Wichterlova B (2000) In: Corma A, Melo FV, Medioroz S et al (eds) 12th International congress in catalysis, vol. 130, p 1463 (Stud Surf Sci Catal)
16. Sobalik Z, Tvaruzkova Z, Wichterlova B (1998) Skeletal T–O–T vibrations as a tool for characterization of divalent cation complexation in ferrierite. J Phys Chem B 102:1077–1085
17. van Saten RA, Vogel DL (1989) Lattice dynamics of zeolites. Adv Solid-State Chem 1:151–224
18. Jacobs WPJH, van Wolput JHMC, van Santen RA (1993) An in situ Fourier transform infrared study of zeolitic vibrations: Dehydration, deammoniation, and reammoniation of ion-exchanged Y zeolites. Zeolites 13:170–182
19. Wichterlova B, Dedecek J, Sobalik Z (2001) Single metal ions in host zeolite matrices. Structure-activity-selectivity-relationships. In: Centi G, Wichterlova B, Bell AT (eds) Catalysis by unique metal ion structures in solid matrices, from science to application, vol 13, NATO Science Series II, Mathematics, Physics and Chemistry, Kluwer, Dordrecht, pp 31–53
20. Brodskii LA, Zhadanov SP (1980) Application of far-infrared spectroscopy for a study of cation positions in zeolites. In: Rees LV (ed) Proceedings of fifth international zeolite conference, Naples, Italy, Heyden, London, pp 234–241
21. Baker MD, Ozin GA, Godber J (1985) Direct probe Fourier transform far-infrared spectroscopy of metal atoms, metal ions, and metal clusters in zeolites. Catal Rev Sci Eng 27:591–651
22. Godber J, Baker MD, Ozin GA (1989) Far-IR spectroscopy of alkali-metal and alkaline-earth cations in faujasite zeolites. J Phys Chem 93:1409–1421
23. Bremard C, Le Maire M (1993) Low-frequency Raman spectra of dehydrated faujasitic zeolites. J Phys Chem 97:9695–9702
24. Yang C, Xu Q, Hu C (2000) Boronation and galliation of zeolites β in an alkaline medium. Mater Chem Phys 63:55–66

25. Uytterhoeven JB, Christner JC, Hall WK (1965) Studies of the hydrogen held by solids. VIII. The Decationated Zeolites. J Phys Chem 69:2117–2126
26. Roessner F, Steinberg KH, Rechenburg S (1987) Infrared spectroscopic study on the location of the hydroxyl groups in erionite. Zeolites 7:488–489
27. Stoecker M, Ernst S, Karge HG et al (1990) ^{29}Si-MAS NMR studies of hydrothermal dealumination of zeolite ZSM-20. Acta Chem Scand 44:519–521
28. Dzwigaj S, Briend M, Shikholeslami A et al (1990) The acidic properties of SAPO-37 compared to faujasites and SAPO-5. Zeolites 10:157–162
29. Hatada K, Ono Y, Ushiki Y (1979) Infrared spectroscopic study of ZSM-5 zeolites. Z Phys Chem NF 117:37–42
30. Barthomeuf D (1980) Structural and physicochemical aspects of acidic catalysis in zeolites. Stud Surf Sci Catal 5:55 (In: Imelik B, Naccache C, Ben Taarit Y et al (eds) Catalysis by zeolites)
31. Jacobs PA, Leuven KU (1982) Acid zeolites: An attempt to develop unifying concepts (P. H. Emmett award address, 1981). Catal Rev Sci Eng 24:415–440
32. Chu CT-W, Chang CD (1985) Isomorphous substitution in zeolite frameworks. 1. Acidity of surface hydroxyls in [B]-, [Fe]-, [Ga]-, and [Al]-ZSM-5. J Phys Chem 89:1569–1571
33. Ward JW (1976) Infrared studies of zeolite surfaces and surface reactions. In: Rabo JA (ed) Zeolite chemistry and catalysis, ACS Monograph 171, American Chemical Society, Washington, DC, pp 118–284
34. Kubelkova L, Beran S, Lercher JA (1989) Determination of proton affinity of zeolites and zeolite-like solids by low-temperature adsorption of carbon monoxide. Zeolites 9:539–543
35. Egerton TA, Hardin AH (1975) The application of Raman spectroscopy to surface chemical studies. Catal Rev Sci Eng 11:71–116
36. Lercher JA, Grundling C, Eder-Mirth G (1996) Infrared studies of the surface acidity of oxides and zeolites using adsorbed probe molecules. Catal Today 27:353–376
37. Knozinger H (1997) Infrared spectroscopy for the characterization of surface acidity and basicity. In: Ertl G, Knozinger H, Weitkamp I (eds) Handbook of heterogenous catalysis. Wiley, Weinheim, pp 707–732
38. Ungureanu A, Hoang TV, Trong On D et al (2005) An investigation of the acid properties of UL-ZSM-5 by FTIR of adsorbed 2, 6-ditertbutylpyridine and aromatic transalkylation test reaction. Appl Catal A: General 294:92–105
39. Sachsenröder H, Brunner E, Koch M et al (1996) Low-temperature ^{1}H MAS NMR and FTIR investigations on the interaction of tetrachloroethylene with surface hydroxyl groups in H-zeolites. Micropor Mater 6:341–347
40. Parry EP (1963) An infrared study of pyridine adsorbed on acidic solids. Characterization of surface acidity. J Catal 2:371–379
41. Kline CH, Turkevich J (1944) The vibrational spectrum of pyridine and the thermodynamic properties of pyridine vapors. J Chem Phys 12:300–309
42. Mortier WJ (1978) Zeolite electronegativity related to physicochemical properties. J Catal 55:138–145
43. Jacobs PA, Mortier WJ, Uytterhoeven JB (1978) Properties of zeolites in relation to their electronegativity: Acidity, carboniogenic activity and strength of interaction in transition metal complexes. J Inorg Nucl Chem 40:1919–1923
44. Sanderson RT (1983) Polar covalence. Academic Press, New York
45. Rabo JA, Gajda GJ (1990) Acid function in zeolites: recent progress. NATO ASI Ser 221:273 (In: Barthomeuf D, Derouane E G, Höelderich W (eds) Guidance for mastering the properties of molecular sieves)
46. Rabo JA, Gajda GJ (1989–90) Acid function in zeolites: Recent progress. Catal Rev Sci Eng 31:385–430
47. Man AJM, Beest BWH, Leslie M et al (1990) Lattice dynamics of zeolitic silica polymorphs, J Phys Chem 94: 2524

48. van Santen RA, de Man AJM, Kramer GJ (1991) Chemical bonding in zeolites. In: Derouane EG, Lemos F, Naccache C et al (eds) Zeolite microporous solids: Synthesis, structure, and reactivity, vol. 352. Kluwer, Dordrecht, p 493 (NATO ASI Ser C)
49. Datka J, Bozar M, Rymarowicz P (1988) Heterogeneity of OH groups in NaH – ZSM-5 zeolite studied by infrared spectroscopy. J Catal 114:368–376
50. Jacobs PA, Martens JA, Wetkamp J et al (1981) Shape-selectivity changes in high-silica zeolitas. Faraday Disc Chem Soc 72:353–369
51. Liu X, Truitt RE, Hodge GD (1998) DRIFTS studies of surface properties of steamed FCC catalysts. J Catal 176:52–60
52. Hathaway PE, Davis ME (1989) Base catalysis by alkali-modified zeolites: I. Catalytic activity. J Catal 116:263–278
53. Kim JC, Li H-X, Chen C-Y et al (1994) Base catalysis by intrazeolitic cesium oxides. Micropor Mater 2:413–423
54. Tsuji H, Yagi F, Hattori H et al (1993) Characterization of basic sites on fine particles of alkali and alkaline earth metal oxides in zeolites. In: Guczi L, Solymosi F, Tetenyi P (eds) New frontier in catalysis, Parts A–C. Elsevier, Amsterdam, vol 75, p 1171
55. Tsuji H, Yagi F, Hattori H (1991) Basic sites on alkali ion-added zeolite. Chem Lett No. 11, 1881–1884
56. Rodriguez L, Cambon H, Brunel D et al (1993) Post-synthetic improvement of the basic character of caesium exchanged X and Y zeolites by occluded caesium oxides. Applications in condensation reactions. In: Guisnet M, Barbier J, Barrault J et al (eds) Heterogeneous catalysis and fine chemicals III, vol 78. Elsevier, Amsterdam, p 623 (Stud Surf Sci Catal)
57. Rodriguez L, Cambon H, Brunel D et al (1998) Activity in the Knoevenagel condensation of encapsulated basic cesium species in faujasite CsNaX or CsNaY. J Mol Catal A: Chem 130:195–202
58. Lasperas M, Cambon H, Brunel D et al (1996) Cesium oxide encapsulation in faujasite zeolites effect of framework composition on the nature and basicity of intrazeolitic species. Micropor Mater 7:61–72
59. Thomas JK (1993) Physical aspects of photochemistry and radiation chemistry of molecules adsorbed on silica, γ-alumina, zeolites, and clays. Chem Rev 93:301–320
60. Edwards PP, Anderson PA, Thomas JM (1996) Dissolved alkali metals in zeolites. Acc Chem Res 29:23–29
61. Barthomeuf D (1991) Acidity and basicity in zeolites. Stud Surf Sci Catal 65:157–170 (In: Öhlmann G, Pfeifer H, Fricke R (eds) Catalysis and adsorption by zeolites)
62. Barthomeuf D (1996) Basic zeolites: Characterization and uses in adsorption and catalysis. Catal Rev Sci Eng 38:521–612
63. Davis RJ (2003) New perspectives on basic zeolites as catalysts and catalyst supports. J Catal 216:396–405
64. Jacobs PA, van Cauwelaert FH, Vansant EF et al (1973) Surface probing of synthetic faujasites by adsorption of carbon dioxide. Part 1. Infra-red study of carbon dioxide adsorbed on Na-Ca-Y and Na-Mg-Y zeolites. J Chem Soc Faraday Trans I 69:1056–1068
65. Jacobs PA, van Cauwelaert FH, Vansant EF (1973) Surface probing of synthetic faujasites by adsorption of carbon dioxide. Part 2. Infra-red study of carbon dioxide adsorbed on X zeolites exchanged with mono- and bi-valent ions. J Chem Soc Faraday Trans I 69:2130–2139
66. Yagi F, Tsuji H, Hattori H (1997) IR and TPD (temperature-programmed desorption) studies of carbon dioxide on basic site active for 1-butene isomerization on alkali-added zeolite X. Micropor Mater 9:237–245
67. Lavalley JC (1996) Infrared spectrometric studies of the surface basicity of metal oxides and zeolites using adsorbed probe molecules. Catal Today 27:377–401
68. Huang M, Kaliaguine S (1992) Zeolite basicity characterized by pyrrole chemisorption: An infrared study. J Chem Soc Faraday Trans 88:751–758
69. Murphy D, Massiani P, Franck R et al (1996) Basic site heterogeneity and location in alkali cation exchanged EMT zeolite. An IR study using adsorbed pyrrole. J Phys Chem 100: 6731–6738

70. Gordymova TA, Davydov AA (1983) Infrared spectra of chloroform adsorbed on γ-Al2O3 and identification of basic surface sites in oxides. React Kinet Catal Lett 23:233–238
71. Uarova EB, Kustov LM, Kazansky VB (1995) Basicity of zeolites: IR-spectroscopic study using adsorbed molecular probes. Stud Surf Sci Catal 94:254 (In: Beyer HK, Karge HG, Kiricsi I et al (eds) Catalysis by microporous materials)
72. Xie J, Huang M, Kaliaguine S (1994) Base and acid sites in alkaline earth cation-exchanged X. Catal Lett 29:281–291
73. de Mallmann A, Barthomeuf D (1988) Change in benzene adsorption with acidobasicity of (Cs, Na) X zeolites studied by i.r. spectroscopy. Zeolites 8:292–301
74. Gramlich V, Meier WM (1971) The crystal structure of hydrated NaA: A detailed refinement of a pseudosymmetric zeolite structure. Zeit Krist 133:134–149
75. Prasad PSR, Prasad KS, Murthy SR (2005) Dehydration of natural stilbite: An in situ FTIR study. Amer Mineralogist 90:1636–1640
76. Gottardi G, Galli E (1985) Natural zeolites. Springer, Berlin
77. Cruciani G, Artioli G, Gualtieri A et al (1997) Dehydration dynamics of stilbite using synchrotron X-ray powder diffraction. Amer Mineralogist 82:729–739
78. Alberti A, Cariati F, Erre L et al (1983) Spectroscopic investigation on the presence of OH in natural barrerite and in its collapsed phases. Phys Chem Minerals 9:189–191
79. Prasad PSR, Sarma LP (2004) A near-infrared spectroscopic study of hydroxyl in natural chondrodite. Amer Mineralogist 89:1056–1060
80. Hanke W, Moller K (1984) Near-infrared study of the dealumination and water desorption from zeolites. Zeolites 4:244–250
81. Cohen de Lara E, Delaval Y (1978) Infrared spectra of nitrogen adsorbed in NaA zeolite. Experimental determination of electrostatic field in the cavities from induced band intensity and comparison with theoretical results. J Chem Soc Faraday Trans 2 74:790–797
82. Cohen de Lara E, Kahn R, Seloudoux R (1985) Effect of an electric field on a methane molecule. I. Infrared analysis of methane (CH4–CD4) adsorbed in NaA zeolite in the temperature range 150–20 K. J Chem Phys 83:2646–2652
83. Busca G, Lorenzelli V, Ramis G et al (1992) FT-IR spectra of ethylene molecularly adsorbed on metal oxides. J Mol Struct 267:315–329
84. Bordiga S, Ricchiardy G, Spoto G et al (1993) Acetylene, methylacetylene and ethylacetylene polymerization on H-ZSM5: A spectroscopic study. J Chem Soc Faraday Trans 89:1843–1855
85. Zecchina A, Buzzoni R, Bordiga S (1995) Host-guest interactions in zeolite cavities. In: Bonneviot L, Kaliaguine S (eds) Zeolites: A refined toll for designing catalytic sites. Elsevier, Amsterdam
86. Kiselev AV, Lygin VI (1975) Infrared spectra of surface compounds. Nauka, Moscow
87. Primet M, Garbowski E, Mathieu M et al (1980) Spectroscopic studies of benzene hydrogenation on platinum-loaded zeolites. Part 1. Benzene adsorption on supports. J Chem Soc Faraday Trans I 76:1942–1952
88. Angell CL, Howell MV (1968) Infrared spectroscopic investigations of zeolites and adsorbed molecules: III. Aromatic hydrocarbons. J Colloid Interface Sci 28:279–287
89. Carter JL, Yates DJC, Luccesi PJ et al (1966) The adsorption of ethylene on a series of near-faujasite zeolites studied by infrared spectroscopy and calorimetry. J Phys Chem 70:1126–1132
90. Liu X, Lampert JK, Arendarskiia DA (2001) FT-IR spectroscopic studies of hydrocarbon trapping in Ag+-ZSM-5 for gasoline engines under cold-start conditions. Appl Catal B Environ 35:125–136
91. Huang YY (1980) Ethylene complexes in copper (I) and silver (I) Y zeolitas. J Catal 61:461–476
92. Zakharieva-Pencheva O, Forster H, Tuzneva M (1994) Normal coordinate analysis of deutero propenes in the free and zeolite-adsorbed state. Spectrochim Acta Part A Mol Spec 50:19–28

93. Zakharieva-Pencheva O, Forster H (1991) Normal coordinate analysis of molecules adsorbed on zeolite surfaces. Part III. cis- and trans-but-2-ene in the gas phase and adsorbed in zeolites A. Vibrational Spectrosc 2:227–238
94. Zakharieva-Pencheva O, Forster H (1992) Quantum chemical studies on the sorption state of hydrocarbons in zeolites. SCC-Xα calculations on propene adsorbed to a calcium occupied six-ring cluster. J Mol Struct: THEOCHEM 258:209–216
95. Kirichi I, Tasi G, Fejes P et al (1989) Adsorption of propene in zeolites possessing sites of different adsorption energies. J Mol Catal 51:341–346
96. Kubes GJ (1988) Molecular hydrogen complexes: coordination of a σ bond to transition metals. Acc Chem Res 21:120–128
97. Grabtree RH (1990) Dihydrogen complexes: Some structural and chemical studies. Acc Chem Res 23:95–101
98. Heinekey DM, Ordham WJ (1993) Coordination chemistry of dihydrogen. Chem Rev 93:913–926
99. Kazansky VB, Pidko EA (2005) A new insight in the unusual adsorption properties of Cu^+ cations in Cu-ZSM-5 zeolite. Catal Today 110:281–293
100. Huang Y, Leech JH, Wang H (2003) An FT-Raman spectroscopic study of the conformational behavior of trans-1,4-dichlorocyclohexane adsorbed in zeolites. J Phys Chem B 107:7632–7639
101. Huang Y, Leech JH (2003) An FT-Raman spectroscopic study of the conformational properties of chlorocyclohexane in zeolites. J Phys Chem B 107:7647–7653
102. Weckhuysen BM (ed) (2004) In-situ spectroscopy of catalysts. American Scientific Publishers, USA
103. Weckhuysen BM (ed) (2004) In-situ spectroscopy of catalysts. In-situ spectroscopy of catalysts. American Scientific Publishers, USA, pp 1–11
104. Ivanov P, Papp H (2001) In situ FT-IR study on the reaction path of skeletal isomerization of n-butene over different zeolites. Appl Surf Sci 179:234–239
105. Mosqueda-Jimenez BI, Jentys A, Seshan K et al (2003) On the surface reactions during NO reduction with propene and propane on Ni-exchanged mordenite. Appl Catal B Environ 46:189–202
106. Traa Y, Burger B, Weitkamp J (1999) Zeolite-based materials for the selective catalytic reduction of NO_x with hydrocarbons. Micropor Mesopor Mater 30:3–41
107. Bando KK, Soga K, Kunimori K et al (1998) Effect of Li additive on CO_2 hydrogenation reactivity of zeolite supported Rh catalysts. Appl Catal A General 175:67–81
108. Wei Y, Zhang D, Liu Z et al (2006) Highly efficient catalytic conversion of chloromethane to light olefins over HSAPO-34 as studied by catalytic testing and in situ FTIR. J Catal 238:46–57
109. Yeom YH, Frei H (2002) Mechanistic study of CH_3OH+O_2 photoredox reaction in a $FeAlPO_4$ sieve by time-resolved FT-IR spectroscopy. J Phys Chem A 106:3350–3355
110. Waugh KC (1992) Methanol synthesis. Catal Today 15:51–75
111. Klier K (1982) Methanol synthesis. Adv Catal 31:243
112. Thibault-Starzyk F, Vimont A, Fernandez C et al (2002) 2D correlation IR spectroscopy of xylene isomerisation on H-MFI zeolite. Chem Commun 1003
113. Chenevarin S, Thibault-Starzyk F (2004) Two-dimensional IR pressure-jump spectroscopy of adsorbed species for zeolites. Angew Chem 9:1175–1178
114. Bonn M, Bakker HJ, Domen K et al (1998) Dynamical studies of zeolitic protons and adsorbates by picosecond infrared spectroscopy. Catal Rev Sci Eng 40:127–173
115. Bonn M, van Santen RA, Lercher JA et al (1997) Picosecond infrared activation of methanol in acid zeolites. Chem Phys Lett 278:213–219
116. Tian H, Ross EI, Wachs IE (2006) Quantitative determination of the speciation of surface vanadium oxides and their catalytic activity. J Phys Chem B 110:9593–9600
117. Banares MA (2005) Operando methodology: Combination of in situ spectroscopy and simultaneous activity measurements under catalytic reaction conditions. Catal Today 100:71–77

Chapter 6
Computational Approach in Zeolite Science

Evgeny A. Pidko and Rutger A. van Santen

Abstract This chapter presents an overview of different computational methods and their application to various fields of zeolite chemistry. We will discuss static lattice methods based on interatomic potentials to predict zeolite structures and topologies, Monte Carlo simulations for the investigation of adsorption phenomena, molecular dynamics technique to model diffusion processes in micropores and electronic structure calculations to study chemical reactivity of zeolitic materials. Various methodologies will be illustrated by the state of the art examples from recent literature.

6.1 Introduction

Computational methods are nowadays widely and extensively applied in the chemical, physical, biomedical, and engineering sciences in assisting the interpretation of experimental data and increasingly in predicting the properties and behavior of matter at the atomic level. They have a long and successful history of application in chemistry and material science for modeling structural and dynamic properties of various compounds, and for understanding chemical reactivity, in which they are playing an increasingly important role. This chapter aims to summarize and illustrate the capabilities of different computational methods and techniques as applied to investigations of zeolites and their properties.

Computational simulation techniques can be divided into two very broad categories. The first is based on the use of interatomic potentials (force fields). These methods usually do not consider explicitly any electron in the system. They are usually empirical and no attempt to solve the Schrödinger equation is made. Instead

E.A. Pidko (✉)
Schuit Institute of Catalysis, Eindhoven University of Technology, NL-5600MB, P. O. Box. 513, Eindhoven, The Netherlands
e-mail: e.a.pidko@tue.nl

A.W. Chester and E.G. Derouane (eds.), *Zeolite Characterization and Catalysis*, DOI 10.1007/978-1-4020-9678-5_6, the system of interest is described with functions (normally analytical), which express its energy as a function of nuclear coordinates. These are then used to calculate structures and energies by means of minimization methods, to calculate ensemble averages using Monte Carlo (MC) simulations, or to model dynamic processes (such as molecular diffusion) via molecular dynamics simulations using classical Newton's law of motion. The first three sections of this chapter describe the application of such methods to modeling structures, molecular adsorption, and diffusion in zeolites.

The second category includes quantum chemical methods based on the calculation of the electronic structure of the system via solution of the Schrödinger equation at some level of approximation. Such methods are the most important for processes that depend on bond breaking or making, which include, of course, catalytic reactions, or for studying the chemical bonding itself. Hartree Fock (HF), Density functional theory (DFT) and post-HF ab initio approaches have been used in modeling zeolites, although DFT methods have predominated in recent applications. The application of quantum chemical methods to zeolite sciences is described in the last section of this chapter.

6.2 Prediction of Structure (Static Lattice Methods)

This section reviews the application of static lattice methods based on interatomic potentials to model zeolite structures as well as to study energetics and stabilities. Such methods can be used to refine approximate models and to predict new structures of zeolites. They remain the most effective and economical approach for modeling structure. Moreover, using quantum chemical methods one can explore and, if necessary, refine the structural models obtained using static lattice methods. Below we will summarize the standard methodologies involved in these calculations and consider their application to modeling structures and energetics as well as to the prediction of new microporous materials.

6.2.1 Methodology

Static lattice methods are based on calculation of an energy term (lattice, surface, defect energy) which is then minimized with respect to structural variables such as positions of nuclei or cell parameters. These methods have been reviewed several times in older and more recent literature [1–4]. To calculate the lattice energy, we need to employ all interatomic interactions, which take place in the material concerned. The overall energy can be thus represented as a sum, first of the electrostatic energies (E_{ELEC}) resulting from interactions of charges on the atoms. These are long range and must be, in principle, summed to infinity in the accurate treatment. On the other hand, the second, the non-Coulomb terms (E_{SR}), which include attractive forces due to dispersion, hydrogen- and covalent bonding as well as Pauli or overlap repulsion are the short-range interactions and, hence, may be safely considered only within a "cut-off" which is typically 15–20Å. Thus, the lattice energy, E_{LAT}, can be written as:

$$E_{LAT} = E_{ELEC} + E_{SR}, \tag{6.1}$$

where the long-range term (E_{ELEC}) is calculated using an Ewald summation of the electrostatic potentials due to point charges, and the energy of short-range interactions (E_{SR}) is represented by:

$$E_{SR} = \frac{1}{2}\sum_{ij} V_{ij}(r_{ij}), \quad r_{ij} < r_{cut}, \tag{6.2}$$

where V_{ij} is the interatomic potential describing the short-range interaction energy between a pair of atoms i and j. r_{ij} is the distance between these atoms. The summation is taken within the cut-off radius r_{cut}. One notes that the interatomic potentials are the crucial input to static lattice calculations, because they represent the energy of the system as a function of atomic coordinates, and therefore, define the accuracy of the simulation.

The potential used depends on the nature of the bonding in the system modeled. For ionic solids usually the Born model is used. In this case the solid is considered as a collection of ions, to which formal or partial charges can be assigned, interacting via short-range potentials which are described using the "Buckinham" (6.3) or "Lenard-Jones" (6.4) potentials:

$$V_{ij} = A_{ij}e^{-r_{ij}/\rho_{ij}} - C_{ij}r_{ij}^{-6}, \tag{6.3}$$

$$V_{ij} = B_{ij}r_{ij}^{-12} - D_{ij}r_{ij}^{-6}, \tag{6.4}$$

where A_{ij}, C_{ij}, ρ_{ij}, and B_{ij}, D_{ij} are the parameters characteristic for the particular short-range interaction. To study zeolites it is essential to introduce in addition the "bond-bending" term:

$$V(\theta) = \frac{1}{2}k_{bend}(\theta - \theta_0)^2, \tag{6.5}$$

where θ is an O–Si–O bond angle, θ_0 is the tetrahedral angle and k_{bend} is the force constant that is a measure of how difficult it is to change the bond angle.

The approximations for the potentials used so far give rise to the rigid-ion model. One can also include ionic polarizabilities, as is done in the shell model originally formulated by Dick and Overhauser [5]. According to this model the ion is thought to consist of a core, in which all the mass of the ion is concentrated, and a massless shell interconnected by a harmonic spring. Short-range interactions thus act between these shells. The displacement of the shell relative to the core results in the formation of a dipole. Therefore, the resulting model includes the necessary coupling between polarizability and short-range repulsion.

On the other hand when the bonding within the system is mainly covalent, different concept should be considered. In this case simple analytic functions (e.g., bond harmonic or Morse) are used for the description of the interaction between chemically bonded atoms. Bond-bending, bond-torsion and non-bonded terms (including both electrostatic and short-range) are also included. Such potentials are usually termed "molecular mechanics" potentials.

When the type of interatomic potential is chosen, one must specify the variable parameters. This can be done either empirically by fitting the parameters to the known experimental parameters (structural, elastic, dielectric, thermodynamic and lattice dynamical properties), or using nonempirical methods by calculation of the interaction within a cluster or periodic array of atoms by ab initio or DFT methods. An interatomic potential function is then fitted to the resulting potential energy surface.

There has been a long debate over the nature of the bonding — ionic versus covalent — in silica and silicates. Nevertheless, the general conclusion is that the bonding in these systems is intermediate in nature, showing characteristics of both covalence and ionicity. Thus, both Born-type and molecular mechanics potentials have been developed for these systems.

Within the Born model, the simple rigid ion pair potentials are the most widely used both for dynamical and static lattice models. On the basis of ab initio calculations Van Beest et al. [6] developed a highly successful parameterization. Partial charges on Si and O atoms and simple pairwise short-range potentials for $Si \cdots O$ and $O \cdots O$ interactions were used in this model. A shell model developed by Sanders et al. [7] included also a bond-bending term. This model was parameterized empirically, while shell model potentials derived by Purton et al. [8] were based on the results of quantum chemical computations.

In addition, there are several "molecular mechanics" potentials available for zeolites. The most widely used are those based on the "cvff" models developed by BIOSYM Inc. For instance the cff91_zeo potential [9] is one of the most widely and successfully used.

In addition to pure static lattice methods, a combined, so-called quantum mechanics/molecular mechanics (QM/MM) is widely applied for studying local structure of zeolites. As it follows from the name, this method represents a combination of the static lattice method, which is used for the description of a larger part of zeolite or even periodically repeated zeolite unit cell, whereas the structure of a smaller part of the system (zeolitic cation site or catalytic active site) is refined using the quantum chemical methods (usually DFT). This approach will be discussed in more detail in Sect. 6.4 of this chapter.

6.2.2 Applications

One of the main areas of application of static lattice methods is simulation of structures and estimation of stabilities of different zeolite structures. As for every force field method, its accuracy is mainly determined by the interaction potential used. When the

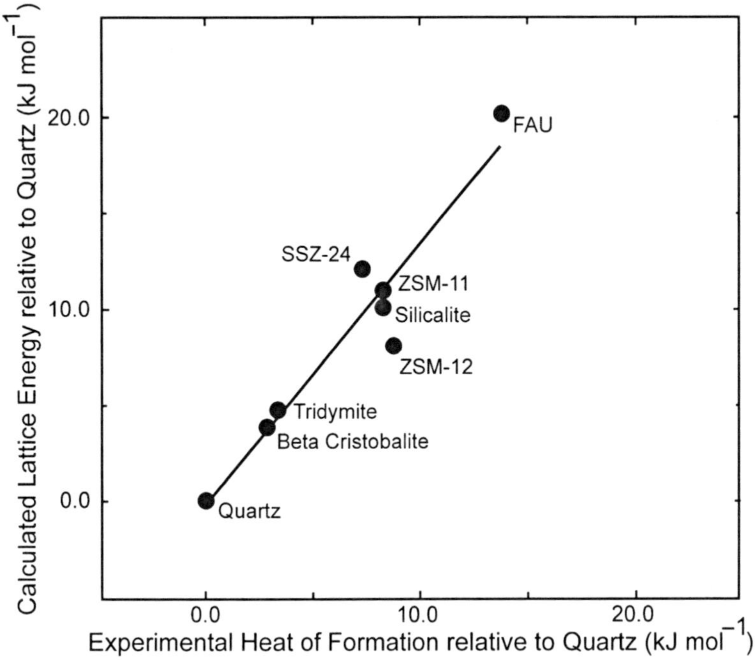

Fig. 6.1 Correlation of the experimental heat of formation with the calculated lattice energy relative to quartz (Reprinted with permission from [12]. © 1994 American Chemical Society)

potential method is appropriate for the system, the results obtained are in a quantitative agreement with the experimental values.

It is known that microporous materials are metastable as compared to the corresponding dense structures. It was established using calorimetric data that the enthalpy difference between pure silica zeolites and quarts is usually in the range 10–20 kJ mol^{-1} [10, 11]. Henson at al. [12] used shell-model potentials of Sanders et al. [7] to calculate energetics and structures of different silica structures including various zeolites. The results obtained are in an excellent agreement with experimental values (Fig. 6.1).

One of the major problems in zeolite science is a lack of knowledge about the distribution of aluminum ions in the framework of high-silica zeolites. Static lattice simulations can be successfully applied to address this question. Ruiz-Salvador et al. [13] used a combination of a Monte Carlo procedure with the minimization of energy using static lattice methods in order to predict Al distribution in Na-clinoptilonote. The Al distribution and cation location in heulandite type zeolites was explored by Channon et al. [14]. Using the static lattice method framework aluminum distribution and proton siting in the synthetic tridirectional large-pore ITQ-7 zeoltie was studied by Sastre et al. [15]. The experimental infrared frequencies of the H-ITQ-7 bringing acid hydroxyl groups were well reproduced by the calculations. All these examples show that the method considered may be succesfully used for the solution of one of the long standing problems in zeolite science — aluminum distribution.

In addition to modeling crystal structures, several successful studies have been reported for local structures of zeolites. For instance, the local structure around framework Fe^{3+} in FeZSM-5 zeolite was investigated by Lewis et al. [16].

Another very interesting and important application of this method is a prediction of new structures of zeolites. Topologically, the zeolite framework can be considered as four connected nets, where each vertex is connected to its four closest neighbors. We are able to imagine an unlimited number of possible zeolitic structures. However, only a limited number of those are of interest due to some important properties with even smaller portion being amenable to synthesis. As it has been mentioned above, zeolites are metastable compounds, and the factors, which govern the synthesis of these materials are not well understood yet. In spite of this, the simulated static lattice energy can be used as an initial criterion of feasibility of the hypothetic microporous structure.

6.3 Molecular Adsorption (Monte Carlo Simulations)

Molecular adsorption is basic to the application of many microporous materials, and therefore, it is very important to know the number and behavior of molecules adsorbed at certain conditions. Many reviews and monographs have been written on this problem [17, 18]. However, adsorption is not well understood yet. In this section the adsorption phenomena in zeolites and application of MC simulations to study them are discussed.

6.3.1 Methodology

The importance of the MC method is that it provides us with a tool for calculating macroscopic quantities (energy, temperature, pressure, etc.) of a system, for which the interatomic potentials are known. Similar to the above-discussed static lattice methods, the reliability of these calculated properties is, therefore, determined by the reliability of the intermolecular potential.

The MC algorithm is based on the proposals of Metropolis et al. [19] and can be characterized by the following three steps: at first a new configuration (random new positions and new orientations) are given to the adsorbed molecules and the energy of this new system is calculated using the methods described in Sect. 2.1. At the next step the difference between energies of the initial and the new structures is calculated $\delta E = E_{\text{new}} - E_{\text{old}}$. Then this new configuration is accepted with a probability proportional to the Boltzmann weight factor at temperature T ($e^{\delta E/kT}$). These steps are repeated in order to obtain a chain of configurations Γ_i ($i = 1,2,\ldots,N$).

Let us assume that we have a system of N particles in a fixed volume V and at a constant temperature T. From statistical mechanics it can be derived that the thermodynamic average of a quantity A can be written as [20]:

$$\langle A \rangle = \frac{1}{Z} \int A(\Gamma) \cdot e^{-E(\Gamma)/kT} \mathrm{d}\Gamma, \tag{6.6}$$

where Γ is the configuration of all particles, hence the integration is taken over all possible configurations; $E(\Gamma)$ and $A(\Gamma)$ are corresponding values of energy and A at configuration Γ; and Z is partition function:

$$Z = \int e^{-E(\Gamma)/kT} \mathrm{d}\Gamma. \tag{6.7}$$

The purpose of the MC method is to calculate the integral as given by (6.6) numerically. One method to achieve this is to generate a set of randomly chosen configurations Γ_i ($i = 1, 2, \ldots, M$). Following this procedure the average of A can be approximated by:

$$\langle A \rangle = \frac{\sum_{i=1}^{M} A(\Gamma_i) \cdot e^{-E(\Gamma_i)/kT}}{\sum_{i=1}^{M} e^{-E(\Gamma_i)/kT}}. \tag{6.8}$$

However, the statistics of this method will be very poor, because most values of Γ_i will be chosen in a region, where $e^{-E(\Gamma_i)/kT}$ is low [20].

To circumvent this problem, Metropolis et al. [19] proposed the method of importance sampling. In this method, the configurations will not be chosen at random but will be selected with a probability $P(\Gamma)$. The average of quantity A for this case can be written as [19, 20]:

$$\langle A \rangle = \frac{\sum_{i=1}^{M} A(\Gamma_i) \cdot P^{-1}(\Gamma_i) \cdot e^{-E(\Gamma_i)/kT}}{\sum_{i=1}^{M} P^{-1}(\Gamma_i) \cdot e^{-E(\Gamma_i)/kT}}. \tag{6.9}$$

If we choose for the distribution of configurations in the chain the equilibrium distribution $P_{eq}(\Gamma)$:

$$P_{eq}(\Gamma) = e^{-E(\Gamma)/kT}, \tag{6.10}$$

the equation (6.9) can be rewritten as:

$$\langle A \rangle = \frac{1}{M} \sum_{i=1}^{M} A(\Gamma_i). \tag{6.11}$$

Generating a chain of configurations ($\Gamma_1 \ldots \Gamma_M$) with a certain distribution can formally be described with the theory of Markov processes [21]. A very important condition to ensure a statistical reliability is that in the course of the simulation it must be in principle possible to reach all accessible configurations.

The adsorption enthalpy directly follows from the averaging over all configurations, which are generated during the simulation (6.11):

$$\langle Q_s \rangle = \frac{1}{M} \sum_{i=1}^{M} A(\Gamma_i). \tag{6.12}$$

In MC simulations, zeolite crystals are allowed to exchange molecules with a reservoir of molecules at a fixed chemical potential in the Grand-Canonical ensemble. In this ensemble the temperature, volume, and chemical potential are constant. Due to equilibrium conditions the temperature and chemical potential of the gas phase inside the zeolite and the external reservoir must be equal. Thus, to determine the equilibrium concentration of molecules in zeolite, we need to know only T and μ of the gas in the reservoir.

It is noticeable that in an MC simulation one does not have to follow the diffusion path of the molecule and can therefore locate a molecule at a random position in zeolite. This technique works very well for small molecules such as carbon monoxide or methane, but has significant drawbacks when studying adsorption of larger molecules. For example, in an MC simulation of methane sorption, from 1,000 attempts to place a CH_4 molecule to a random position within a zeolite 999 attempts will be rejected due to the overlap of the methane molecule with the lattice of the zeolite. In the case of ethane approximately only one attempt from 1,000,000 will be successful. One can see that the strategy of randomly inserting molecules to a zeolite will not be effective for larger molecules. In order to circumvent this problem, recently a so-called configuration-bias Monte Carlo (CBMC) method was developed by Frenkel and Smit [20, 22, 23]. The principle idea of this technique is to grow an alkane chain, atom by atom, within a zeolite matrix instead of random inserting.

6.3.2 *Applications*

For small molecules such as water, methane, CO_2 or noble gases, conventional grand-canonical Monte Carlo (GCMC) simulations can be used to calculate adsorption isotherms in various zeolites [24, 25]. Fig. 6.2 illustrates a comparison of the experimental adsorption isotherms for carbon dioxide in faujasites with those calculated by means of the GCMC simulations [26]. One can see that the computer simulation provides results which very well agree with the experiment.

Knowledge of the adsorption of alkanes in zeolites is very important both from the practical and scientific point of view. As it has been mentioned above,

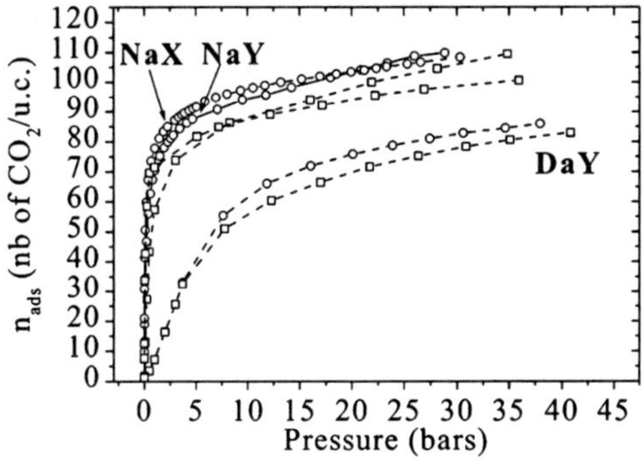

Fig. 6.2 Absolute isotherms for CO_2 on DAY, NaY and NaLSX at 300 K; (□) simulation; (○) experiment (Reprinted with permission from [26]. ©2005 American Chemical Society)

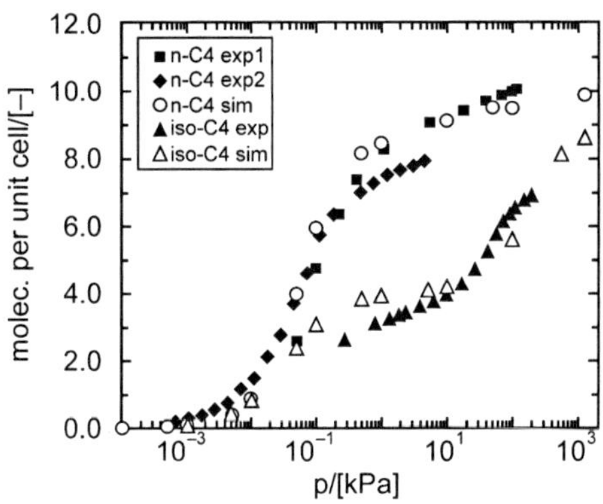

Fig. 6.3 Sorption isotherms for normal and isobutene at 300 K. Comparison of CBMC simulations with experiments. (Reprinted with permission from [27]. ©1998 American Chemical Society)

the GCMC technique fails in simulation of longer hydrocarbons, whereas the CBMC method is preferred. To illustrate the capability of the latter method, Fig. 6.3 shows a comparison of simulated and experimental adsorption isotherms of linear and branched alkanes in the silicalite [27]. The simulations give a nearly quantitative description of the experimental adsorption isotherms. For other alkanes in silicalite a similar agreement has been obtained [28]. The simulations reproduce all qualitative features observed in the experiments, and more important, provide a deeper insight into sorption behavior than available from the experiment. For example, the adsorption isotherms of branched alkanes in silicalite show a step for a given number of molecules per unit cell (Fig. 6.3), which is not observed for linear isomers. Using CBMC simulations it was shown that due to a preferential sitting of the branched molecules at the intersection of the linear and zig-zag channels, these adsorption sites are being occupied at first. Once all intersections are occupied, the next molecule has to adsorb in between two intersections. Since these sites are less favorable for the bulky branched molecules, this requires much higher pressure before they are occupied.

Usually experimental techniques to determine adsorption isotherms are based on measuring the weight increase of the zeolite due to the adsorption of molecules. For pure gases this directly relates to the number of molecules adsorbed, whereas for mixtures additional experiments are required. As a consequence the experimental data on the adsorption behavior of gas mixtures are scarce. Molecular simulation techniques, as a kind of "computer experiments," offer an attractive method for data acquisition of multicomponent adsorption. For instance, Du et al. [29] and Macedonia et al. [30] successfully simulated the adsorption of short alkanes (methane, ethane, propane, and butane) and their binary mixtures in silicalite using CBMC or by GCMC techniques.

It is noticeable that most simulation studies use a rigid zeolite structure. For molecules that do not have a tight fit in zeolite pores, this appears to be a good assumption. However, in the case of more rigid molecules, such as aromatic compounds, the fit can be very tight and adsorption of such molecules can even induce phase transitions. Clark and Snurr [31] have shown that the adsorption isotherms of benzene in silicalite are very sensitive to small changes in the zeolite structure. Therefore, this requires simulation with a model of a zeolite with accurate flexible lattice potentials.

Another very important application of the MC technique is simulation of the interaction between templates and zeolite framework. Typically, a combination of molecular dynamics (MD), MC and energy minimization techniques is used in this case. An MD trajectory for the template molecule in the gas phase is used to generate a library of sorbate conformations, each of which is then inserted randomly into the framework using an MC procedure. It was shown by Lewis et al. [32] that on the basis of these calculations it is possible to select templates for a given zeolite framework.

6.4 Diffusion (Molecular Dynamics Simulations)

Diffusion of adsorbed molecules in zeolites plays an important role in the use of this type of materials as adsorbents in separation processes and in shape-selective catalysis. Computational methods now are in a stage where diffusion phenomena can be treated at a high level of accuracy. In this section the application of classical MD and also kinetic MC (KMC) approaches for the study of diffusion in zeolites will be discussed.

6.4.1 Methodology

It has been shown above that classical MC simulations deal with systems, which are in equilibrium, and thus, static thermodynamic properties such as adsorption heats and average sitting behavior can be derived according to (6.11). In the MC there is no time scale involved and successive configurations are chosen at random. In contrast, MD explicitly simulate time evolution. Therefore, using MD simulations one can evaluate not only static thermodynamic properties, but also diffusivities and site residence times of molecules adsorbed in zeolitic pores. These dynamic properties are derived by generating a system trajectory followed by averaging over time. Thus, for any thermodynamic property A, the average value of A following from an MD simulation is given by:

$$\langle A \rangle = \frac{1}{t}\sum_{i=1}^{M} A \cdot \delta t_i, \tag{6.13}$$

where M is the number of MD sampling points, t is the total time of the simulation, δt_i is the time step length, and A is a value of property A at a specific time along the trajectory.

In MD, successive configurations of the system are generated by integration of classical Newton's law of motion. The result is a trajectory that specifies how the position and velocities of the particles in the system vary with time. Let us consider an ensemble of particles moving in a force field. The number of particles equals N, the force on particle i ($i = 1,\ldots,N$) at time t is given by $F_i(t)$, and the position of the particle at time t is represented by $r_i(t)$. A Taylor expansion of $r_i(t)$ around t results in:

$$r_i(t+\delta t) = r_i(t) + \left(\frac{\mathrm{d}r_i(t)}{\mathrm{d}t}\right)_t \delta t + \frac{1}{2}\left(\frac{\mathrm{d}^2 r_i(t)}{\mathrm{d}^2 t}\right)_t (\delta t)^2, \tag{6.14}$$

$$r_i(t-\delta t) = r_i(t) - \left(\frac{\mathrm{d}r_i(t)}{\mathrm{d}t}\right)_t \delta t + \frac{1}{2}\left(\frac{\mathrm{d}^2 r_i(t)}{\mathrm{d}^2 t}\right)_t (\delta t)^2. \tag{6.15}$$

Combination of (6.14) and (6.15) yields:

$$\begin{aligned} r_i(t+\delta t) &= 2r_i(t) - r_i(t-\delta t) + \left(\frac{\mathrm{d}^2 r_i(t)}{\mathrm{d}^2 t}\right)_t (\delta t)^2 \\ &= 2r_i(t) - r_i(t+\delta t) + \frac{F_i(t)}{m_i}(\delta t)^2, \end{aligned} \tag{6.16}$$

and

$$\frac{\mathrm{d}r_i(t)}{\mathrm{d}t} = \frac{r_i(t+\delta t) - r_i(t-\delta t)}{2\delta t} = v_i(t), \tag{6.17}$$

where m_i is the mass of particle i and $v_i(t)$ is its velocity at time t.

Equation (6.16) clearly shows that, given the forces on the particles, one is able to generate a trajectory in time without using any knowledge about the velocities of the particles. However, determination of the velocities might be useful for the evaluation of the particle ensemble temperature from the kinetic energy of the system:

$$E_{\mathrm{kin}}(t) = \frac{1}{2}\sum_{i=1}^{N} m_i v_i^2(t), \tag{6.18}$$

and hence,

$$T(t) = \frac{2Nk}{3}E_{\mathrm{kin}}(t) = \frac{Nk}{3}\sum_{i=1}^{N} m_i v_i^2(t), \tag{6.19}$$

where k is the Boltzman constant.

MD simulations rest on the (6.16)–(6.19). As follows from (6.19) the temperature of the particle ensemble is not constant but fluctuates around average temperature, and therefore, the kinetic energy of the system is also not a constant. However, it is noticeable that the total energy of the system is conserved during the simulation and is given by:

$$E_{\mathrm{tot}} = E_{\mathrm{pot}} + E_{\mathrm{kin}}, \tag{6.20}$$

where the potential energy (E_{pot}) is related to the force via

$$F = -\mathrm{d}E_{\mathrm{pot}}/\mathrm{d}r. \tag{6.21}$$

The potential energy at each time step can be calculated as a sum of all interaction potentials (Sect. 2) considered in the simulation (e.g., interaction with zeolite walls, adsorbate–adsorbate interactions, and for molecular species bond, bond-bending etc. potentials). Then using (6.21) one can get a force field for molecules migrating in the zeolite pores. This force field can be directly used in the MD algorithm. From this diffusivities are easily calculated using the Einstein relation [33]:

$$D = \frac{1}{6}\frac{\langle r^2\rangle}{t}, \tag{6.22}$$

where $\langle r\rangle^2$ is the mean square distance traveled by the adsorbate molecules during period t.

It has been shown in Sect. 2 that at every step of the classical MC simulation, molecules do not follow the diffusion paths and are located at random positions in zeolite. As a result, the trajectory of molecules and properties over the time, which are necessary to derive diffusivity, cannot be investigated. This problem is solved in the so-called KMC approach. In such simulations of diffusion, the pore space, through which molecules diffuse, is simplified to a lattice, consisting of a grid of coarse-grained adsorption sites, which are connected by bonds. Molecules occupy a certain fraction of the sites, and are assumed to hop from site to site, normally along bonds connecting neighboring sites. Their movement is described by the set of Markovian master equations:

$$dP_j/dt = \sum_i (W_{ji}P_i - W_{ij}P_j), \tag{6.23}$$

where P_j is the probability for a system to be in a certain configuration j and W_{ij} is the transition probability per unit time to move from configuration j to i.

Another very useful approach for simulation of diffusion in zeolites is transition-state theory (TST). When the sluggish, activated diffusion of tightly fitting or strongly adsorbed molecules in zeolites is considered, the motion can be accurately described by a succession of hops between the adsorption sites. Once these sites have been localized, knowledge of the jump or transition rates between them allows to calculate the diffusivity using KMC method or in the harmonic approximation by normal-mode analysis [34]. The thus obtained transition rate is then corrected for multi-step transitions or recrossings of the transition state. The corresponding dynamic correction factor is usually evaluated by short local MD simulations.

6.4.2 Applications

Since the first application of MD technique to guest molecules loaded into zeolite in 1986 [35], this subject has attracted growing interest. Recently, several comprehensive reviews [34, 36–38] on modeling of diffusion in zeolites have been published. Since these reviews cover virtually all applications of the MD, KMC and TST methods, below we will discuss very briefly the most common applications of these techniques.

Most MD calculations are performed to generate trajectories for studying adsorbate self-diffusion. In this case the simulated trajectories must be long enough to compute a statistically converged self-diffusion coefficient. Auerbach et al. [39] demonstrated that the diffusivity, below which the adsorbate moves too slow for MD, is around $5 \times 10^{-10} m^2 s^{-1}$. This value is higher than most measured diffusivities in cationic zeolites, explaining why MD studies usually focus on hydrocarbons in all-silica zeolite analogs or on diffusion of relatively small molecules such as CH_4. In these cases, when the diffusion is sufficiently rapid, MD provides quantitative information on dynamic behavior of pure gases as well as of multicomponent mixtures in zeolites.

However, for many systems the molecules diffuse very slowly in the zeolite pores due to "diffusion barriers" such as strong adsorption to some specific sites (exchangeable cations) or narrow windows between zeolite cages. As a consequence the total simulation time needed to compute a diffusion coefficient can become prohibitively large. In such cases diffusion can be considered a hopping process. If the molecule has sufficiently high kinetic energy, it may overcome the diffusion barrier and "jump" from one adsorption site to another. These hopping rates can be successfully computed using transition-state theory (TST).

Snurr, Bell, and Theodorou [40] applied harmonic TST to benzene diffusion in silicalite-1, assuming that both the C_6H_6 molecule and the zeolite lattice remain rigid upon diffusion. Their results underestimate experimental diffusivities by one to two orders of magnitude, which is most likely due to the assumption of rigid zeolite rather than due to the failure of the harmonic TST method. Subsequently, it was shown by Forester and Smith [41] that lattice flexibility indeed influences substantially the hopping rates. Diffusivities for benzene in silicalite-1 obtained from the flexible framework simulations ($3.36 \times 10^{-14} m^2 s^{-1}$ [41]) are in excellent accord with the experiment ($2.2 \times 10^{-14} m^2 s^{-1}$ [42]).

KMC simulations demand much less computer power but do not possess the predictive power of MD simulations. High efficiency of this method is usually achieved by several assumptions. This method is applied by coarse-graining the molecular motions, keeping only their diffusive character, while zeolite structure is approximated by lattice models. In contrast to the above-discussed methods based on the atomistic representation of adsorbate and zeolite, in this case the zeolite lattice is represented as a numerical grid. The grid points have important physical meaning for dynamics in zeolites, as shown schematically in Fig. 6.4.

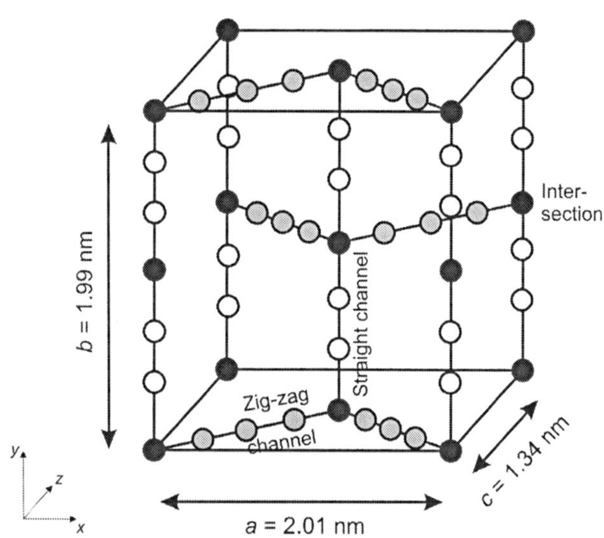

Fig. 6.4 Unit cell of MFI lattice topology with a total of 24 sites per unit cell. Sites are indicated as black circles (intersections), white circles (straight channel) and grey circles (zig-zag channel). Long-chain alkanes can only be located at the intersections. The jump frequencies along straight or zig-zag channels can be estimated using TST. (Reprinted with permission from [45]. © 2005 Wiley)

The KMC method models diffusion on a lattice as a random walk composed of uncorrelated, single-molecule jumps from one adsorption site (grid point) to another, providing a stochastic solution to the master (6.21) associated with the lattice model. The jump frequencies of molecules need to be supplied as inputs to the KMC simulations. These can be estimated from the TST theory [43] or MD simulations [34, 37, 44]. Recently Krishna and van Baten [45] have shown that the accuracy of the KMC simulations can be remarkably improved by a proper tuning in order to reproduce either MD or experimental diffusivities. As a result, KMC simulations can be used to probe the mixture diffusion characteristics with considerably less computer power as compared to that needed for the corresponding MD simulations.

6.5 Reactivity and Chemical Properties (Quantum Chemical Methods)

Most of the properties of zeolites such as chemical reactivity, bonding, properties of the exchangeable cations or different species encapsulated into the zeolite cages are directly related to the properties and behavior of electrons of the chemical system, and moreover, can be described only when the electron structure is taken into account. In all of the methods discussed above the properties of the system depend only on the positions of the atoms, which are either structureless (the Born model) or have a structure approximated using a very simplistic model ("shell model"). Obviously, such approaches cannot be applied to study processes that depend on bond breaking or making, i.e., reactivity, and for studying the chemical bonding itself. Here quantum mechanical methods have a pivotal role to play as the only means of directly addressing such questions and providing a framework for the interpretation of experimental data. In this section application and limitations of different modern methods based on electronic structure calculations as applied to study of either chemical reactivity or properties of different molecules loaded into zeolite will be discussed.

6.5.1 Methodology

The goal of quantum chemical methods is to predict the structure, energy and properties of a system containing many particles. In contrast to the methods discussed above here both electrons and nuclei are considered. The energy of the system is a direct function of the exact position of all atoms and the forces that act upon the electrons and the nuclei of each atom. In order to calculate the electronic states of the system quantum chemical methods attempt to solve Schrödinger's equation, $\hat{H}\Psi = E\Psi$, where Ψ is the wavefunction and E is the energy of the N-particle system. $\hat{H}$ is the Hamiltonian operator, which is comprised of the kinetic and potential energy operators acting on the overall wavefunction of the system. The exact solution for this equation can be found only for a very limited number of systems and thus, a number of simplifying approximations resulting in different quantum chemical methods are required to solve it for larger systems. Below we present a simple overview of the most important approximations and on the important limitations of the resulting quantum chemical methods as applied to the zeolite chemistry. More detail and in-depth discussion on the electronic structure calculations can be found in a number of very good references [46–51].

There are two important approximations to solve the Schrödinger equation for a multi-body system. The first is the Born-Oppenheimer approximation [52], which assumes uncoupling the electron motion from the nuclear motion since the mass of the latter species is much greater than the electron mass. The electronic wavefunction can then be solved separately for the nuclear one for a fixed set of nuclear positions. The second approximation is usually done in order to take into account repulsion between electrons. Since the electron-electron interaction cannot be directly solved, it was proposed to consider each electron moving in a field of the other electrons of the system. The solution then requires convergence of the electronic structure via an iterative scheme. This is known as the self-consisted field approximation [46–51].

The basic structure of nearly all quantum chemical codes is shown in Fig. 6.5, where there are two primary nested loops. The inner loop requires the convergence of the electronic structure for a fixed set of atoms, whereas the outer loop performs the geometry optimization in order to minimize the forces upon each atom and to achieve the geometry corresponding to the lowest energy of the system.

Electronic structure methods can be categorized as ab initio wavefunctions-based, ab initio density functional or semiempirical methods. All of them can be applied for the solution of different problems of zeolite chemistry. Wavefunction methods start with the Hartree-Fock (HF) solution and have well-prescribed methods that can be used to increase its accuracy. One of the deficiencies of the HF theory is that it does not treat dynamic electron correlation, which refers to the fact that the motion of electrons is correlated so as to avoid one another. The neglect of this effect can cause very serious errors in the calculated energies, geometries, vibrational, and other molecular properties.

There are numerous so-called post Hartree-Fock (post-HF) methods for treating correlated motion between electrons. One of the most widely used approaches is based on the definition of the correlation energy as a perturbation. In other words,

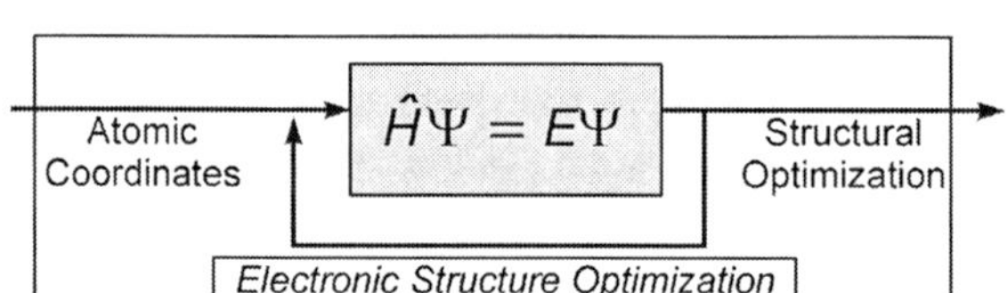

Fig. 6.5 The lowest level hierarchical structure for most quantum chemical algorithms

configurational interactions are treated as small perturbations to the Hamiltonian. Using this expansion the HF energy is equal to the sum of the zero and first order terms, whereas correlation energy appears only as a second order term. The second order Møller-Plesset perturbation theory (MP2) typically recovers 80–90% of the correlation energy, while MP4 provides a reliably accurate solution to most system.

Another approach to treat the electron correlation is the configuration interactions method (CI). The general solution strategy for it is to construct a trial wavefunction that is comprised of a linear combination of the ground-state wavefunction and excited-state wavefunctions. The trial wavefunction can include the exchange of 1, 2 or 3 electrons from the valence band into unoccupied orbitals; these are known as CI singles (CIS), CI doubles (CID) and CI triples and allow for single, single/double and single/double/triple excitations.

Coupled cluster (CC) methods differ from perturbation theory in that they include specific corrections to the wavefunction for a particular type to an infinite order. CC theory therefore must be truncated. The lowest level of truncation is usually at double excitations (CCSD) since the single excitations do not extend the HF solution. CC theory can improve the accuracy for thermochemical calculations to within 1 kcal mol^{-1}.

Despite using these methods a very high accuracy can be achieved, almost all of the post-Hartree-Fock methods are prohibitive for calculation of reliable models of zeolites due to very high computational costs. Only the computationally cheapest post-HF methods can be currently used for studying zeolites. Thus, the application of post-HF methods in zeolite chemistry is limited to MP2 method, which still can be applied to calculations of only rather small zeolite models. One notes however that when the resolution of identity (RI) approximation is used [53], the resulting RI-MP2 method can be applied for calculations of systems containing more than hundred of atoms.

A more attractive method is density functional theory (DFT). DFT is "ab initio" in the sense that it is derived from first principles and does not usually require adjustable parameters. These methods formally scale with increase in the number of basis functions (electrons) as N^3, and thus permit more realistic models compared to the higher-level post-HF methods, which usually scale as N^5 for MP2 and up to N^7 for MP4 and CCSD(T). On the other hand theoretical accuracy of DFT is not as high as the higher level ab initio wavefunction methods.

The DFT is attributed to work of Hohenberg and Kohn [54], who formally proved that the ground-state energy for a system is a unique functional of its electron density. Kohn and Sham [55] extended the theory to practice by showing how the energy can be portioned into kinetic energy of the motion of the electrons, potential energy of the nuclear-electron attraction, electron-electron repulsion, which involves with Coulomb as well as self interactions, and exchange correlation, which covers all other electron-electron interactions. The energy of an N-particle system can then be written as

$$E[\rho] = T[\rho] + U[\rho] + E_{XC}[\rho]. \tag{6.24}$$

Kohn and Sham demonstrated that the N-particle system can be written as a set of n-electron problems (similar to the molecular orbitals in wavefunction methods) that could be solved self-consistently in a manner that is similar to the SCF [55].

Although the DFT is in principle an exact approach, unfortunately, due to the fact that the exact expression for the electron density functional as well as for the exchange correlation energy are not known, lots of assumptions and approximations are usually made. The most basic one is local density approximation (LDA) which assumes that the exchange-correlation per electron is equivalent to that in a homogeneous electron gas, which has the same electron density at a specific point r. The LDA is obviously an oversimplification of the actual density distribution and usually leads to overestimation of calculated bond and binding energies.

The non-local gradient corrections to LDA functional improve the description of electron density. In this case the correlation and exchange energies are functionals of both the density and its gradient. The gradient corrections take on various different functionals such as P86 [56], B88 [57], PW91 [58], PBE [59], etc. However, the accuracy of those is typically less than what can be expected from high level ab initio methods.

One notes that HF theory provides a more exact match of the exchange energy for single determinant systems. Thus, numerous hybrid functionals have been developed, where the exchange functional is a linear combination of the exact HF exchange and the correlation (and exchange) calculated from LDA theory. The geometry and energetics calculated within this approach (B3LYP and B3PW91 [60], MPW1PW91 [61], PBE0 functional [62] etc.) are usually in a good agreement with the experimental results and with those obtained using the post-HF methods. On the other hand, hybrid functionals still fail in describing of chemical effects mainly based on electron-electron correlation such as dispersion and other weak interactions [63, 64].

As it was mentioned above, the energy in the DFT methods is formally a function of the electron density. However, in practice the density of the system $\rho(r)$ is written as a sum of squares of the Kohn-Sham orbitals:

$$\rho(r) = \sum |\psi_i(r)|^2. \tag{6.25}$$

This leads to another approximation usually done both in DFT and wavefunction-based methods. It consists in representation of each molecular orbital by a specific orthonormal basis set. The true electronic structure of a system can be in principle mathematically represented by an infinite number of basis functions. However, due to computational limitations, these functions are truncated in practice and described by a finite number of basis sets resulting in some potential loss in accuracy. A wide range of different basis sets currently exists and the choice of a certain one strongly depends on the solution method used, the type of the problem considered and the accuracy required in the particular case. These functions can take on one of several forms, including Slater-type functions, Gaussian functions, and plane waves. The basis sets build using these functions can be either full-electron basis sets, describing both core and valent electrons of the atoms, or so-called effective core potential (ECP) ones. In the latter case it is assumed that core electrons do not significantly influence the electronic structure and properties of atoms and therefore, replaced with an approximate pseudopotential. Such simplification is useful for the description of heavy atoms, because it decreases the number of basis functions and, correspondingly, the required computational time without dramatic loss of accuracy.

Although the increase of the number of basis functions leads to a more accurate mathematical treatment of the electron structure; the computational time significantly increases in parallel. Therefore, a reasonable compromise must be found between the accuracy needed and computational costs taking into account both the accuracy of the quantum chemical method and the completeness of the basis set. One notes that the better method is used, the larger basis set is needed. It is generally accepted that the basis set used in DFT calculations should be at least a valence-double-ζ quality augmented with polarization functions. The basis set requirements for the post-HF methods are usually higher.

The approximations done in order to solve Schrödinger equation by different quantum chemical methods as well as the use of finite basis set for the description of molecular orbitals are not the only factors leading to limited accuracy. When modeling zeolites, one can seldom take into account all the atoms of the system. Typically, a limited subset of the atoms of the zeolite is used to construct an atomistic model. The size of the model used to describe reaction environment can be critical for obtaining reliable results. Indeed, although the coordination of small molecules such as CO, CH_4, H_2O to isolated cations can be studied very accurately using even CC theory, obviously the results thus obtained can hardly represent adsorption to the exchangeable cations in zeolites. This is an example of a competition of "model" versus "method" accuracy.

The current progress in computational chemistry also made it possible to use rather efficiently periodic boundary conditions in DFT calculations of solids. This allows theoretical DFT studies of structure and properties of some zeolites with relatively small unit cells using an experimental crystal structure of the zeolite as a model. However, the periodic DFT calculations as applied to zeolites are mostly limited to the use of LDA and GGA density functionals.

Another method for the zeolite modeling is so-called cluster approach. Here only a part of the zeolite containing finite number of atoms is considered while the influence of the rest atoms of the zeolite lattice is neglected. A minimum requirement to the zeolite model is that it involves the reactive site or the adsorption site together with its environment. In this case the model is built of several TO_4 units to mimic the local structure of a part of the zeolite framework. Although this approach results in some loss of "model" accuracy, it can be very useful for the analysis of different local properties of zeolites such as elementary reaction steps, adsorption, etc. In addition, in the case of cluster modeling the higher level ab initio methods as well as hybrid density functionals can be successfully used.

Recently quantum chemical embedding schemes became very popular for quantum chemical modeling of zeolites. They allow the combination of two or more computational techniques in one calculation and make it possible to investigate the chemistry of such systems like zeolites with high precision. The region of the system where chemical process takes place (similar to that used for cluster modeling) is treated with an appropriately accurate method, while the remainder of the system is treated at a lower level. The main difficulty comes in linking two different regions together. The link region is usually defined in order to provide an adequate transfer of information between the inner and outer regions.

The energy for this system is then calculated as

$$E_{\text{hybrid}}(\text{System}) = E_{\text{high}}(\text{Model}) + E_{low}(\text{System}) - E_{\text{low}}(\text{Model}), \quad (6.26)$$

where E_{high} (Model) refers to the energy calculated at higher level for the inner core region only. E_{low} (System) $-$ E_{low} (Model) refers to the difference in energy between the full system and the core region both calculated at the low level of theory. Although the most common case is when the quantum chemical methods (QM) are used for the description of the core region, while the rest atoms of the system are treated by molecular mechanics, it can be any lower level method which is faster than that used for the core. Therefore, using cluster embedding one can rather accurately investigate the local chemistry of zeolites and also take into account longer range effects.

6.5.2 Applications

As was shown above the quantum chemical methods are based on the calculation of the electronic structure of a chemical system. This peculiarity reflects on the main area of application of them, that is calculation and prediction of chemical properties and reactivity of zeolites. Indeed, a chemical reaction can be considered as a rearrangement of electrons within the interacting species, and therefore, quantum chemistry is the only method, which can directly address questions related to such processes.

Chemically active sites in zeolites are usually local defects in zeolite structure such as exchanged cations, organometallic complexes embedded onto zeolitic cages or Brønsted acid sites. Usually these species can be present in zeolite in various forms, which differ one from another with different local surrounding, oxidation state, etc. Unfortunately, there is no direct experimental technique to find out what active site is responsible for the particular chemical reaction. Moreover, the local structure of the zeolite itself in the vicinity the active as well as the structures of these local defects are often unknown.

The direct comparison between the experimentally (X-ray diffraction (XRD)) determined and the computed structures of zeolites is usually complicated. Indeed, XRD provides averaged values for bond lengths and angles and does not distinguish between framework silicon and aluminum atoms, while the local zeolite structure can be significantly influenced by the presence of Al ions. It is however possible to assume that this effect is minor in the case of high-silica zeolites. The high-silica zeolite structures obtained by means of periodic DFT calculations with GGA functionals are usually in a good agreement with the respective experimental data [65, 66].

Quantum chemical methods are also widely used to provide an insight in the location of extra-framework species in zeolites. This application is particularly important for high-silica aluminosilicates for which such information cannot be obtained from the experiment. The problem similar to the one discussed above arises when one compares the calculated results for the local structure of the extra-framework species in zeolites with those derived from EXAFS data. The interatomic distances from EXAFS correspond to the values averaged over all of the existing sites of the cation-exchanged zeolites. One must note that stabilization of a metal cation at different zeolite sites can results in very different coordination and bond lengths. On the other hand quantum-chemical methods can be directly applied to address the problem of local structure of zeolites such as localization of exchangeable cations.

The structure and coordination of extra-framework cations in zeolites were investigated for many zeolites using various quantum chemical techniques. Very successful studies have been presented by Nachtigall and co-workers [67–70] on the coordination of copper ions in high-silica zeolites. The embedded cluster approach (QM/MM) with periodic boundary conditions was used. The computed results provided a detailed understanding of the experimentally observed photo-luminescence spectra [68] as well as information on the preferential location and properties of Cu^{+} ions in ZSM-5 zeolite. Periodic planewave DFT calculations were used in [71, 72] to elucidate the location of Zn and Pt cations in high-silica mordenite.

Besides the structural problem, quantum chemical methods are widely and extensively used to address the reactivity of zeolites. Here both periodic and cluster modeling approach can be successfully applied. It should be however noted that it is rather difficult to assess the accuracy of methods for calculation of activation barriers. It is generally accepted that the GGA functional underestimate barriers by tens kJ mol^{-1}. This error can be significantly reduced when hybrid functionals are used [73]. The calculations of activation barriers for reactions in zeolites are very challenging and at the same time extremely difficult. The difficulties arise not only from the inaccuracies of the computational techniques, but also from the features of the zeolite chemistry. Indeed, in many cases there are multiple reaction paths existing for a particular chemical process, most of which show very similar energetics. In addition, as it was mentioned above, there is often a lack of information on the structure of the active site. Thus, numerous reaction paths over different possible sites must be compared.

Using cluster modeling approach one can efficiently model the local structure of different possible zeolitic active sites and then compare their reactivity, i.e., energy profile of a chemical reaction over these species. On the basis of these computations one can, obviously, not only decide on the most probable catalytically active species, but also draw a conclusion on the mechanism of the catalytic process.

Recently a comprehensive DFT study of various reaction paths of methanol over a simple 5 T cluster representing a Brønsted acid site of ZSM-5 zeolite have been reported [74]. It has been shown that every imaginable combination of direct reactions in this case fails completely in the formation of the initial carbon-carbon bonds.

The reactivity of cation-exchanged zeolites in various important chemical processes has been also investigated by means of cluster modeling approach. Different reaction paths for ethane dehydrogenation over various possible extra-framework metal-containing species in Zn- and Ga-exchanged zeolites have been recently reported [75–77]. The obtained results allowed rationalization of the available experimental results and provided information on the most probable active sites as well as the reaction mechanism. Similarly, various reaction paths for the nitrous oxide decomposition over mono- and binuclear iron-containing sites in ZSM-5 zeolite have been investigated in detail in [78–80]

Thus, the cluster approach can be very useful for the analysis of chemical processes, which are dominantly controlled by the electronic properties of the system studied. One should also realize that the activation barriers obtained using this method usually cannot be directly used for evaluation of kinetic parameters of the process, but only for qualitative comparison of different possibilities (reaction paths or active sites). This is not only due to the deficiencies of the DFT methodology but also due to a rather poor "model accuracy" of cluster models in many cases. This approach leads to the neglect of zeolite framework electrostatic contribution and steric constrains due to the shape of the zeolitic micropores. Thus, the cluster approximation fails to describe a very important shape selectivity effect of the zeolite. To address this problem, the periodic approach must be applied.

It has been shown by Rozanska et al. [81–83] that electrostatic stabilization in zeolite pores can strongly influence the energetics of the chemical reactions within it. Although elementary reaction steps remain in essence the same for periodic and cluster calculations, the exact values of the activation energies can be lowered by 40 and in some cases even by ~100 kJ mol^{-1} for those calculated within the periodic approach. In order to illustrate this effect, Fig. 6.6 shows comparison of reaction diagrams for toluene isomerization calculated at DFT level using a cluster containing 4 T atoms representing zeolitic Brønsted acid site [81] and infinite lattice of acidic mordenite [82].

The examples discussed so far are mostly the results of pure DFT calculations. The DFT method significantly underestimates dispersive interactions, which can occur between the reaction intermediates as well as the transition state structures and zeolitic walls. For larger hydrocarbon molecules these interactions can significantly contribute the adsorption energies, enthalpies and activation energies of chemical reactions. The calculated energetic values can be numerically corrected for van der Waals interactions via estimation of their contribution using Buckingham or Lenard-Jones potentials (see Sect. 2) with appropriate coefficients [82, 83]. On the other hand, when the embedded cluster approach (QM/MM) is used, such correction is applied at every step of calculations. This was very well illustrated by Yoshi and Thomson [84] with the hexadiene cyclization reaction over HZSM-5 zeolite.

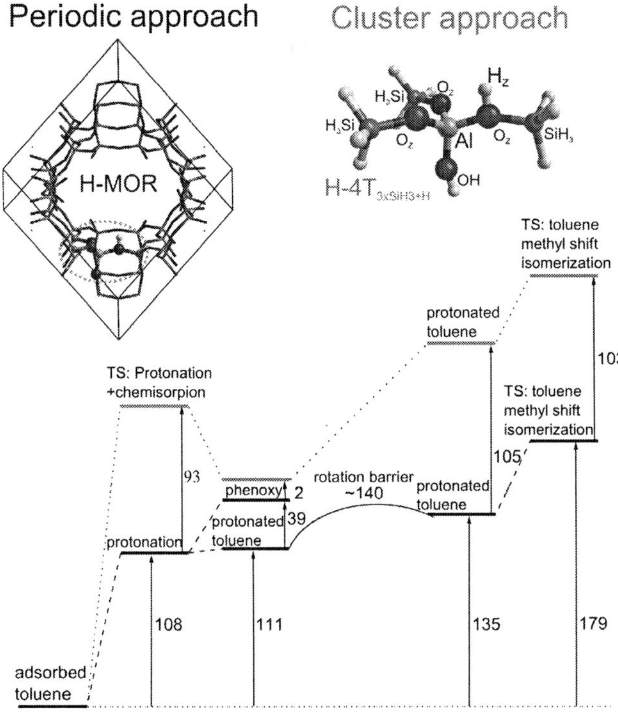

Fig. 6.6 Reaction energy diagrams of toluene isomerization catalyzed by acidic mordenite (energy values are in kJ mol^{-1}). Gray lines correspond to the results obtained using cluster modeling (H-$4T_{3\times SiH3+H}$ cluster model), whereas black lines refer to periodic approach (H-MOR model represents a monoclinic unit cell of a hydrogen form of high-silica mordenite). (The data is compiled from [81, 82])

A more accurate treatment of the dispersive interactions can be obtained when the post-HF methods are used. The simplest quantum chemical method for electron correlation is the second-order Møller-Plesset perturbation theory (MP2). It is far more expensive than DFT calculations. Even the most efficient parallel implementation of RI-MP2 [85] and local approaches [86, 87] do not allow studies of large chemical systems. However, Tuma and Sauer [88] have recently proposed use of a hybrid method, which combines MP2 calculations for the reaction site represented by a cluster model with DFT calculations applied for a remaining part of the zeolite under periodic boundary conditions. This approach allows of reliable calculations of chemical processes in zeolites with a large contribution of dispersion effects such as catalytic transformations of hydrocarbons over acidic zeolites.

One should note that the energy data (adsorption heats, activation barriers, etc.) are frequently absent for most of reactions and processes over well defined zeolites, and thus, direct comparison of the calculated parameters with the experimental ones must be performed very carefully and mostly, qualitatively.

Another related to the issue above and very important application of quantum chemical methods is interpretation of experimental spectroscopic data.

The properties of local defects of zeolite structure are frequently studied by IR spectroscopy of adsorbed probe molecules. The vibrational parameters calculated at the DFT level are usually in a good agreement with the experiment, and therefore, can be directly compared. For example, a thorough interpretation of the vibrational spectra of light alkanes adsorbed on magnesium- and calcium-exchanged zeolites Y have been performed by means of hybrid DFT calculations using cluster modeling approach in [89]. Interaction of CO with K-FER zeolite was investigated by a combination of IR spectroscopy and plane wave DFT (GGA) calculations using a periodic zeolite model by Garrone et al. [90]. The calculated CO stretching vibrations for the adsorbed carbon monoxide were shown to coincide with those observed experimentally with the accuracy of 2 cm^{-1}.

On the basis of spectroscopic data Kazansky et al. [91, 92] suggested that unusual adsorption properties and high reactivity of bivalent zinc cations in high-silica zeolites are due to the specific location of exchangeable cations in the zeolite matrix. A model was suggested where the bivalent cation is located in the vicinity of one $[AlO_2]^-$ framework unit with another charge-compensating anion site located distantly. Such localization results in only partially compensation of the positive charge of Zn^{2+}, and hence, determines their enhanced reactivity and adsorption properties. This hypothesis was subsequently supported by cluster DFT calculations [75, 93, 94].

Combined QM/QM approach with DFT applied to the core region containing the adsorption site and HF used for the description of the surrounding zeolitic cage has been shown to provide a very good prediction for the values of ^{13}C NMR shifts for organic molecules adsorbed to zeolites [95].

One can see that different experimentally observed properties of adsorbed molecules can be quantitatively predicted on the basis of quantum chemical calculations. Moreover, these methods can be used for understanding the nature of bonding within the adsorption complexes in zeolite. A comprehensive review has been published by Frenking and Fröhlich on the nature of the bonding in transition-metal compounds [96]. Almost all the quantum chemical methods for analyzing the chemical bond are described there, and thus, we will limit ourselves to several very recent examples on application of those methods.

Zeolites modified with copper attract attention of scientists not only due to high activity in NO_x decomposition reactions, but also due to their specific adsorption properties. It was found that molecular adsorption of H_2 to univalent copper-exchanged ZSM-5 results in about 1,000 cm^{-1} red shift of H-H stretching vibration. This reflects an extremely strong perturbation of the adsorbed H_2 molecules [97]. A similar effect was found for the adsorption of light alkanes, where the C-H stretching bands are shifted due to interaction with copper ions by more than 400 cm^{-1} [98]. Such perturbations have never been observed before for hydrogen or alkanes adsorbed to any other form of zeolites. Solans-Monfort et al. [97] performed an analysis of the H_2 adsorption energy using partition scheme implemented in ADF program [99], while Pidko and Kazansky [98] used NBO analysis [100] for investigation of bonding within ethane/Cu(I)-ZSM-5 complexes. Both studies showed that the very strong perturbations of the adsorption molecules are due to specific donor-acceptor interactions between the adsorbate and the adsorption site.

Using the Quantum Theory of Atoms in Molecules (QTAIM) theory proposed by Bader [101] one can reveal all of the chemical bonds (including hydrogen and van der Waals bonds) between the interacting species; determine their properties and relative contribution to the total adsorption energy. The application of QT AIM to various chemical problems has been recently thorough reviewed in [102]. This approach is based on the topological analysis of the electron density distribution function that is always provided by quantum chemical calculations (6.25). Using this method an unusual localization of ferrocene molecule in NaY zeolite was explained in [103]. Experimentally it was found that against expectations, the Fe $(C_5H_5)_2$ molecule in the supercage of NaY is located about 1 Å above the line connecting the exchanged cations, although it results in less effective interaction of the π-systems of the cyclopentadienyl with the exchanged cations. QTAIM analysis revealed numerous weak van der Waals bonds between H atoms of the ferrocene molecule and basic oxygens of the zeolite. It was concluded that these very weak bonds are responsible for the way a $Fe(C_5H_5)_2$ molecule is held in a zeolite. Similar effect has been recently reported in [89, 104] for the adsorption of light alkanes to alkaline-earth-exchanged Y zeolites based on MP2 calculations and topological analysis of the resulting electron density distribution functions. The adsorption energy decomposition analysis based on QTAIM was applied [105]. It has been shown that despite the rather small value of the energy of an individual CHO contact, these interactions significantly contribute to the adsorption energy even in the case of light alkane adsorption on cation-exchanged low-silica zeolites.

In addition to the investigation of properties of zeolites, quantum chemical calculations play nowadays an important role for understanding the mechanism of the zeolite formation. Recently, several successful DFT studies [106, 107] devoted to the investigation of the mechanism of silica oligomerization as well as the factors, which control formation of zeolite structures, were reported. The influence of the water solvent media on the computed reaction energies and activation barriers was taken into account by using the continuum solvation COSMO method [108].

References

1. Catlow CRA, Price GD (1990) Nature 347:243
2. Catlow CRA, Bell RG, Gale JD (1994) J Mater Chem 4:781
3. Watson G, Tschaufeser P, Wall A, Jackson RA, Parker SC (1997) In: Catlow CRA (ed) Computer Modelling in Inorganic Crystallography, Academic Press, San Diego, Boston Chapter 3
4. Catlow CRA, Bell RG, Slater B (2004) In: Catlow CRA, van Santen RA, Smith B (eds) Computer Modelling of Microporous Materials, Elsevier, Amsterdam, Chapter 1
5. Dick BG, Overhauser AW (1958) Phys Rev 112:90
6. van Beest BWH, Kramer GJ, van Santen RA (1990) Phys Rev Lett 60:1955

7. Sanders MJ, Leslie M, Catlow CRA (1984) J Chem Soc Chem Commun 1271
8. Purton J, Jones R, Catlow CRA, Leslie M (1993) Phys Chem Miner 19:392
9. Hill JR, Sauer J (1994) J Phys Chem 98:1238
10. Petrovic I, Navrotsky A, Davis ME, Zones SI (1993) Chem Mater 5:1805
11. Piccione PM, Laberty C, Yang SY, Camblor MA, Navrotsky A, Davis ME (2000) J Phys Chem 104:10001
12. Henson NJ, Cheetham AK, Gale JD (1994) Chem Mater 6:1647
13. Ruiz-Salvador AR, Lewis DW, Rubayo-Soneira J, Rodriguez-Fuentes G, Sierra LR, Catlow CRA (1998) J Phys Chem B 102:8417
14. Cannon YM, Catlow CRA, Jackson RA, Owens SL (1998) Micropor Mesopor Mater 24:153
15. Sastre G, Fornes V, Corma A (2002) J Phys Chem B 106:701
16. Lewis DW, Carr S, Sankar G, Catlow CRA (1995) J Phys Chem 99:2377
17. Stach H, Lohse U, Thamm H, Schirmer W (1986) Zeolites 6:74
18. Ruthven DM (2000) Ind Eng Chem Res 39:2127
19. Metropolis N, Rosenbluth AW, Rosenbluth MN, Teller AH, Teller E (1953) J Chem Phys 21:1081
20. Frenkel D, Smit B (1996) Understanding molecular simulation: from algorithms to applications. Academic Press, San Diego
21. van Kampen NG (1980) Stochastic processes in physics and chemistry. North Holland, Amsterdam
22. Smit B (1995) Mol Phys 85:153
23. Smit B (1995) J Phys Chem 99:5597
24. Smit B, Krishna R (2003) Chem Eng Sci 58:557
25. Fuchs AH, Cheetham AK (2001) J Phys Chem B 105:7375
26. Maurin G, Llewellyn PL, Bell RG (2005) J Phys Chem B 109:16084
27. Vlugt TJH, Zhu W, Kapteijn F, Moulijn JA, Smit B, Krishna R (1998) J Am Chem Soc 120:5599
28. Vlugt TJH, Martin MG, Smit B (1999) J Phys Chem B 103:1102
29. Du ZM, Manos G, Vlugt TJH, Smit B (1998) AIChE J 44:1756
30. Macedonia MD, Maginn EJ (1999) Fluid Phase Equilib 158:19
31. Clark LA, Snurr RQ (1999) Chem Phys Lett 308:155
32. Lewis DW, Freeman CM, Catlow CRA (1995) J Phys Chem 99:11194
33. Crank J (1975) The mathematics of diffusion. Clarendon Press, Oxford
34. Auerbach SM (2000) Int Rev Phys Chem 19:155
35. Demontis P, Suffrinti GB, Alberti A, Quartieri S, Fois ES, Gamba A (1986) Gazz Chim Ital 116:459
36. Krishna R, Baur R (2003) Sep Purif Technol 33:213
37. Keil FJ, Krishna R, Coppens MO (2000) Rev Chem Eng 16:71
38. Demontis P, Suffritty GB (1997) Chem Rev 97:2845
39. Auerbach SM, Jousse F, Vercauteren DP (2004) In: Catlow CRA, van Santen RA, Smith B (eds) Computer modelling of microporous materials. Elsevier, Amsterdam, Chapter 3
40. Snurr RQ, Bell AT, Theodorou DN (1994) J Phys Chem 98:11948
41. Forester TR, Smith W (1997) J Chem Soc Faraday Trans 93:3249
42. Snurr RQ, Bell AT, Theodorou DN (1994) J Phys Chem 98:5111
43. Tunca C, Ford DM (1999) J Chem Phys 111:2751
44. Paschek D, Krishna R (2000) Phys Chem Chem Phys 2:2389
45. Krishna R, van Baten JM (2005) Chem Eng Technol 28:160
46. Jensen F (1999) Introduction to computational chemistry. Wiley, New York
47. Levine IN (1983) Quantum chemistry. Allyn and Bacon, Boston
48. Leach AR (1996) Molecular modeling: principles and applications. Pearson Education, Harlow

49. Foresman JB (1996) Frish A: exploring chemistry with electronic structure, 2nd edn. Pittsburg, PA Gaussian
50. Parr RG, Yang W (1989) Density functional theory of atoms in molecules. Oxford University Press, New York
51. Young DC (2001) Computational chemistry: a practical guide for applying techniques to real-world problems. Wiley, New York
52. Born B, Oppenheimer JR (1927) Ann Phys 79:361
53. Feyereisen M, Fitzgerald G, Komornicki A (1993) Chem Phys Lett 208:359
54. Hohenberg P, Kohn W (1964) Phys Rev 136:B864
55. Kohn W, Sham L (1965) Phys Rev 140:A1133
56. Perdew JP (1986) Phys Rev B 33:8822
57. Becke AD (1988) Phys Rev A 38:3098
58. Perdew JP, Chevary JA, Vosko SH, Jackson KA, Pederson MR, Singh DJ, Fiolhais C (1992) Phys Rev B 46:6671
59. Perdew JP, Burke K, Ernzerhof M (1996) Phys Rev Lett 77:3865
60. Becke AD (1993) J Chem Phys 98:5648
61. Adamo C, Barone V (1998) J Chem Phys 108:664
62. Adamo C, Barone V (1999) J Chem Phys 110:6158
63. Johnson ER, DiLabio GA (2006) Chem Phys Lett 419:333
64. Zhao Y, Truhlar DG (2005) J Chem Theory Comput 1:415
65. Benco L, Demuth T, Hafner J, Hutschka F (2001) Micropor Mesopor Mater 42:1
66. Campana L, Selloni A, Weber J, Goursot A (1995) J Phys Chem 99:16351
67. Spuhler P, Holthausen MC, Nachtigallová D, Nachtigall P, Sauer J (2002) Chem Eur J 8:2099
68. Nachtigall P, Nachtigallová D, Sauer J (2000) J Phys Chem B 104:1738
69. Nachtigallová D, Nachtigall P, Sierka M, Sauer J (1999) Phys Chem Chem Phys 1:2019
70. Nachtigallová D, Nachtigall P, Sauer J (2001) Phys Chem Chem Phys 3:1552
71. Grybos R, Hafner J, Benco L, Toulhoat H (2007) J Phys Chem C 111:6454
72. Bucko T, Benco L, Hafner J, Toulhoat H (2005) J Phys Chem B 109:20361
73. Johnson BG, Gonzales CA, Gill PMW, Pople JA (1994) Chem Phys Lett 221:100
74. Lesthaeghe D, van Speybroeck V, Marin GB, Waroquier M (2006) Angew Chem Int Ed 45:1714
75. Pidko EA, van Santen RA (2007) J Phys Chem C 111:2643
76. Pidko EA, Kazansky VB, Hensen EJM, van Santen RA (2006) J Catal 240:73
77. Pidko EA, Hensen EJM, van Santen RA (2007) J Phys Chem C 111:13068
78. Heyden A, Peters B, Bell AT, Keil FJ (2005) J Phys Chem B 109:1857
79. Hansen N, Heyden A, Bell AT, Keil FJ (2007) J Phys Chem C 111:2092
80. Hansen N, Heyden A, Bell AT, Keil FJ (2007) J Catal 248:213
81. Rozanska X, Saintigny X, van Santen RA, Hutschka F (2001) J Catal 202:141
82. Rozanska X, van Santen RA, Hutschka F, Hafner J (2001) J Am Chem Soc 123:7655
83. Vos AM, Rozanska X, Schoonheydt RA, van Santen RA, Hutschka F, Hafner J (2001) J Am Chem Soc 123:2799
84. Joshi YV, Thomson KT (2005) J Catal 230:440
85. Hättig C, Hellweg A, Köhn A (2006) Phys Chem Chem Phys 8:1159
86. Schütz M, Hetzer G, Werner HJ (1999) J Chem Phys 111:5691
87. Lee MS, Maslen PE, Head-Gordon M (2000) J Chem Phys 112:3592
88. Tuma C, Sauer J (2006) Phys Chem Chem Phys 8:3955
89. Pidko EA, Xu J, Mojet BL, Lefferts L, Subbotina IR, Kazansky VB, van Santen RA (2006) J Phys Chem B 110:22618
90. Garrone E, Bulánek R, Frolich K, Otero Areán C, Rodrígues Delgado M, Turnes Palomino G, Nachtigallová D, Nachtigall P (2006) J Phys Chem B 110:22542
91. Kazansky VB, Pidko EA (2005) J Phys Chem B 109:2103
92. Kazansky VB, Serykh AI, Pidko EA (2004) J Catal 225:369

93. Zhidomirov GM, Shubin AA, Kazansky VB, van Santen RA (2005) Theor Chem Acc 114:90
94. Zhidomirov GM, Shubin AA, Kazansky VB, van Santen RA (2004) Int J Quant Chem 100:489
95. Zheng A, Chen L, Yang J, Tue Y, Ye C, Lu X, Deng F (2005) Chem Commun 2474
96. Frenking G, Fröhlich N (2000) Chem Rev 100:717
97. Solans-Monfort X, Branchadell V, Sodupe M, Zicovich-Wilson CM, Gribov E, Spoto G, Busco C, Ugliengo P (2004) J Phys Chem B 108:8278
98. Pidko E, Kazansky V (2005) Phys Chem Chem Phys 7:1939
99. te Velde G, Bickelhaupt FM, Baerends EJ, Fonseca Guerra C, van Gisbergen SJA, Snijders JG, Ziegler T (2001) J Comput Chem 22:931
100. Reed AA, Curtiss LA, Weinhold F (1988) Chem Rev 88:899
101. Bader RFW (1990) Atoms in molecules a quantum theory. Clareondon Press, Oxford
102. Matta C, Boyd RJ (eds) (2007) The quantum theory of atoms in molecules. Wiley, Weinheim
103. Kemner E, de Schepper IM, Kearley GJ (2001) Chem Commun 2466
104. Pidko EA, van Santen RA (2006) Chem Phys Chem 7:1657
105. Espinosa E, Molins E, Lecomte C (1998) Chem Phys Lett 285:170
106. Trinh TT, Jansen APJ, van Santen RA (2006) J Phys Chem B 110:23099
107. Mora-Fonz MJ, Catlow CRA, Lewis DW (2007) J Phys Chem C 111:18155
108. Baldridge K, Klamt A (1997) J Chem Phys 106:6622

Chapter 7
Reactions of C_1 Building Blocks

Michael Stöcker

Abstract The chapter "Reactions of C_1 Building Blocks" covers the direct conversion of methane to aromatics, the methanol-to-hydrocarbons (MTHC) conversion with respect to gasoline (methanol to gasoline) and olefins (methanol to olefins, methanol-to-propene) as well as some combinations like the TIGAS and Mobil's olefin-to-gasoline and distillate processes. The main focus within this chapter will be on the industrial processes, especially concerning the MTHC reactions – including catalytic systems, reaction conditions, process – and to a minor extent related to the mechanistic aspects and kinetic considerations.

7.1 Dehydroaromatization of Methane

7.1.1 Introduction/Background

The direct conversion of methane to aromatics has been a demanding challenge in catalysis research during the last two decades due to the requirements of effective activation. In addition to the oxidative activation, like the coupling of methane to C_2^+, the non-oxidative activation of methane to aromatics over metal modified microporous materials has been in the focus during this period [1].

Since the crude oil reserves are declining, the energy supply is changing from mainly being oil based to becoming more natural gas based. Methane is the superior component of most of the natural gas reserves as well as coal mine release gases. In addition, methane is formed in rather large amounts as undesired by-product in

M. Stöcker
SINTEF Materials and Chemistry, P. O. Box 124, Blindern, N-0314 Oslo, Norway
e-mail: michael.stocker@sintef.no

A.W. Chester and E.G. Derouane (eds.), *Zeolite Characterization and Catalysis*,
DOI 10.1007/978-1-4020-9678-5_7,

a number of chemical processes and contributes as greenhouse gas to the pollution of the atmosphere. Therefore, the catalytic conversion of methane to valuable chemical products or fuels is an important topic for both the research and development within the field of catalysis as well as for the chemical and petrochemical industry. Methane conversion with respect to a better utilization of natural gas and a decreasing contribution as greenhouse gas will definitely contribute to an improvement of the global climate and the quality of life [1, 2].

However, methane is the most stable and symmetric organic molecule consisting of four C–H bonds with a bond energy of 435 kJ mol^{-1}, consequently, the effective activation of methane in connection with its conversion represents a demanding task for this sub-field of catalysis research and development [1, 2]. Lunsford has summarized the different routes with respect to the catalytic conversion of methane to valuable chemicals and fuels, which can be grouped into two main categories [3]. The indirect route is described by the formation of syngas either by steam reforming or partial oxidation and subsequent Fischer-Tropsch processing. The direct route is based on the conversion of methane to, for example ethene, methanol, formaldehyde or aromatics, directly. During the last 25 years, a strong focus has been concentrated on the oxidative coupling of methane as an alternative route for the direct methane conversion; however, no catalysts so far have been able to increase the C_2 yield beyond 25% and the C_2 selectivity higher than 80%, thus resulting in limited industrial application [2].

So far, the largest potential for progress with respect to the methane conversion lies within the direct conversion dedicated to the production of formaldehyde and methanol. Anyway, the transformation of methane to aromatics is thermodynamically more favorable than the conversion of methane to ethene, and therefore, a lot of efforts have been made to investigate the direct transformation along this line. A number of different groups tried to convert methane to higher hydrocarbons in the absence of oxygen and using different multi-component catalysts, like supported Pt and Pt-CrO_3/H-ZSM-5 and others. In 1993, Wang and co-workers presented the conversion of methane to aromatics under non-oxidative conditions applying transitions metal ions supported H-ZSM-5 catalysts [4]. The H-ZSM-5 is a zeolite with strong acidity, and the mechanism of the methane conversion over transition metal ion containing H-ZSM-5 zeolites has been investigated in detail. Since that time, a number of scientists have devoted their interest to the activation of methane under non-oxidative conditions in order to form aromatics, with the main focus on the following topics:

1. Catalyst preparation with respect to the pore architecture and transition metals.
2. Investigations of the active phases and mechanistic studies.
3. Optimization of reaction conditions.
4. Synergetic effects between the transition metals and the zeolitic supports.
5. Coke formation and its role for the conversion.

A lot of interest has been devoted to the zeolite supported catalysts, due to the fact that their lattice can stabilize the transitions metal ions, especially when these ions migrate into intra-crystalline zeolite cavities and/or channels [1, 2, 5–8].

7.1.2 Thermodynamics/Reaction Mechanism/Active Sites

As already mentioned, the conversion of methane under non-oxidative conditions is, from a thermodynamic point of view, more favorable to aromatic compounds than to olefins, as demonstrated in Fig. 7.1.

A number of investigations have been conducted with respect to the dehydro-aromatization of methane leading to aromatic compounds (mainly benzene) and hydrogen, mainly using bi-functional catalysts like Mo/H-ZSM-5 or Mo/H-MCM-22 (however, other metals (besides Mo) like Re and W have been studied as well). Detailed investigations on the title reaction revealed that the location and oxidation state of the metal species as well as the pore architecture and acidity of the microporous materials applied are important items with respect to the catalytic behavior of the systems used. A number of different mechanisms have been suggested, like the carbenium ion mechanism by Wang et al. [4], with Mo^{6+} species or protonic sites as hydride acceptor. Xu et al. [9] suggested a possible mechanism in which both an intermediate Mo-carbide like complex and the heterolytic splitting of methane were important. The formation of free methyl radicals followed by the formation of ethane and ethene and subsequent aromatization has been proposed by Chen et al. [10]. Mirodatos and co-workers performed a series of transient kinetic experiments intending to understand the nature of the methane activation.

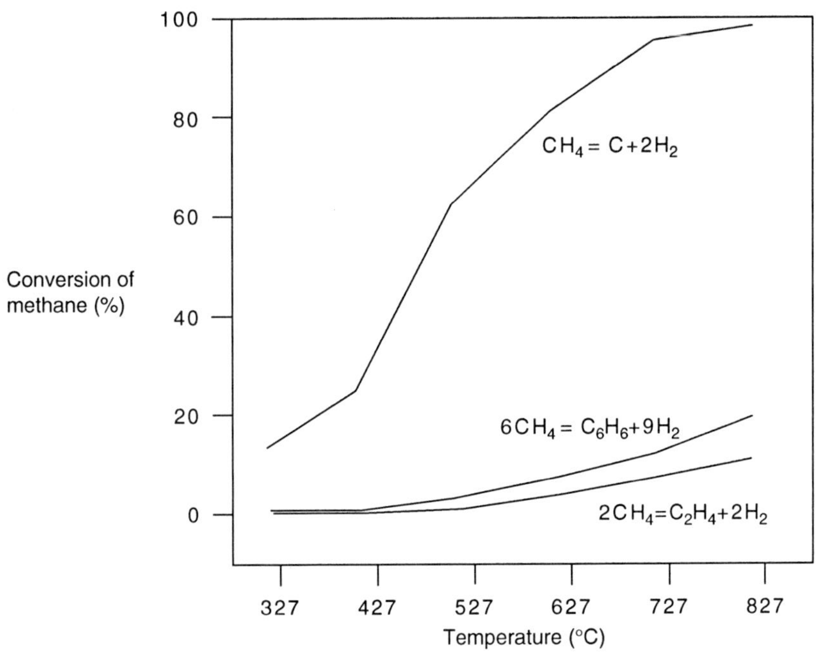

Fig. 7.1 Direct transformation of methane under non-oxidative conditions [13] (reproduced by permission of Elsevier, Amsterdam)

They suggested that several types of sites could be involved in the dehydroaromatization of methane [11]. Iglesia and co-workers studied the Mo species in Mo-exchanged H-ZSM-5 for the methane reaction via the kinetics of formation and their stoichiometry, among others by using in situ Raman and XAS spectroscopy and the isotopic exchange of deuterium with OH groups in the zeolite [12].

All the different mechanisms suggested are speculative since direct experimental evidence for the proposed mechanisms is currently not available [1]; however, important aspects of the conversion can be summarized as follows [13]:

1. An induction period at the early stage of the conversion has been observed, probably reducing the MoO_x species by methane into Mo-carbide and/or other Mo-species containing oxygen and carbon.
2. Coke deposits are formed during the conversion leading to severe catalyst deactivation.
3. Different interactions are influencing the catalytic behavior, like interactions between the Mo species and the acid sites as well as framework aluminum.
4. The final product distribution is governed by the shape-selective effect given through the pore architecture of the zeolite applied.

In conclusion, it seems that there are close correlations between the activity of the benzene formation in connection with the dehydroaromatization of methane and the pore architecture as well as the acidity of the catalyst under study. Possibly, the Mo carbide species are highly dispersed on the outer surface and the partially reduced Mo species are located in the channels of the zeolite. These findings suggest that methane is dissociated on the Mo carbide cluster supported on the catalyst having optimum Brønsted acidity to form C_2-species as primary intermediates, which are then subsequently oligomerized to aromatic compounds at the interface of the Mo carbide and the catalysts – for example H-ZSM-5 – having the proper Brønsted acidity [1, 2, 14]. The main transformation steps of this process could be outlined as shown in Fig. 7.2.

In contrast to the Mo carbide as the active species for the dehydroaromatization of methane over the Mo/H-ZSM-5 catalyst, it was proposed that the metallic Re on Re/H-ZSM-5 is responsible for the catalytic conversion of methane with respect to the formation of C_2 hydrocarbons and aromatic compounds [2].

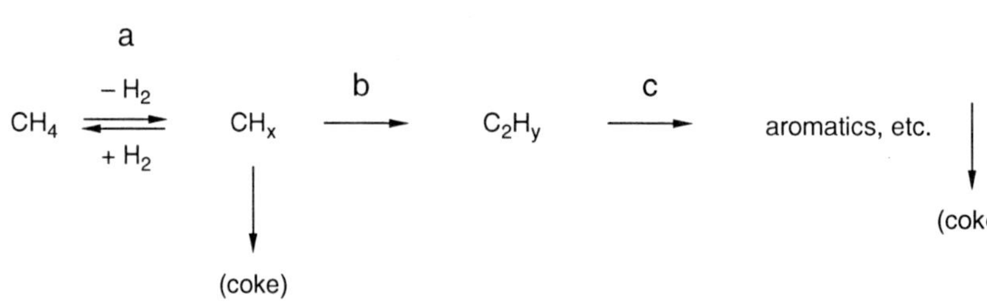

Fig. 7.2 Transformation steps of the dehydroaromatization of methane. (**a**) and (**b**) are on the Mo species, whereas (**c**) occurs on the zeolite [2] (reproduced by permission of Elsevier, Amsterdam)

7.1.3 Catalysts/Effects of Porosity and Surface Dealumination/ Interaction Between Transition Metal Ions and the Zeolite Support/Synergetic Effects

Quite a large number of research groups have investigated series of different catalysts, both porous and non-porous supports, however, mostly containing Mo as the active metal species. Besides Mo, other metals like Re, Fe, V, Cr, W, Ni, Pt, Cu and Zn have been evaluated as well. Mo impregnated on H-ZSM-5 has been claimed to be one of the best catalysts for the dehydroaromatization of methane. It was concluded that both the framework and the acidity of H-ZSM-5 zeolite, as well as the oxidation state and the location of the Mo species are crucial factors with respect to the catalytic behavior [9]. Zhang et al evaluated the methane conversion to aromatics over Mo-modified porous carriers and they ranged the activities of the various Mo-based catalysts as follows [15]:

Mo/H-ZSM-11 > Mo/H-ZSM-5 > Mo/H-ZSM-8 > Mo/H-BEA > Mo/H-MCM-41 > Mo/H-SAPO-34 > Mo/H-MOR ~ Mo/H-X > Mo/H-Y > Mo/H-SAPO-5 > Mo/H-SAPO-11.

The group correlated the catalytic performance to the structure of the porous materials and concluded that the 2 D zeolites with a pore diameter close to the kinetic diameter of benzene are the best supports of Mo-modified catalysts with respect to the conversion of methane to aromatics. The modification with Re, Pt, W, Zr, Co, Fe and Cr can enhance the reactivity and/or stability of the catalysts, whereas the addition of V, Li and P has the opposite effect. Furthermore, the synthesis method applied has also an influence on the catalytic performance. Finally, the synergetic effect between the Mo species and the Brønsted acid sites of the H-ZSM-5 zeolite has been regarded to be responsible for the bifunctionality of the Mo/H-ZSM-5 catalyst – a Mo/[H+]-ratio per unit cell around one was found to be an optimum concerning the activation of methane and the following transformation of the intermediates into aromatics [13, 16].

Post-steam treatment has been regarded as an effective method for improving the catalytic performance of Mo/H-ZSM-5 catalysts in methane dehydroaromatization under non-oxidative conditions. The treatment enhanced the stability of the catalyst and resulted in a higher yield of light aromatics, as well as in a decrease in the formation rate of carbonaceous deposits. Zhang et al studied the catalytic performance of Mo-supported micro-sized and nano-sized H-ZSM-5 zeolites. The catalytic behavior of Mo/micro-sized H-ZSM-5 was observed to be better than that of Mo/nano-sized H-ZSM-5 [2, 17].

Tetraethylorthosilicate (TEOS) was used to modify H-ZSM-5 zeolite by CVD. An Mo/H-ZSM-5 catalyst prepared by using the TEOS modified zeolite showed higher activity and stability than those synthesized by using non-modified H-ZSM-5 zeolite [18, 19].

The catalytic performances of H-ZSM-5 modified by either hexagonally packed beta-Mo-carbide and face-centered cubic alpha-Mo-carbide have been compared, concluding that the latter showed superior performance like higher activity and higher selectivity to benzene as well as a higher stability as a function of time on stream within the dehydroaromatization of methane [20].

Re/H-MCM-22 was investigated as a catalyst for the conversion of methane to aromatics and compared with the results from Re and Mo-supported H-MCM-22 and H-ZSM-5. Re/H-MCM-22 was similarly active and selective for the reaction as Mo/H-ZSM-5, Re/H-ZSM-5 and Mo/H-MCM-22. However, the catalytic performances of H-MCM-22 supported Re and Mo catalysts were characteristic of higher benzene selectivities as well as higher catalytic stability than those supported on H-ZSM-5. The differences between these two zeolites were related to their different channel structures, where H-MCM-22 has large cavities and slit-like pore openings [21].

Mo containing H-MCM-22 zeolite with reduced surface aluminum concentration seems to be a promising catalyst for the dehydroaromatization of methane. The removal of Al atoms (acid sites) from the surface of H-MCM-22 zeolite by ammonium hexafluorosilicate increased its selectivity to benzene and its stability by reducing coke formation – and, consequently, less pore blockage [22]. A similar beneficial effect has been observed for a surface dealuminated, Mo-supported H-ZSM-5 system [23]. Furthermore, alkali treatment of MCM-22 and ZSM-5 created some mesopores on the parent zeolites, resulting in the formation of new porosities and pore architectures, revealing a coexistence of micro- and mesopores, beneficial for the diffusion of the reactants and products, and thus increasing the stability of the title reaction over the Mo-supported zeolites [24, 25].

As already discussed, the catalyst for dehydroaromatization of methane consists of two components, the metal species (usually Mo) and the zeolite support. However, since the catalyst is calcined at high temperatures and the reaction is carried out at about 700°C, various kinds of interactions may happen during the catalyst preparation and reaction; for example, the interaction between Mo species and acid sites as well as framework aluminum. The strong interaction between Mo species and framework Al can lead to the formation of MoO_4^{2-} species, which are detrimental to methane activation. Another interaction between the Mo species and the zeolite resulted in a high dispersion of the Mo species in the channel of the zeolites applied. This kind of interaction increases with increasing Mo loading and/or calcination temperature [1].

7.1.4 Reaction Conditions/Nature of Coke and its Role for the Reaction

The non-oxidative dehydroaromatization of methane is usually carried out at 700°C (or slightly higher) applying plug-flow hydrodynamics in a continuous flow system under atmospheric pressure (or slightly higher) [26–28]. However, carbonaceous deposits are formed during the conversion of methane forming aromatic compounds, which represents a major obstacle with respect to a better understanding of the reaction and process development. The chemical nature of the carbonaceous deposits has been identified as carbidic carbon in Mo carbide, Mo associated coke and aromatic type coke on acid sites. The carbidic coke, which is formed during the induction period, is possibly the active species for the methane activation, whereas the coke on acid sites is active for the formation of aromatics from ethane [1, 2, 13].

Recently, it was reported that treatment of the parent zeolite by steam dealumination, dealumination by acid solutions or silanization with Si-containing compounds could significantly reduce the rate of coke formation and enhance the durability of the Mo-containing H-ZSM-5 and H-ZSM-22 catalysts. The suppression of carbonaceous deposits during the methane aromatization could also be achieved by increasing the methane pressure and the addition of CO and CO_2 to the methane feed, with an enhanced catalyst stability as a beneficial effect in addition. Further, hydrogen may react with the coke which is burnt off at high temperatures, whereas CO_2 can eliminate the carbonaceous deposits, which are burnt off at low and high temperatures [1, 2, 13, 29].

7.1.5 Use of Membrane Technology

The removal of hydrogen is one of the most efficient ways to overcome the thermodynamic and kinetic limits with respect to the methane dehydroaromatization. Therefore, investigation of the membrane reactor technology for hydrogen removal is highly demanded. Ichikawa and co-workers observed a marked enhancement of the methane conversion towards benzene using an effective Pd catalytic membrane reactor with Mo/ZSM-5 (3 wt.% Mo). The formation rate of aromatics and hydrogen was enhanced up to 10 times by removal of hydrogen using a Pd membrane for 100 h at 610°C. About 50–60% of hydrogen from the total hydrogen produced during the methane dehydroaromatization was selectively removed by the Pd membrane, resulting in a shift of the equilibrium towards the product side [30].

7.1.6 Future Perspectives

Energy based on fossil fuels will, in the future, be replaced by energy based on hydrogen. Therefore, the direct conversion of methane under non-oxidative conditions has a strong importance for the production of hydrogen and/or for the hydrogen formation accompanied by basic chemicals. The development of methane dehydroaromatization is a potentially important process for the hydrogen production – besides the formation of aromatics. Strong efforts have been devoted to the development of active and selective catalysts for the title reaction as well as the understanding of the bifunctional behavior of the Mo species supported on the zeolites, mainly applied H-ZSM-5 and H-ZSM-22. In addition, a lot of work has been dedicated to find out the nature and properties of the carbonaceous deposits formed during the coarse of the reaction. However, a real breakthrough with respect to new active and selective catalysts as well as the complete understanding of the reaction mechanism has still not yet been achieved. On the other hand, our general understanding of the conversion of methane to aromatics has improved to a large extent, and further research dedicated to this sub-field of catalysis will certainly enlarge our knowledge of the direct activation and conversion of methane and lead, finally, to the development of a clean, profitable and efficient process [13].

7.2 Methanol to Hydrocarbons (MTHC)

7.2.1 Introduction/Background

The known worldwide natural gas resources vs. crude oil reserves are quite impressive – the proven natural gas reserves amount to about 141 trillion m^3; however, roughly 60% of these amounts are recovered at remote gas field sites. These facts have been a driving force with respect to the conversion of natural gas into liquids and higher value-added products either via the direct route "gas to liquids (GTL)" – based on Fischer-Tropsch technology or via the indirect route, the production of methanol from synthesis gas, which is made by steam-reforming of natural gas or gasification of coal, and the consecutive formation of hydrocarbons.

These technologies are alternatives to the chemical conversion of methane, either via direct coupling, which is thermodynamically not favorable, or via oxidative coupling, a route *not* successful so far from an industrial point of view. The interest to manufacture fuels and petrochemicals from natural gas is driven by the desire to apply this technology directly, for example at remote natural gas field sites, in order to minimize transportation costs and gas burning at the recovery sites.

To begin with, the methanol to hydrocarbons (MTHC) technology was primarily regarded as a powerful method to convert coal into high-octane gasoline. This concept has been expanded since, not only with respect to the formation of other fuels, but also to chemicals in general. Of course, light olefins are important components for the petrochemical industry, and the demand of high-quality gasoline is increasing as well. In fact, with this new technology, one can make almost anything out of coal or natural gas that can be made out of crude oil. Methanol is made from synthetic gas (a mixture of carbondioxide and hydrogen) which is formed by steam reforming from either natural gas or coal. The methanol is then converted to an equilibrium mixture of methanol, dimethylether (DME) and water, which can be processed catalytically to either gasoline (methanol to gasoline (MTG)) or olefins (methanol to olefins (MTO)), depending on the catalyst and/or the process operation conditions. Although methanol itself is a potential motor fuel or can be blended with gasoline, it would require large investments to overcome the technical problems connected with it.

The commercial MTG reaction runs at temperatures around 400°C at a methanol pressure of several atmospheres and uses a ZSM-5 catalyst. These are the optimal conditions for converting the olefins that form within the catalyst into paraffins and aromatics. However, at one point in the MTG reaction, the product mixture consists of about 40% light olefins. The importance of light olefins as intermediates in the conversion of MTG was recognized early. Consequently, a number of attempts were made to selectively form light olefins from methanol on not only medium-pore zeolites but also on small-pore zeolites and SAPO type molecular sieves. If one interrupts the reaction at the point of about 40% light olefin formation, one could harvest these C_2–C_4 olefins. By adjusting the reaction conditions (like for example raising the temperature to 500°C) as well as the catalyst applied, one can increase dramatically the olefin yield. This discovery led to the development of the MTO process, which generates mostly propene and butene, with high-octane gasoline as a by-product. But the catalyst can be modified to selectively produce more ethene [31].

The discovery of the MTG reaction was an accident. One group at Mobil was trying to convert methanol to other oxygen-containing compounds over a ZSM-5 catalyst. Instead, they received unwanted hydrocarbons. Somewhat later, another Mobil group, working independently, was trying to alkylate isobutane with methanol over ZSM-5 and identified a mixture of paraffins and aromatics boiling in the gasoline range – all coming from methanol. Although the discovery of MTG was accidental, it occurred due to a balanced effort in catalysis over many years. The MTO reaction seems to benefit from this development, although independent research has been performed since. The evolution of the MTHC technology, from their discovery to the demonstrative and/or commercial realization, has been accompanied by extensive research related to the basic question of the mechanism of formation of the initial C–C bond [31].

7.2.2 *The Chemistry of MTHC: Mechanistic and Kinetic Considerations*

The main reaction steps of the methanol conversion to hydrocarbons can be summarized as follows:

Methanol is first dehydrated to DME. The equilibrium mixture, thus formed, consisting of methanol, DME and water is then converted to light olefins. In the last step of this scheme, the light olefins react to form paraffins, aromatics, naphthenes and higher olefins by alkylation and polycondensation (see Fig. 7.3).

There is a consensus that the intermediate in the dehydration of methanol to DME (step 1 in (7.1))

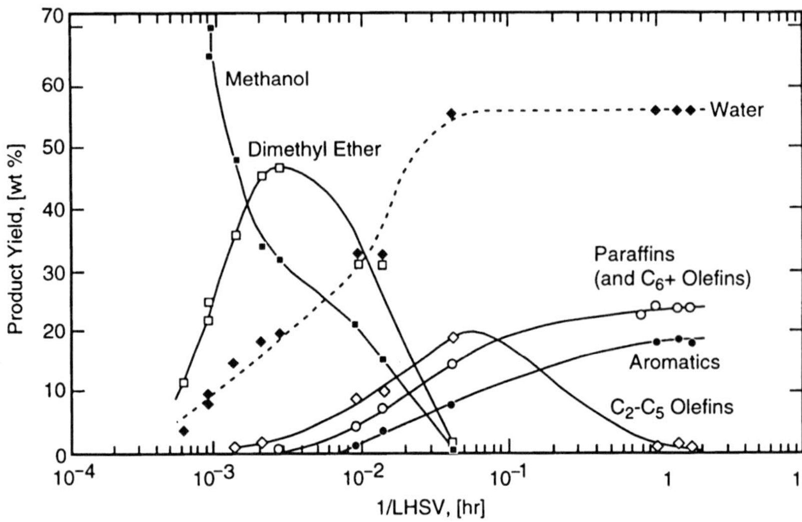

Fig. 7.3 Methanol-to-hydrocarbons reaction path [31] (reproduced by permission of Elsevier, Amsterdam)

$$2\ CH_3OH \underset{+H_2O}{\overset{-H_2O}{\longleftrightarrow}} CH_3OCH_3 \overset{-H_2O}{\longrightarrow} \text{light olefins} \longrightarrow \text{n/iso-parafins higher olefins aromatics napthenses} \tag{7.1}$$

over solid acid catalysts is a protonated surface methoxyl, which is subject to a nucleophilic attack by methanol. The subsequent conversion of light olefins to paraffins, aromatics, naphthenes and higher olefins (step 3 in (7.1)), which proceeds via classical carbenium ion mechanisms with concurrent hydrogen transfer, is well known from hydrocarbon chemistry in acid media. However, the second step in (7.1), which represents the initial C–C bond formation from the C_1 reactants, has been the topic of an extensive discussion throughout the years. A large number of papers are available presenting more than 20 possible mechanistic proposals for the formation of the first C–C bond. For a complete overview the interested reader may consult reference [31] and the references cited therein. Of all the direct mechanisms, the oxonium ylide proposal has received the strongest experimental attention. However, recent research has suggested that direct mechanisms might not be responsible for the sudden boost in hydrocarbon formation after all [32]. Nowadays, however, the importance of the initial C–C bond formation is considered to be minor and the so-called "hydrocarbon-pool" mechanism (a parallel type mechanism) is gaining general acceptance [33]. The "hydrocarbon-pool" mechanism was suggested by Dahl et al [34, 35], who studied the MTHC conversion applying SAPO-34 (a silico-alumino-phosphate material with the chabasite (CHA) structure)

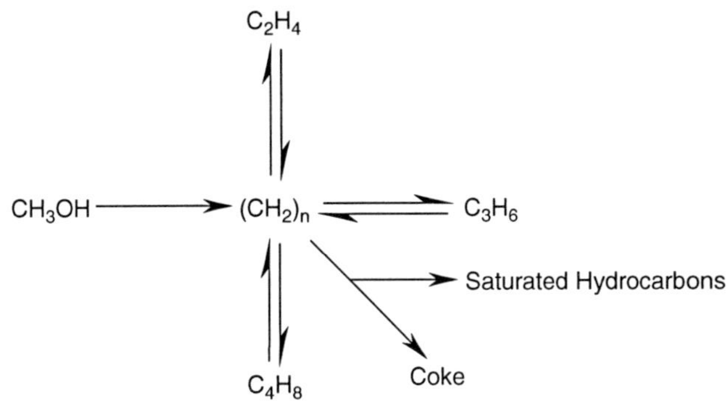

Scheme 7.1 Hydrocarbon pool mechanism [31] (reproduced by permission of Elsevier, Amsterdam)

as catalyst and ^{13}C-labeled methanol as feed [and (^{12}C) ethene (made in situ from ethanol)] (see Scheme 7.1).

The "hydrocarbon-pool"=$(CH_2)_n$ represents an adsorbate, which may have many characteristics in common with ordinary coke, and which might easily contain less hydrogen than indicated. It would perhaps be better represented by $(CH_x)_n$ with $0<x<2^{40}$. Once the C–C bond formation has started, an induction period for olefins formation is observed, since oligocyclization has to produce a significant amount of aromatics ("hydrocarbon pool"), a pre-requisite for a closed loop, in which they behave as catalysts [36]. When using SAPO-34 in the title reaction, the product pattern is much simpler than, e.g., ZSM-5, where a much wider range of products was found. Therefore, it might be easier to obtain a picture of the reaction pathway using SAPO-34, and Dahl et al showed that the consecutive mechanism, insofar propene formation was concerned, did not turn out to be valid. They concluded that only a minor part of the propene molecules may have been formed by addition of methanol to ethene since this would imply a ^{12}C/^{13}C ratio larger than one. The majority of the propene molecules were formed directly from methanol [34, 35].

During the last few years, it has become clear that methylbenzenes play central roles in the "hydrocarbon pool" mechanism and they are regarded as important species taking part during the course of the reaction. For example, isotopic scrambling in all polymethylbenzenes has been found after switching from a ^{12}C-methanol feed to a ^{13}C-methanol feed in the organic material trapped inside a SAPO-34 catalyst [33]. Further, the role of polymethylbenzenes as the major "hydrocarbon pool" species appears to be independent of the microporous catalyst used. Detailed studies on the reactivity of polymethylbenzenes have been carried out over H-BEA zeolite, since this 12-MR zeolite allows direct feeding of polymethylbenzenes. Reacting ^{13}C-methanol and ^{12}C-benzene over zeolite H-BEA gained the formation of methylbenzenes consisting of a ^{12}C-benzene ring and ^{13}C-methyl groups, indicating a "paring" type mechanism (alkyl side chain growth by ring contraction/expansion) responsible for the olefin formation over this zeolite. Studies performed by Svelle et al dealing with the methylation of ^{12}C-ethene, -propene and -butene with

^{13}C-methanol over a H-ZSM-5 catalyst, confirmed that methylation, oligomerization, polymethylbenzene formation (hydrocarbon pool) and cracking reactions took place in parallel, again underlining the complex reaction pathways of the MTHC conversion [33, 37].

Kinetic investigations related to the MTHC conversion normally regard the methanol/DME mixture as a single species. This is confirmed by the observation that the DME formation is much faster than the subsequent reactions, so that the oxygenates are at equilibrium [38, 39]. Based on the autocatalytic nature of the methanol conversion over ZSM-5, Chen and coworkers applied a kinetic model, which takes into account that the rate of disappearance of oxygenates is accelerated by their reaction with olefins. At low conversions (<50%) the conversion of olefins into aromatics can be neglected and a reasonable fit was obtained, meaning that the autocatalytic step is much faster than the formation of the first olefin from the oxygenates. Fitting experimental data obtained on H-ZSM-5 with varying acidity showed a linear correlation between the rate constant of the reaction of oxygenates with olefins and the intrinsic acid activity of the catalyst [38, 40].

The model of Chen et al. was modified by Chang through adding a bimolecular step accounting for the carbene insertion in the primary olefins [41]. By assuming all reactions to be first order and the carbenes to be in steady state, his model predicts a lower rate constant for the reaction of the oxygenates with the olefins than for the carbene insertion. This was found to be in line with the observation that carbene insertion into a C–H bond is slower than carbene addition to double bonds. The comparison with experimental data was acceptable. Chang's description of the carbene insertion in olefins predicts a redistribution of the olefinic species, but not a net increase in the production of olefins.

A kinetic model for the methanol conversion to olefins with respect to methane formation at low conversion numbers was discussed in the literature [42]. The study of Hutchings et al revealed that at low conversion, hydrogen formation from methanol decomposition and the water gas shift reaction was significant over the zeolite ZSM-5, and therefore, these reactions should not be included in the derivation of kinetic models for this reaction. In view of this, the authors considered it as unlikely that methane formation predominantly proceeds via reaction of carbene intermediate with molecular hydrogen [42].

7.2.3 *Methanol to Olefins (MTO)*

7.2.3.1 Catalysts and Reaction Conditions/Deactivation

Catalysts mainly applied for the MTO reaction are ZSM-5 and SAPO-34. Whereas strong acid sites are those mainly responsible for aromatization reactions, moderate acidity is required for the MTO process. ZSM-5 and SAPO-34 have different properties and behavior due to their different compositions and topologies. ZSM-5 is an alumino-silicate possessing a three-dimensional pore structure

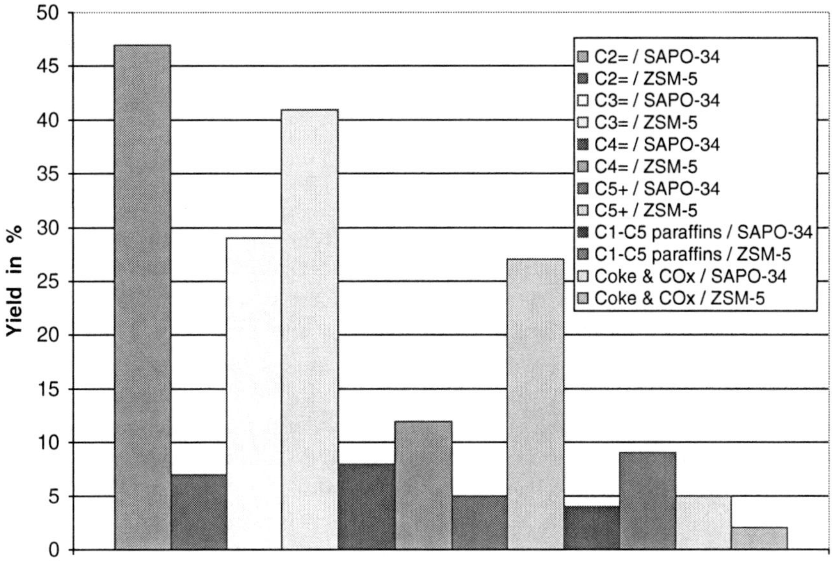

Fig. 7.4 MTO product yields for SAPO-34 and ZSM-5 [43] (reproduced by permission of DGMK, Hamburg)

consisting of 10-membered ring pore openings (medium pore zeolite), whereas SAPO-34 is a small pore silico-alumino-phosphate consisting of an 8-membered ring system with a cage structure. The SAPO-34 catalyzed MTO reaction yields both ethene and propene in variable amounts with very low formation of heavier by-products. Further, the MTO process can be designed for an ethene to propene ratio between 0.75 and 1.5 – at nearly complete methanol conversion and with ethene formation favored at higher severity. Catalyst deactivation is quite fast for SAPO-34 producing mainly aromatic coke. In addition, SAPO-34 is more stable than corresponding zeolite structures towards coking, whereas ZSM-5 reveals slower deactivation by aromatic coke. ZSM-5 produces propene as the major olefin, however, significant higher amounts of C_5^+ and aromatic by-products are formed, compared to SAPO-34 (see Fig. 7.4) [36, 43].

The high selectivity to ethene gives SAPO-34 a significant advantage over other types of catalyst systems, like ZSM-5 or SSZ-13 (synthetic alumino-silicate with CHA structure). In addition, SAPO-34 has a significantly better stability due to a lower rate of coke formation than the other catalytic systems with comparable acid site densities. Further, the catalyst operates best at a certain equilibrium level of coke. Besides coke, heavier olefins are formed via cationic polymerization. Coking, which is responsible for non-irreversible catalyst deactivation, is many times faster during the induction period, then it decreases to a constant rate, roughly with a linear dependence to space velocity. Due to this fact, the catalyst life time is inversely related to the total methanol fed. Not recoverable deactivation has been reported as well, mainly due to water formation (dealumination). The deactivated catalyst can be regenerated by combustion with air. The vapor phase reaction is carried out at temperatures between 425 and 500°C within a pressure range of 1–3 bar [36].

7.2.3.2 Process Technology

Since the MTO conversion is an exothermic reaction, the need to remove the high exothermic heat as well as the need for frequent regeneration led to a fluidized-bed reactor and regenerator design – last but not the least to cope with the fast catalyst decay. The heat of the reaction is controlled by steam generation. The catalyst is sent continuously to the regenerator, where the coke is burned off and steam is generated to remove the heat resulting from burning. After heat removal, the reactor effluent is cooled, and some of the water is condensed. After compression, the effluent passes through a caustic scrubber to remove carbon dioxide and to a dryer to remove water [44]. The selectivity to ethene (48%) and propene (33%) is about 81% (90% if butenes are included as well). SAPO-34 is extremely selective towards ethene and propene formation with the flexibility of altering the ratio between the two olefins by varying the reactor conditions [31]. A 0.5 ton per day demo unit operated by Norsk Hydro in Porsgrunn (Norway) has verified the olefin yields and catalyst performance (see Fig. 7.5).

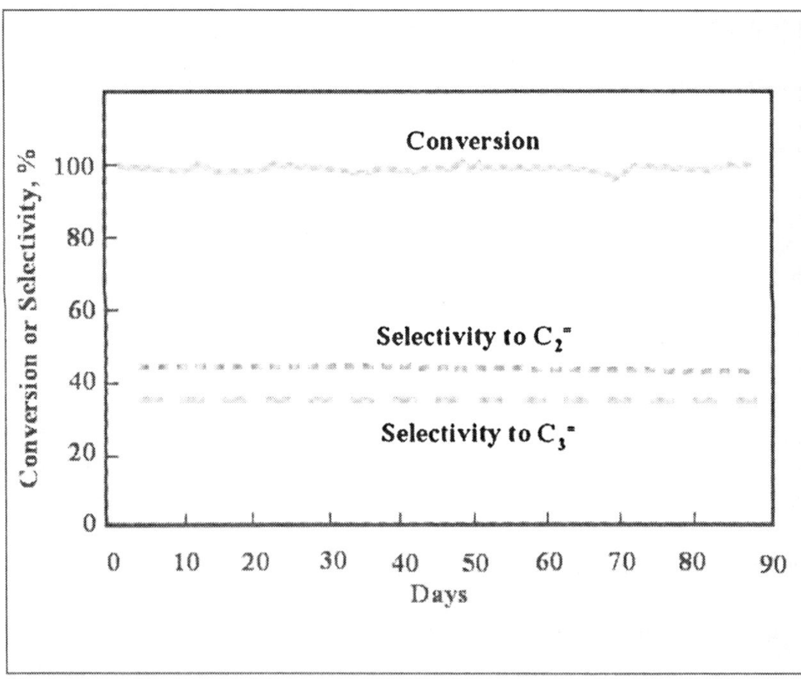

Fig. 7.5 Stability of the MTO process using methanol from a demo plant [43] (reproduced by permission of DGMK, Hamburg)

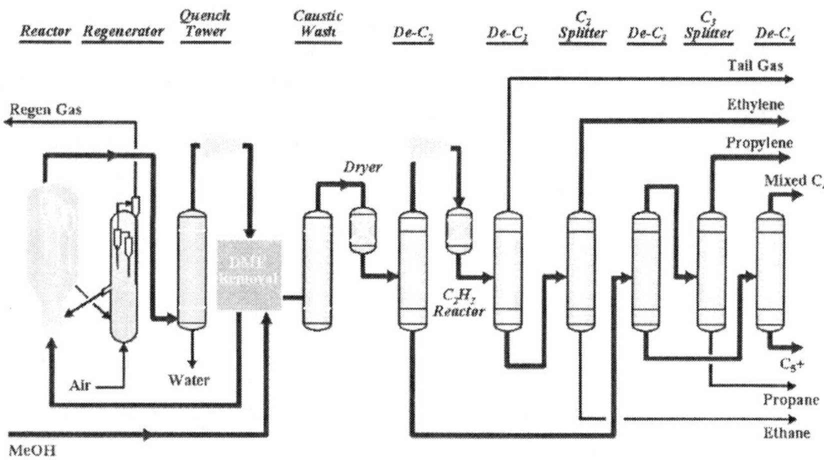

Fig. 7.6 Process flow scheme of the UOP/Norsk Hydro MTO process [43] (reproduced by permission of DGMK, Hamburg)

The fluidized-bed technology provides all advantages in terms of increased product yield, better quality and a very efficient heat recovery. UOP and Norsk Hydro have commercially manufactured the MTO catalyst (MTO-100^{TM}), based on SAPO-34, which has shown the type of attrition-resistance and stability suitable to handle multiple regeneration steps and fluidized-bed conditions. The MTO technology seems now to be ready for commercial use. UOP has, in cooperation with Norsk Hydro, announced in 1996, their SAPO-34 based MTO process to be realized for construction of a 250.000 ton per year plant using a natural gas feedstock for production of ethane [31]. The first commercial plant has been scheduled for 2006 in Nigeria [36, 43]. The process flow scheme of the UOP/Norsk Hydro MTO process is shown in Fig. 7.6.

Finally, Mobil Corp., Union Rheinische Braunkohlen Kraftstoff AG and Uhde GmbH jointly designed, engineered and operated a 4,000 ton per year demo plant at Wesseling (Germany) in order to prove the fluidized-bed MTG process, however, this concept was demonstrated for the MTO process as well, applying a ZSM-5 as catalyst [44].

7.2.3.3 Commercial Aspects/Economic Impact

There is a strong industrial interest for the methanol-to-olefin technology: Besides the MTO process owned by UOP/Norsk Hydro, ExxonMobil has a strong patent position. Further, the Casale Group has rights to a Romanian MTO catalyst and Atofina/UOP have certain perspectives concerning the gas-to-olefins (GTO) in the Middle East. In addition, Lurgi has developed a methanol-to-propene (MTP) process (Chapt. 7.2.6), demonstrated successfully in a pilot unit.

An economical evaluation between an MTO unit and a naphtha cracker – based on a 500 kilo metric tons per annum (kMTA) production of ethene – has been performed: Based on a methanol price of 90 USD/ton, the breakeven point between a naphtha cracker and an MTO unit is around a naphtha price of 150 USD/ton, corresponding to a crude oil price of about 15–16 USD/barrel. This can be taken as an indication of the competitiveness of an MTO unit with a naphtha cracker – with a delivered methanol price in the order of 90 USD/ton [45].

7.2.3.4 Future Perspectives

The MTO process provides an economically attractive alternative to naphtha cracking for the production of ethene and propene, and the corresponding polyolefins (gas-to-polyolefins – GTP) from methanol that can be produced at low cost and in large quantities from natural gas or other hydrocarbon sources. The UOP/Norsk Hydro MTO process and catalyst have been successfully demonstrated and are available for license. MTO/GTO – GTO/GTP is the most economical gas monetization alternative and the plant capacity is flexible with respect to the gas field size. MTO is very flexible with respect to the ethene/propene ratio and the increased future propene demand can be met without an excessive co-production of ethene. The olefin market is large, has a healthy growth and can consume large amounts of gas without being disturbed [43].

The UOP/Norsk Hydro MTO process can be economically viable in different scenarios:

1. Production of methanol at a remote gas field site (less sensitive to the oil price) and transportation of the methanol to an MTO plant or other downstream facilities located at the olefins user's site.
2. An integrated GTO complex at the gas field site and transportation of olefins or polyolefin products to customers.
3. Increased olefins production and feedstock flexibility at an existing naphtha or ethane-propane cracker facility by installing an MTO reactor section and feeding into a revamped cracker fractionation section.
4. A smaller MTO unit using methanol produced in a single-train methanol plant to meet the local demand for olefins or polyolefins, or both [43, 44].

Furthermore, MTO is also considered as a valuable option for the valorization of stranded gas reserves [36].

Finally, the market for ethene and propene is large compared to that of methanol and, therefore, the impact of MTO/GTO on methanol could be strong. Currently, several MTO alternatives are being studied. The abundance of natural gas and the economy of large methanol plants lead to the conclusion that the MTO technology is an attractive alternative to naphtha cracking for the production of ethene and propene.

7.2.4 *Methanol to Gasoline (MTG)*

7.2.4.1 Catalysts and Reaction Conditions/Deactivation

In 1979, the New Zealand government selected the MTG process over the Fischer-Tropsch (SASOL process) for converting natural gas from their extensive Maui and Kapuni fields to gasoline. At that time, Mobil's fixed-bed MTG process was unproven commercially, whereas the SASOL technology was already commercialized. The New Zealand plant started to produce about 14.500 barrels per day gasoline from April 1986, supplying one-third of the nation's gasoline demand. The gasoline production part of the factory was later closed down, due to the price available for gasoline vs. the price of methanol; however, the methanol production part is still in operation [31].

The commercial MTG reaction runs at temperatures around 400°C at a methanol pressure of several atmospheres and uses a ZSM-5 catalyst. The gasoline produced is fully compatible with conventional gasoline. The conversion of MTHC and water is virtually complete and essentially stoichiometric. The reaction is exothermic. A simplified block diagram of the MTG process is shown in Fig. 7.7 [44].

During the MTG process, the ZSM-5 catalyst becomes less effective due to three forms of catalyst deactivation:

1. the deposition of a carbonaceous residue on and in the ZSM-5 catalyst
2. an irreversible loss of activity due to the effect of steam on the zeolite structure and
3. during regeneration, when coke is removed by oxidation, the high temperatures involved may effect the zeolite structure

These processes can be controlled by minimizing the regeneration temperature, the reaction temperature and partial pressure of steam, and adjusting operating procedures to minimize coke formation. The total number of cycles before a new ZSM-5 catalyst charge is required for the MTG process has not yet been reported, but catalyst life appears to be longer than 2 years, the period once anticipated [31].

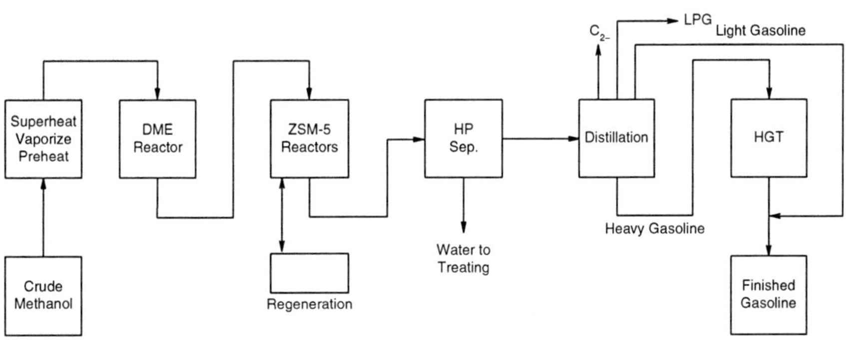

Fig. 7.7 Block diagram of the fixed-bed MTG process [44] (reproduced by permission of Elsevier, Amsterdam)

7.2.4.2 Process Technology

Bench-scale studies of the MTG process have been performed both with fixed-bed and fluidized-bed reactors. Consequently, the major objective of the demo plant was to verify the bench-scale results. The only different variable between a bench-scale unit and a commercial-size reactor is the linear velocity of the reactants. Further, the fluidized-bed MTG process has been proved in the 4,000 ton per year demo plant at Wesseling (Germany) [44]. Finally, Mobil's fixed-bed MTG process was in operation from April 1986 at New Zealand, as mentioned earlier–with a production of about 14.500 barrels per day gasoline to begin with.

7.2.5 *Conclusion*

A broad variety of well-proven technologies for the production of hydrocarbons from methanol is established; however, their future perspectives are depending strongly on the gasoline, methanol and natural gas prices [44].

7.2.6 *Methanol to Propene (MTP)*

The future demand of propene is much larger than that of ethene, and this gap can probably only be bridged by the application of the MTP technology. Currently, worldwide about 70% of the propene is produced via steam cracking, followed by fluidized catalytic cracking (FFC) with about 28% (see Fig. 7.8).

However, a ZSM-5 based catalyst developed by Süd-Chemie AG is used in Lurgi's MTP process. The slower deactivation rate of this catalyst allows the application of a fixed-bed reactor technology, which is adiabatically operated. After a DME production intermediate step, the DME/methanol/water mixture enters the MTP section, which includes three reactors in series with intermediate cooling (see Fig. 7.9).

The process operates at about 425°C and a low pressure of about 1.5 bar, with propene as the main product (about 70% yield and 97% selectivity) and gasoline-range compounds as by-products – which are recycled. The low coke formation allows long cycle times and regeneration in situ at the reaction temperature. The process has been optimized with more than 8,000 h pilot plant operation and has been demonstrated at Tjeldbergodden, Norway in co-operation with Statoil [36].

7.2.7 *TIGAS Process*

The MTG plant in New Zealand has been combined with a methane steam-reforming unit for production of synthesis gas and a methanol plant to produce

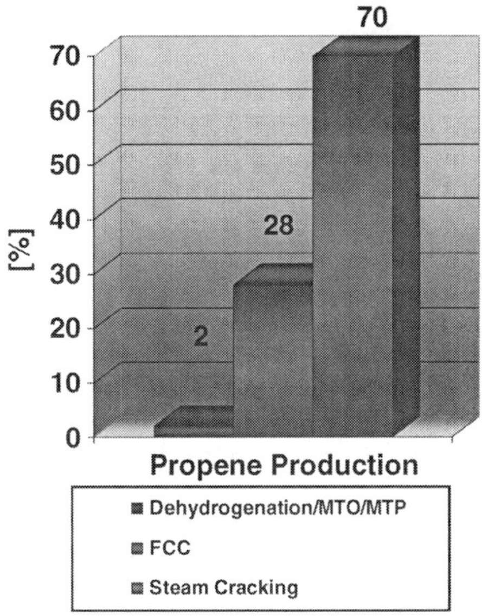

Fig. 7.8 Current propene production figures (reproduced by permission of Statoil and Lurgi)

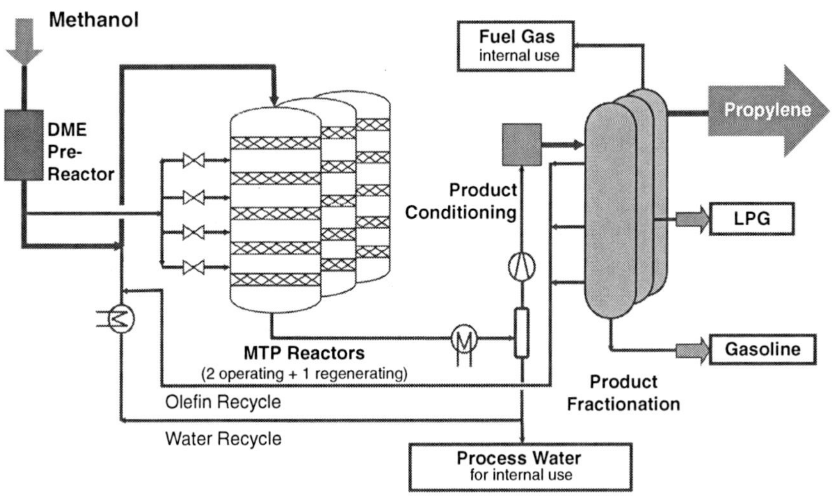

Fig. 7.9 MTP process: simplified process flow diagram (reproduced by permission of Statoil and Lurgi)

gasoline from natural gas. In this so-called TIGAS process (developed by Haldor Topsøe AS) the two process steps – the methanol synthesis and the MTG process – are integrated into one single synthesis loop without isolation of methanol as an intermediate (see Fig. 7.10).

The process economics has been improved by a clever combination and close integration of the different steps. The purpose of the process development work on

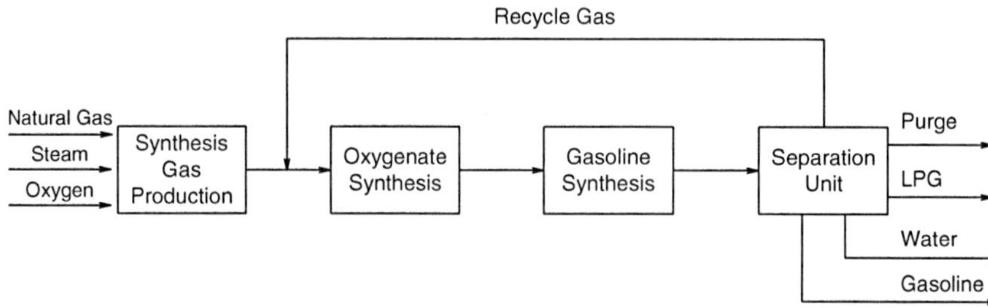

Fig. 7.10 Haldor Topsøe's TIGAS process [46] (reproduced by permission of Elsevier, Amsterdam)

the integrated gasoline synthesis was to modify the three process steps: synthesis gas production, oxygenate synthesis and the MTG conversion – in order to be able to operate all steps at the same procedure and the last two steps in one single synthesis loop.

By selecting combined steam reforming and autothermal reforming for the synthesis gas production, and by using a multifunctional catalyst system (not based on zeolites or related microporous materials), producing a mixture of oxygenates instead of only methanol, the processes can operate at the same pressure of about 20 bar.

The TIGAS process was developed due to the fact that future synthetic fuel plants will be built in remote areas where the price of natural gas is very low and not related to gasoline [44, 46–48].

7.2.8 Mobil's Olefin-to-Gasoline and Distillate Process (MOGD)

7.2.8.1 Introduction/Background

1973 marked the beginning of the energy crisis, and the new interest in synfuels and other chemicals favored the continuation of the methanol-to-hydrocarbon research. Already the MTG and MTO processes represent a sort of chemical factory, to be brought on stream as the technological and/or economic demands arise. One can go a step further and convert the olefins to an entire spectrum of products, through yet another ZSM-5 based process: Mobil's olefin-to-gasoline and distillate process (MOGD), originally developed as a refinery process, which works well coupled with the MTO process. In the MOGD reaction, ZSM-5 oligomerizes light olefins from either refinery streams or MTO into higher-molecular-weight olefins that fall into the gasoline, distillate/diesel and lubricant range (see also Fig. 7.11) [31].

7.2.8.2 Catalyst and Process Operation

Both processes use ZSM-5 as catalyst and the combined MOGD process offers a high valuable gasoline (C_5 to C_{10}) and distillate/diesel (C_{10} to C_{20}) yield in various

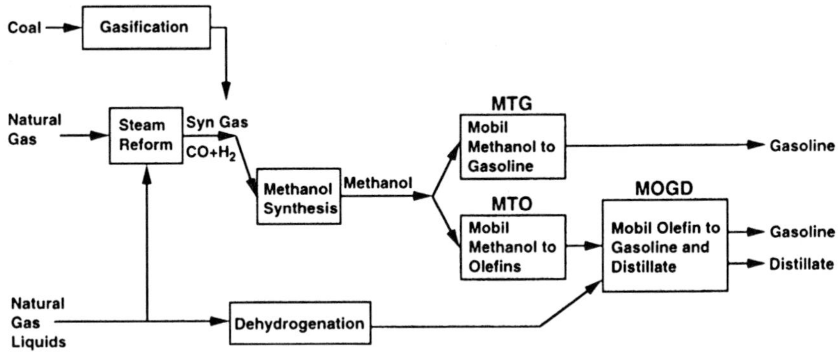

Fig. 7.11 Gasoline and distillate production via methanol and Mobil's ZSM-5 technology [31] (reproduced by permission of Elsevier, Amsterdam)

proportions – starting from light olefins (C_3 to C_4). This process is conducted by a product pattern constrained by both the shape selectivity of the zeolite catalyst as well as the thermodynamics governing the oligomerization reaction. In the MOGD process, the gasoline and distillate/diesel selectivity is larger than 95% of the olefins in the feed and the gasoline/distillate product ratios range from 0.2 to $>$ 100. The distillate product is mostly iso-paraffinic and is an exceptionally good blending stock due to its high cetane index, low pour point and negligible sulfur content. The physical properties, such as flash point, boiling range and viscosity, are comparable with conventional distillate/diesel fuels. While the ZSM-5 pore architecture determines the product pattern (branched hydrocarbons), the conditions of the reaction (temperature, pressure, space velocity) have an influence on the product molecular weight. In general, the reaction window for olefins over ZSM-5 is large, with reactions having been demonstrated for temperatures as low as 40°C and pressures as low as 7 kPa and as high as 14,000 kPa. There is no upper limit concerning the reaction temperature, however, equilibrium constrains become significant at about 330°C [44, 49, 50].

Commercial interest with respect to this reaction is mainly under conditions where there is essentially a complete conversion, and the primary effect of the process variables is to alter the average product molecular weight [50].

7.2.8.3 Thermodynamic Considerations

The conditions which can be used for condensation of light olefins overlap those for cracking of large paraffins or olefins. Thus, the consideration of chemical equilibrium can clarify the effects of process conditions on yields of heavy hydrocarbons obtainable. The problem in calculating equilibrium yields is that extremely large numbers of compounds are involved and very few of the required free energy data are available. The computations can be greatly simplified by the fact that when a group of isomers are in equilibrium with each other they can be treated as a single compound in calculating their equilibrium with other compounds [50].

7.2.8.4 Technical Process

The transformation of these olefin reactions into a large-scale process requires that quite a number of technical aspects have to be taken into account. The two main concerns for the MOGD process are the heat control of the reaction and maximizing the yield of either the gasoline or the distillate/diesel range products. The solution chosen consists of four fixed-bed reactors – three on-line and one in the regeneration mode – during the course of technical operation. The three on-line reactors are operated in series with interstage cooling and condensed liquid recycle to control the heat of reaction. The olefins feed is mixed with a gasoline recycle stream and passed, after heating, through the three reactors. The conceptual design allows that both the maximum gasoline as well as the maximum diesel modes can be achieved by suitably changing the reactor temperature and recycle composition. In order to generate a gasoline-rich stream for recycle to the reactors, a fractionation is applied. The recycle improves the distillate/diesel selectivity as well [44, 50].

References

1. Xu Y, Lin L (1999) Recent advances in methane dehydro-aromatization over transition metal ion-modified zeolite catalysts under non-oxidative conditions. Appl Catal A: Gen 188:53–67
2. Shu Y, Ichikawa M (2001) Catalytic dehydrocondensation of methane towards benzene and naphthalene on transition metal supported zeolite catalysts: Templating role of zeolite micropores and characterization of active metallic sites. Catal Today 71:55–67
3. Lundsford JH (2000) Catalytic conversion of methane to more useful chemicals and fuels: A challenge for the 21st century. Catal Today 63:165–174
4. Wang L, Tao L, Xie M, Xu G, Huang J, Xu Y (1993) Dehydrogenation and aromatization of methane under non-oxidizing conditions. Catal Lett 21:35–41
5. Johns JR, Howe RF (1985) Preparation of molybdenum mordenite from $MoCl_5$. Zeolites 5:251–256
6. Huang M, Yao J, Xu S, Meng C (1992) E.P.R. evidence of the unusual coordinational environment for Mo^{5+} in SAPO-5 molecular sieves. Zeolites 12:810–814
7. Agudo AL, Benitez A, Fierro JLG, Palacios JM, Neira J, Cid R (1992) Surface properties of molybdenum-impregnated ZSM-5 catalysts. J Chem Soc Faraday Trans 88:385–390
8. Lee CW, Saint-Pierre T, Azuma N, Kevan L (1993) Electron spin resonance and electron spin echo modulation studies of oxomolybdenum species in thermally reduced molybdenum-doped H-SAPO-5 and MoH-SAPO-11 silicoaluminophosphate molecular sieves: Comparison of adsorbate coordination with copper-doped H-SAPO-5 and CuH-SAPO-11. J Phys Chem 97:11811–11814
9. Xu Y, Liu S, Wang L, Xie M, Guo X (1995) Methane activation without using oxidants over Mo/HZSM-5 zeolite catalysts. Catal Lett 30:135–149
10. Chen LY, Lin LW, Xu ZS, Li XS, Zhang T (1995) Dehydro-oligomerization of methane to ethylene and aromatics over molybdenum/HZSM-5 catalyst. J Catal 157:190–200

11. Schuurman Y, Mirodatos C (1997) Uses of transient kinetics for methane activation studies. Appl Catal A: Gen 151:305–331
12. Ding W, Li S, Meitzner GD, Iglesia E (2001) Methane conversion to aromatics on Mo/H-ZSM5: Structure of molybdenum species in working catalysts. J Phys Chem B 105:506–513
13. Xu Y, Bao X, Lin L (2003) Direct conversion of methane under non-oxidative conditions. J Catal 216:386–395
14. Wang DJ, Lunsford JH, Rosynek MP (1996) Catalytic conversion of methane to benzene over Mo/ZSM-5. Topics Catal 3:289–297
15. Zhang C-L, Li S, Yuan Y, Zhang W-X, Wu T-H, Lin L-W (1998) Aromatization of methane in the absence of oxygen over Mo-based catalysts supported on different types of zeolites. Catal Lett 56:207–213
16. Su L, Xu Y, Bao X (2002) Study on bifunctionality of Mo/HZSM-5 catalysts for methane dehydro-aromatization under non-oxidative condition. J Nat Gas Chem 11:18–27
17. Zhang W, Ma D, Han X, Liu X, Bao X, Guo X, Wang X (1999) Methane dehydro-aromatization over Mo/HZSM-5 in the absence of oxygen: A multinuclear solid-state NMR study of the interaction between supported Mo species and HZSM-5 zeolite with different crystal sizes. J Catal 188:393–402
18. Liu H, Li Y, Shen W, Bao X, Xu Y (2004) Methane dehydroaromatization over Mo/HZSM-5 catalysts in the absence of oxygen: Effects of silanation in HZSM-5 zeolite. Catal Today 93–95:65–73
19. Malinowski A, Ohnishi R, Ichikawa M (2004) CVD synthesis in static mode of Mo/H-ZSM-5 catalyst for the methane dehydroaromatization reaction to benzene. Catal Lett 96:141–146
20. Bouchy C, Schmidt I, Anderson JR, Jacobsen CJH, Derouane EG, Derouane-Abd Hamid SB (2000) Metastable fcc α-MoC_{1-x} supported on HZSM-5: Preparation and catalytic performance for the non-oxidative conversion of methane to aromatic compounds. J Mol Catal A: Chem 163:283–296
21. Shu Y, Ohnishi R, Ichikawa M (2003) Improved methane dehydrocondensation reaction on HMCM-22 and HZSM-5 supported rhenium and molybdenum catalysts. Appl Catal A: Gen 252:315–329
22. Rodrigues ACC, Monteiro JLF (2004) Dehydro-aromatization of methane under non-oxidative conditions over MoMCM-22: Effect of surface aluminium removal. Stud Surf Sci Catal 147:709–714 (In: Bao X, Xu Y (eds) Natural gas conversion VII)
23. Sarioglan A, Erdem-Senatalar A, Savasci ÖT, Ben Taarit Y (2004) The effect of dealumination on the apparent and actual rates of aromatization of methane over MFI-supported molybdenum catalysts. J Catal 226:210–214
24. Su L-L, Li Y-G, Shen W-J, Xu Y-D, Bao X-H (2004) Methane dehydroaromatization over alkali-treated MCM-22 supported Mo catalysts: Effect of porosity. Stud Surf Sci Catal 147:595–600 (In: Bao X, Xu Y (eds) Natural gas conversion VII)
25. Su L, Liu L, Zhuang J, Wang H, Li Y, Shen W, Xu Y, Bao X (2003) Creating mesopores in ZSM-5 zeolite by alkali treatment: A new way to enhance the catalytic performance of methane dehydroaromatization on Mo/HZSM-5 catalysts. Catal Lett 91:155–167
26. Liu B, Yang Y, Sayari A (2001) Non-oxidative dehydroaromatization of methane over Ga-promoted Mo/HZSM-5-based catalysts. Appl Catal A: Gen 214:95–102
27. Liu S, Wang L, Ohnishi R, Ichikawa M (1999) Bifunctional catalysis of Mo/HZSM-5 in the dehydroaromatization of methane to benzene and naphthalene: XAFS/TG/DTA/MASS/FTIR characterization and supporting effects. J Catal 181:175–188
28. Chu W, Qiu F (2003) Remarkable promotion of benzene formation in methane aromatization with ethane addition. Topics Catal 22:131–134
29. Liu H, Li T, Tian B, Xu Y (2001) Study of the carbonaceous deposits formed on a Mo/HZSM-5 catalyst in methane dehydro-aromatization by using TG and temperature-programmed techniques. Appl Catal A: Gen 213:103–112

30. Kinage AK, Ohnishi R, Ichikawa M (2003) Marked enhancement of the methane dehydroaromatization toward benzene using effective Pd catalytic membrane reactor with Mo/ZSM-5. Catal Lett 88:199–202
31. Stöcker M (1999) Methanol-to-hydrocarbons: Catalytic materials and their behaviour. Micropor Mesopor Mater 29:3–48
32. Lesthaeghe D, Van Speybroeck V, Marin GB, Waroquier M (2005) What role do oxonium ions and oxonium ylides play in the ZSM-5 catalysed methanol-to-olefin process? Chem Phys Lett 417:310–316
33. Olsbye U, Bjørgen M, Svelle S, Lillerud K-P, Kolboe S (2005) Mechanistic insight into the methanol-to-hydrocarbon reaction. Catal Today 106:108–111
34. Dahl IM, Kolboe S (1994) On the reaction mechanism for hydrocarbon formation from methanol over SAPO-34: I. Isotopic labeling studies of the co-reaction of ethene and methanol. J Catal 149:458–464
35. Dahl IM, Kolboe S (1993) On the reaction mechanism for propene formation in the MTO reaction over SAPO-34. Catal Lett 20:329–336
36. Bellussi G, Pollesel P (2005) Industrial applications of zeolite catalysts: Production and uses of light olefins. Stud Surf Sci Catal 158:1201–1212 (In: Cejka J, Zilková N, Nachtigall P (eds) Molecular sieves: From basic research to industrial applications)
37. Svelle S, Rønning PO, Kolboe S (2004) Kinetic studies of zeolite-catalyzed methylation reactions: 1. Co-reaction of [^{12}C]ethene and [^{13}C]methanol. J Catal 224:115–123
38. Froment GF, Dehertog WJH, Marchi AJ (1992) Chapter 1 in "A review of the literature": Zeolite catalysis in the conversion of methanol into olefins. Catalysis 9:1–64
39. Chang CD, Silvestri AJ (1977) The conversion of methanol and other O-compounds to hydrocarbons over zeolite catalysts. J Catal 47:249–259
40. Chen NY, Reagan WJ (1979) Evidence of autocatalysis in methanol to hydrocarbon reactions over zeolite catalysts. J Catal 59:123–129
41. Chang CD (1980) A kinetic model for methanol conversion to hydrocarbons. Chem Eng Sci 35:619–622
42. Hutchings GJ, Gottschalk F, Hunter R (1987) Comments on "Kinetic model for methanol conversion to olefins" with respect to methane formation at low conversion. Ind Eng Chem Res 26:635–637
43. Kvisle S, Nilsen HR, Fuglerud T, Grønvold A, Vora BV, Pujado PR, Barger PT, Andersen JM (2002) Methanol to olefins: state of the art and perspectives. Erdöl Erdgas Kohle 118:361–365
44. Keil FJ (1999) Methanol-to-hydrocarbons: Process technology. Micropor Mesopor Mater 29:49–66
45. Heber J (2000) 3rd Asia olefins and polyolefins markets conference 2000, Bangkok, Thailand, 20–21 January 2000
46. Topp-Jørgensen J (1988) Topsøe integrated gasoline synthesis – The TIGAS process. Stud Surf Sci Catal 36:293–305 (In: Bibby DM, Chang CD, Howe RF, Yurchak S (eds) Methane conversion)
47. Maxwell IE, Stork WHJ (2001) Hydrocarbon processing with zeolites. Stud Surf Sci Catal 137:747–819 (In: van Bekkum H, Flanigen EM, Jacobs PA, Jansen JC (eds) Introduction to zeolite science and practice, 2nd edn)
48. Stöcker M (2005) Gas phase catalysis by zeolites. Micropor Mesopor Mater 82:257–292
49. Avidan AA (1988) Gasoline and distillate fuels from methanol. Stud Surf Sci Catal 36:307–323 (In: Bibby DM, Chang CD, Howe RF, Yurchak S (eds) Methane conversion)
50. Tabak SA, Krambeck FJ, Garwood WE (1986) Conversion of propylene and butylene over ZSM-5 catalyst. AIChE J 32:1526–1531

Chapter 8
Applications in Synthesis of Commodities and Fine Chemicals

Michel Guisnet and Matteo Guidotti

Abstract Channels and cages of microporous and mesoporous materials can be considered as nanoreactors whose properties (topology, morphology, nature of the active site and chemical surroundings) can be finely tuned and fitted to the synthesis of functionalised high-added value molecules, such as commodities and fine chemicals.

This text summarises a selected list of examples and reactions where zeolites, zeotypes and ordered mesoporous inorganic molecular sieves can be successfully employed as catalysts with enhanced performance with respect to conventional systems. These applications encompass electrophilic substitution of aromatics (alkylation, acylation, nitration, halogenation), acid-catalysed rearrangement of olefins, epoxides, oximes and arylesters, synthesis of amines and pyridines, cycloaddition, epoxidation, ammoximation, oxidation of C–H bonds and alcohols as well as stereoselective reactions (diastereo- and enantioselective systems) and cascade one-pot transformations over multifunctional solid catalysts.

8.1 Introduction

Microporous molecular sieves (zeolites and zeotypes) showed to be active and selective catalysts in an incredibly large number of transformations of

M. Guisnet
Faculté des Sciences Fondamentales et Appliquées, Université de Poitiers, UMR CNRS 6503, 40 av. du Recteur Pineau, 86022 Poitiers, France
Departamento de Engenharia Quimica, IST-Technical University of Lisbon, Av. Rovisco Pais, 1, 1049-001 Lisboa, Portugal
e-mail: michel.guisnet@univ-poitiers.fr

M. Guidotti
CNR-Istituto di Scienze e Tecnologie Molecolari, IDECAT-CNR Unit, Dip. Chimica IMA, "L. Malatesta", via G. Venezian, 21 20133 Milano, Italy
e-mail: m.guidotti@istm.cnr.it

A.W. Chester and E.G. Derouane (eds.), *Zeolite Characterization and Catalysis*,
DOI 10.1007/978-1-4020-9678-5_8, © Springer Science+Business Media B.V 2009

hydrocarbons and functionalised compounds. This broad applicability of zeolites in catalysis is due to three main factors:

- The rich variety of their active sites (protonic and Lewis acid sites, basic sites, redox-active sites) which may have very different physicochemical properties (strength, density, etc.) and which can act independently or cooperatively for catalysing bi- and even multi-functional processes [1].
- Their so-called "shape selectivity" related to the molecular size of the pores (channels, cages, etc.) where the catalytic reactions occur [2, 3]. Zeolite catalysts can be considered as a network of "nanoreactors" whose characteristics (their shape and size and those of their apertures) can affect positively not only the selectivity, but also the rate and stability of the reactions.
- The properties of both the active sites and the pore systems can be finely tuned and fitted to the desired reaction by established and well-mastered methods [1, 4].

When considering commercial processes, zeolites present additional advantages over conventional systems, especially for acid-catalysed transformations. Indeed, metal chlorides, such as $AlCl_3$, or Brønsted acids, such as H_2SO_4, which were firstly used as catalysts are corrosive, cannot be easily recovered from the reaction mixture and are generally responsible for large amounts of wastes. However, this is not the case with acidic zeolites which, in addition can be easily used in continuous processes.

Thanks to their remarkable properties, zeolite catalysts play a major role in the large-scale processes of refining and petrochemical industries (FCC, hydrocracking, C_5–C_6 alkane isomerisation, production of *p*-xylene, ethylbenzene and cumene, etc.). Conversely, they play only a limited role in the synthesis of fine chemicals. Several practical and fundamental reasons can be found [5]:

- The relatively small size of commercial units and hence the relatively low absolute amount of wastes, even for high values of E-factors, i.e., the weight ratio between the amount of wastes and the amount of desired products [6, 7].
- The pressure of time that leads industry to prefer often well-established stoichiometric processes (or homogeneously catalysed ones) to innovative heterogeneous pathways. Moreover, owing to large differences in national environmental rules, these polluting processes can be still carried out in countries with less demanding restrictions.
- The completely independent historic evolution of heterogeneous catalysis (closely related for a long time with gas-phase refining and petrochemical processes) and of organic synthesis (often in liquid phase), that causes difficult communication and reciprocal understanding between experts in these fields.

Since the development of synthetic methods in 1990s, mesoporous molecular sieves (among them some widely used materials, e.g., MCM-41 or SBA-15) have found a rapid and broad success in academia, as demonstrated by the exponentially increasing number of papers in literature, but then they have a limited use in industry. This is mainly due to the fact that the synthetic procedures of these solids are more expensive and more difficult compared to microporous sieves and that there is a shorter tradition of development and application of these relatively new catalysts. Actually, ordered mesoporous materials can find an industrial use as sorbents and for separation purposes (thanks to their well-defined pore-size distribution) and several patents report their potential or practical application as catalysts for the transformation of bulky chemicals, but currently no large-scale industrial synthetic processes are based on this kind of catalysts. Nevertheless, since they possess very wide channels and cavities, they are particularly suitable as solids for the catalytic conversion of bulky and high molecular-mass molecules, as fine chemicals are. Furthermore, the high added-value of the products can justify the use of catalysts prepared by expensive and time-consuming routes.

Selected examples of application of microporous and mesoporous catalysts for the commercial synthesis of relatively complex and functionalised compounds are presented here. The focus is placed on the conditions that have to be satisfied in order to replace the conventional homogeneous processes by environmentally friendly processes based on zeolite catalysts.

8.2 Electrophilic Substitution of Aromatics

The mechanism of the acid-catalysed electrophilic substitution of aromatics (shown in Fig. 8.1 in the case of benzene as substrate) implies the attack of the aromatic ring by an activated electrophile E^+ formed on protonic sites, followed by elimination of

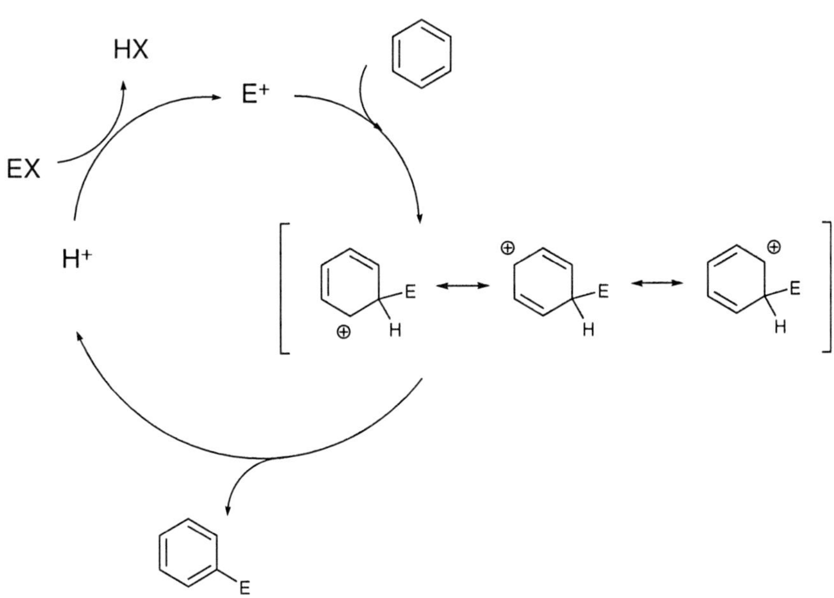

Fig. 8.1 Electrophilic substitution of benzene

a proton located on the sp^3 carbon. The attack of the aromatic ring is generally considered as the kinetic limiting step [8].

The aromatic ring of the product can undergo additional electrophilic attacks (by the same electrophile or by another electrophilic species) leading to bi- then to poly-substituted products. The rate of this poly substitution depends on the activation or deactivation of the aromatic ring by the E substituent. The electron donor or acceptor character of the E group can be due to inductive effect, to resonance effect or to both of them. With electron donor E substituents, such as alkyl groups, the poly substitution is favoured, whereas electron acceptor substituents such as acyl, nitro, halogen groups inhibit further substitutions. The regioselectivity of poly substitution depends also on the E substituent. The substituents which activate the aromatic ring, either by inductive (e.g., alkyl groups) or by resonance effect (e.g., NH_2 groups) orient the substitution in *ortho* and *para* positions. Those which deactivate the ring (e.g., nitro, acyl groups) orient the substitution in *meta* positions. As an exception, however, halogen substituents, though deactivating the aromatic ring, orient the substitution in *ortho* and *para* positions [8].

8.2.1 Alkylation of Aromatics

The know how for the alkylation of aromatic compounds over zeolitic catalysts, which allows the large scale production of petrochemicals such as ethylbenzene and cumene, is also applied in the synthesis of alkyl binuclear aromatics. Such compounds are precursors of fine chemicals and intermediates [3, 9–11]: for instance, vitamin K is commercially produced from 2-methylnaphthalene (MN); 2,6-dialkylnaphthalene (26DAN) are precursors of polyethylene naphthalate (PEN), of polyisobutylene naphthalate (PBN) and liquid crystal polymers; 4,4′-dialkylbiphenyl (44′DABPH) can be oxidized into 4,4′-biphenyldicarboxylic acid, which is a monomer for high-performance polymers and liquid crystal applications.

Producing selectively the desired compound is the main problem to be solved. Considering the Friedel-Crafts alkylation mechanism, a selective production is not likely with conventional catalysts. Indeed alkyl groups activate the aromatic ring and, hence, favour the formation of polyalkylated products. Moreover, over conventional acid catalysts, the regioselectivity of alkylation is not very high and many of the numerous isomers (e.g., there are two monoalkyl and ten dialkyl naphthalene isomers against only one mono and three dialkyl benzene isomers) appear among the reaction products. In addition, the very high cost for the separation of the desired isomer as well as the necessity of isomerisation - recycling steps make this route economically unfeasible. The search for selective catalysts was naturally focussed on shape selective zeolites which typically limit polyalkylation and orient the alkylation to the less bulky isomers, which are the desired products, such as, for example, 2,6-dialkylnaphtalene (26DAN) and 4,4′-dialkylbiphenyl (44′DABBPH). The shape-selective effect of zeolites is more pronounced with bulky alkyl substituents and, thus, ethylation and especially isopropylation is more selective than

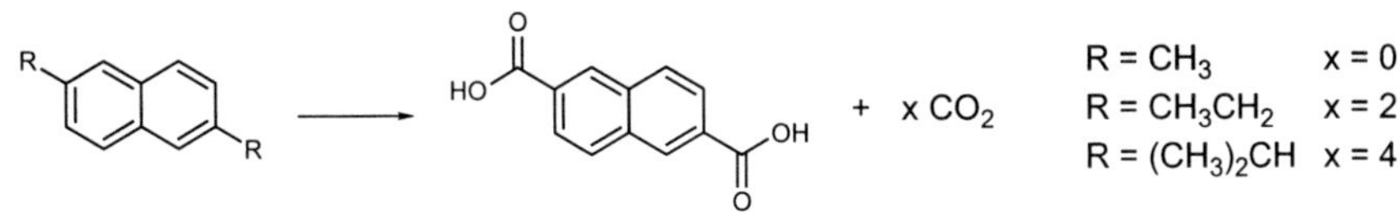

Fig. 8.2 Oxidation of 2,6-dialkylnaphthalenes

methylation reactions. However, this approach may be unfavourable in terms of atom economy. In fact, 2 or 4 carbon atoms are lost per molecule during the oxidation of diethyl and diisopropyl aromatics, respectively, into dicarboxylic acid polymer precursors (Fig. 8.2).

8.2.1.1 Selective Synthesis of 2,6-Dimethyl Naphthalene

Two main types of processes were developed for the production of 2,6-dimethyl-naphthalene (2,6-DMN): the first based on Friedel-Crafts methylation and trans methylation of naphthalene (N), the second on the formation of the naphthalene skeleton from toluene, *ortho* or *para* xylenes (Fig. 8.3). Most of the routes [12] involve one or two zeolite catalysed steps.

The Teijin [13] and Mobil-Kobe [14] processes based on route 1 imply two steps: (a) transmethylation between DMN and N to form MN with simultaneous DMN isomerisation to increase the amount in 2,6 isomer and (b) alkylation with methanol of MN into DMN. The main problem is due to the poor selectivity to 2,6-DMN (in both steps, the ten isomers of DMN are formed) and therefore to the costly separation of the desired isomer. The separation of 2,6-DMN from 2,7-DMN is particularly difficult. The two isomers cannot be separated by distillation, because their boiling points differ by 0.3°C only. Moreover, as they form eutectic crystals with a 2,6-DMN/2,7-DMN ratio of 0.7, their separation by crystallization is possible only from mixtures with much higher ratios. Such optimal conditions can be satisfied by an adequate choice of the zeolite pore structure and of the operating parameters. Thus, good results were obtained by Mobil [15] by using the MWW zeolite (MCM-22) employed for ethylbenzene and cumene synthesis. 2,6-DMN/2,7-DMN ratios of 1.2–1.5 were reached under usual operating conditions and even values of 2.3 were recorded with 2-MN alkylation with methanol. Better results were obtained over a one-dimensional large-pore zeolite (MTW) having puckered channels with 0.56×0.60 nm apertures, only if polymethylbenzenes (e.g., trimethylbenzenes, TMB) are added to the feed. [178] explains the excellent performance of MTW as due to a restricted transition-state shape selective transmethylation of MN with polymethylbenzenes.

The high cost of 2,6-DMN separation explains why synthesis schemes requiring the formation of the naphthalene skeleton were explored (routes 2-1 and 2-2, Fig. 8.3). Such routes present the great advantage of avoiding the formation of 2,7-DMN. Up to date, only the multistep BP-Amoco process (route 2-2) is

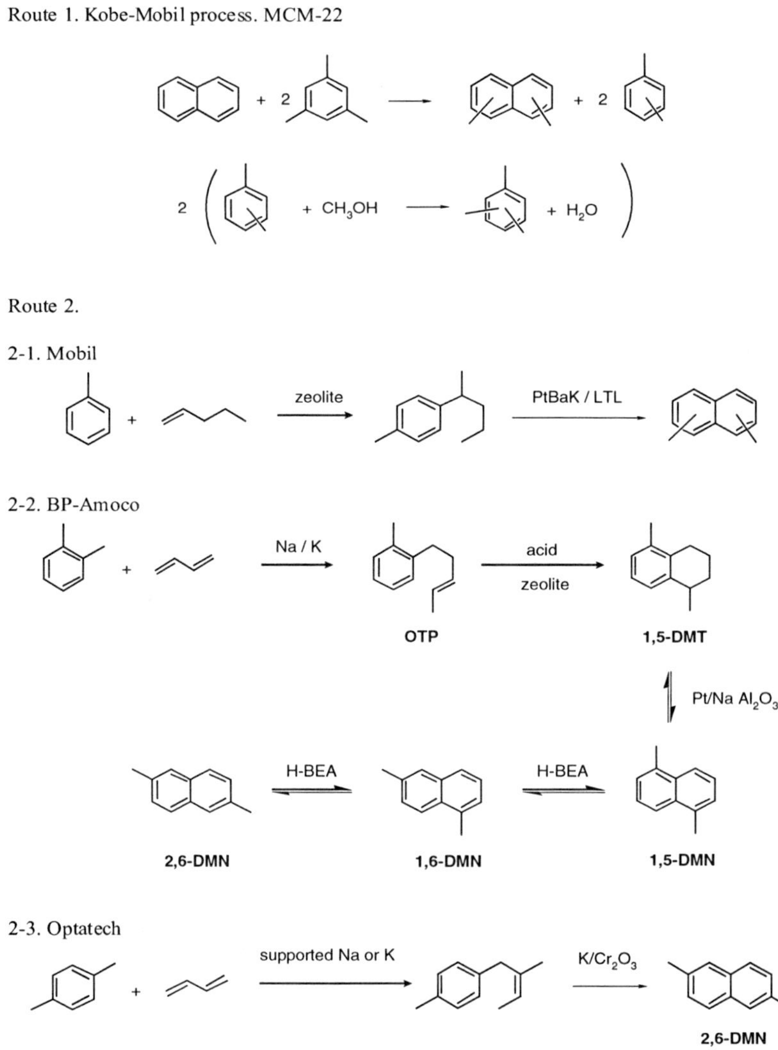

Fig. 8.3 Synthesis of 2,6-dimethylnaphthalene. Main commercial processes Route 1. Kobe-Mobil process. MCM-22 Route 2. 2-1. Mobil 2-2. BP-Amoco 2-3. Optatech

operating at large scale (30 kton/year). However, the more simple Cataryl process (route 2-3) developed by Optatech [16] with a first step similar to that of the BP-Amoco process, but from *p*-instead of *o*-xylene seems promising. The BP-Amoco process comprises four successive catalytic reaction steps: alkenylation of *o*-xylene with butadiene over a basic catalyst, cyclization of the OTP product over an acid zeolite, dehydrogenation of 1,5-DMT into 1,5-DMN over Pt/Na-Al_2O_3 and isomerisation of 1,5-DMN into the 2,6-, 1,5-, 1,5-mixture over a H-BEA zeolite (route 2-2, Fig. 8.3). Under the chosen operating conditions (in particular, moderate temperature), such isomerisation is very selective and this facilitates significantly the separation and purification steps to obtain the desired 2,6-DMN. The high selectivity is due to the fact that methyl shifts from α to β position on the same ring are highly probable, whereas those from β to β position, or from one ring to another, are unlikely. As the isomerisation by methyl shift proceeds through a successive scheme:

$$1,5-\mathrm{DMN}\rightleftarrows 1,6-\mathrm{DMN}\rightleftarrows 2,6-\mathrm{DMN}$$

The optimal 2,6-DMN yield is obtained at the thermodynamic equilibrium within the above triad. Furthermore, to catalyse this reaction, the zeolite micropores should be large enough to allow the diffusion of the most hindered 1,5-DMN isomer. The values of the energy diffusion barriers show that large-pore zeolites, such as H-BEA (selected for BP-Amoco process) are suitable for such isomerisation, while medium-pore zeolites are not. Recent papers [17–19] show that alkyl benzenic compounds, used as solvents in 1,5-DMN isomerisation over H-BEA and as desorbents in the adsorptive separation of 2,6-DMN from the triad, play a positive role in both processes. Toluene, which is not transformed under the conditions of isomerisation is the most appropriate additive for producing high purity 2,6-DMN.

8.2.1.2 Selective Synthesis of 2,6-Diisopropylnaphthalene (2,6-DIPN) and of 4,4′-Diisopropylbiphenyl (44′-DIBPH)

The shape selectivity effect of zeolites can be strongly increased by choosing alkyl groups more hindered than methyl groups. The isopropyl group is the best compromise between steric hindrance and carbon atom economy and various alkylating agents are used for this purpose: propene, isopropanol, isopropyl bromide and diisopropylbenzene, higher selectivity values being obtained with the two first reactants.

Molecular graphics screening of some potential 2,6-DIPN selective catalysts [20] indicate that the one-dimensional large-pore MOR zeolite was a promising candidate, which was experimentally verified [21]. Furthermore, this zeolite was also claimed as highly selective catalyst for biphenyl diisopropylation [3]. Various treatments, namely, dealumination, passivation of the external surface, etc. improve the selectivity and the stability of the zeolite. Dealumination creates mesopores and allows a quasi three-dimensional diffusion of the organic molecules limiting the formation and deposition of heavy products (the so-called "coke") within the channels as well as their deactivating effect [22]. Furthermore, all the treatments eliminate the acidic sites of the external surface responsible for unselective diisopropylation and for secondary isomerisation of the desired product.

Diisopropylation of N over MOR catalysts occurs with a high selectivity (>90%) to β,β'-diisopropyl naphthalenes (DIPN). The desired 2,6-DIPN product is largely favoured and values of the 2,6-DIPN/2,7-DIPN molar ratio close to 3 can be

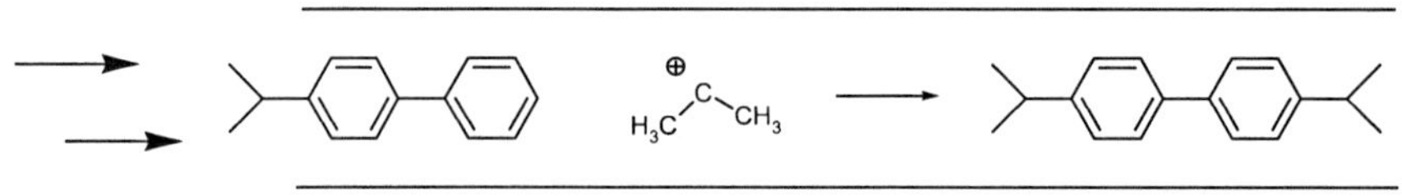

Fig. 8.4 Isopropylation of isopropylbiphenyl over H-MOR

obtained. This latter observation can be related to significant energy barriers in the diffusion of the slightly bulkier 2,7-DIPN molecules (0.662 nm × 0.726 nm × 1.376 nm) along the monodimensional channels of MOR and no limitation in the diffusion of the slightly smaller 2,6-DIPN molecules (0.661 nm × 0.661 nm × 1.423 nm). Therefore, in these channels, the attack of 2-MIPN by the isopropyl group preferentially occurs at the position 6, with formation of 2,6-DIPN. The same explanation was proposed to justify the high selectivity obtained over H-MOR to the less hindered diisopropyl biphenyl isomer (44′-DIBPH) (Fig. 8.4). An additional explanation (electronic transition-state selectivity) based on a computational analysis of the frontier electron density at ring carbons of MN, was advanced by Song et al. [3]. In 2-MN, the frontier electron density was found to be higher at position 6 than at position 7, with therefore a higher reactivity towards electrophilic substitution, hence a preferential formation of 2,6-DIPN.

Unfortunately, most of the reported data have been obtained in batch reactors, hence cannot be easily used to develop industrial processes which are generally operated under flow conditions. Moreover, no valuable information on catalyst deactivation by coking, which is the major problem to be solved can be drawn from batch experiments.

8.2.2 Acylation of Aromatics

Aromatic acylation is the main route for the synthesis of aromatic ketones that are widely used for the production of pharmaceuticals, pesticides, flavours, perfumes, UV absorbers and dyes [11, 23, 24]. Industrially (e.g., Ibuprofen synthesis), this reaction is often carried out in batch reactor with acyl chlorides as acylating agents and using Lewis acid metal chlorides ($AlCl_3$, $ZnCl_2$, etc.) as catalysts, with as a consequence major environmental problems. Indeed, as a very stable 1:1 molar complex is formed between the metal chloride and the arylketone, an over-stoichiometric amount of metal chloride has to be used and the following hydrolysis of the complex leads to the production of huge quantities of hydrochloric acid and corrosive effluents to be neutralised, treated, and properly disposed of. For these reasons, great efforts were taken for the liquid-phase acylation of aromatics and heteroaromatics over acid zeolite catalysts with anhydrides as acylating agents, which could constitute a sustainable and eco-friendly alternative.

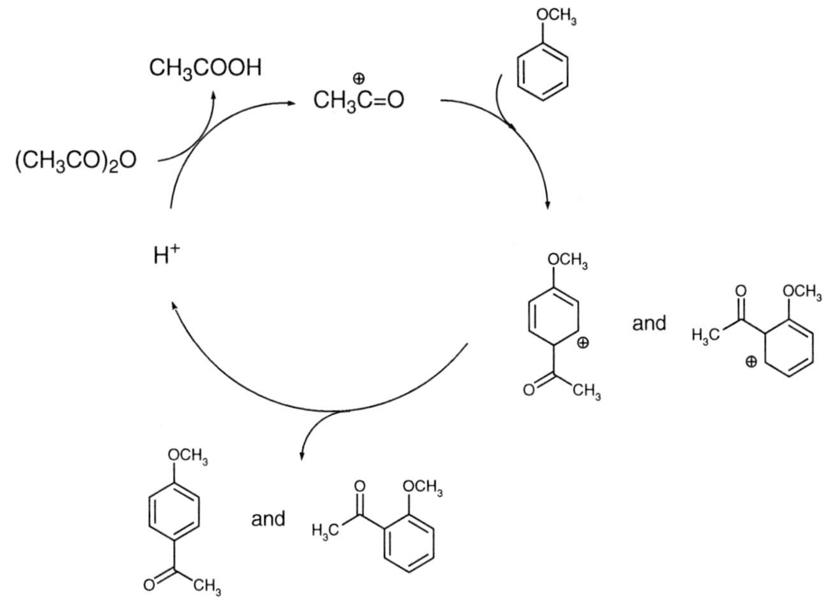

Fig. 8.5 Acetylation of anisole with acetic anhydride. Reaction mechanism

Most of the works deal with the acetylation over large-pore H-FAU and H-BEA zeolites and the medium-pore H-MFI (H-ZSM-5), often in batch reactors and in the presence of a solvent [24, 25]. The acylium cation resulting from acetic anhydride protonation is generally considered as the acylating species. Ketene, which was shown to be formed under the operating conditions [26], was also proposed. However, its participation in the reaction was recently rejected thanks to the use of perdeuterated acetic anhydride [27]. As could be expected from the reaction mechanism (Fig. 8.5), the acylation rate strongly depends on the degree of activation of the aromatic ring by substituents and the polyacylation of the aromatic ring is inhibited by the deactivating effect of the acyl group. Kinetic studies and adsorption experiments [28–30] demonstrate a strong effect of the competition among reactant, solvent and product molecules for occupancy of the zeolite micropores and for chemisorption on the protonic acid sites. As a consequence, an inhibition of acetylation by the acetylated products (i.e., an autoinhibition), which are more polar and bulkier than the substrates, and also by the very polar acetic acid product is generally observed. This inhibition is responsible for a rapid decrease in reaction rate with as a result, a plateau in the yield in acetylated products after short times under batch conditions (Fig. 8.6). The larger the difference in polarity between substrate and product molecules, the more pronounced is the effect [31, 32]. This constitutes a severe limitation for the acylation of hydrophobic substrate molecules such as methyl- and fluoro-substituted aromatics, which are weakly activated, and hence poorly reactive.

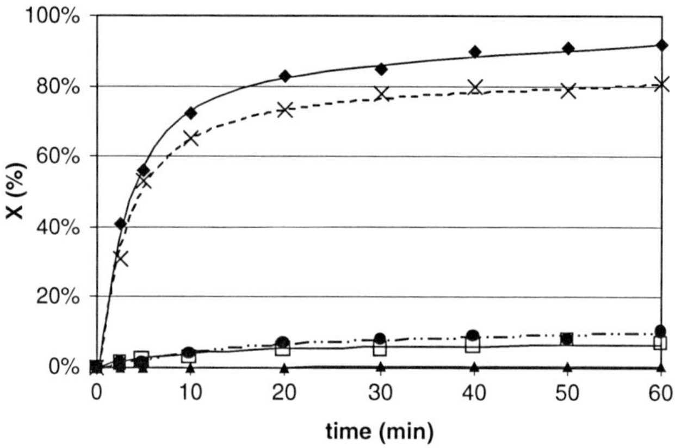

Fig. 8.6 Acetylation in batch reactor of various substrates: anisole (◆), 2-methoxynaphthalene (×), toluene (●), 2-methyl naphthalene (□) and fluorobenzene (▲) over a H-FAU zeolite. Conversion X (%) vs. reaction time. Reproduced (adjusted) from [32], with permission from IUPAC

Moreover, owing to their long residence-time within the zeolite micropores, the polar product molecules undergo secondary condensation reactions leading to bulkier and generally more polar products, which are consequently more strongly retained and, even, sterically blocked within the pores. Actually, in the micropores of a H-BEA zeolite used for long reaction times for anisole acetylation, di- and tri-acetylated anisoles (resulting from additional acetylation of the side chain of the desired *p*-methoxyacetophenone) and various other heavy side products were found. These bulky and polar-trapped compounds block the access of nitrogen, as well as of the reactants, to the micropores. The pore blockage is the main cause for zeolite deactivation during acylation reactions [25].

As the acetylation of poorly activated substrates is limited by both their low reactivity and the strong inhibiting effect of the polar product molecules, most of the studies dealt with the acetylation of aromatic ethers (anisole, veratrole, methoxynaphthalene) and heterocyclic compounds (furans, thiophenes, pyrroles). Indeed, the liquid-phase acetylation of such activated substrates can be efficiently and selectively carried out over zeolite catalysts under mild conditions. The key point is to favour desorption of the bulky and polar desired products. This can be enhanced by:

1. choosing the adequate zeolite type and tuning the physico-chemical properties: namely, use of large pore three-dimensional zeolite under the form of nanosized crystals and with hydrophobic properties
2. using a flow reactor (which favours the removal of the products by sweeping) rather than a batch reactor
3. choosing the proper operating conditions: high substrate to acetic anhydride molar ratio; eventual use of a suitable solvent; temperature high enough to help desorption, but low enough to avoid extensive secondary reactions; high space velocity values to promote the sweeping of the products out of the catalyst

Table 8.1 Acetylation of anisole. Comparison of the characteristics of the conventional process and the zeolite-based one

	Conventional process ($AlCl_3$)	Novel process (H-BEA)
Yield to anisole	95%	95%
Chemical+Physical steps	8	2
Aqueous effluents (t/t anisole)	4.5	0.035
Effluent composition (wt.%)	68.7	99
H_2O	5	0
Al^{3+}	24	0
Cl^-	2.3	1
Organics		

These rules are applied in the industrial processes developed by Rhodia for the acetylation of anisole and veratrole. They are also valid for aromatic acylation with bulkier acylating agents (benzoylation, etc.) and their strict application can allow the acetylation of poorly activated substrates such as toluene.

Table 8.1 shows the dramatic improvement brought by the substitution of the conventional technology of anisole acetylation with $AlCl_3$ catalyst by the new technology with a H-BEA zeolite catalyst. The new process is much more environmentally friendly and more economical than the previous one, i.e., it is typically a green process.

8.2.3 Nitration of Aromatics

Nitroaromatic compounds are important intermediates for the synthesis of drugs, pesticides, dyes, polymers and explosives. The commercial nitration process, developed nearly 150 years ago, uses a mixture of nitric and sulphuric acids as nitrating agent. This strongly exothermic reaction, firstly performed in batch reactors, is now carried out in adiabatic flow reactors. High yields relative to aromatics (typically of 95–98 wt.%) are obtained for the mononitration of benzene, toluene and mononitrotoluene (MNT). The operating conditions (temperature, composition of the acid mixture, etc.) are milder for activated aromatic substrates, such as toluene, and more severe for deactivated substrates such as MNT. Sulphuric acid has to be regularly replaced owing to a gradual deactivation by water generated by nitration.

The increase in severity of the environmental legislation has led to the development of improved methods for the recovering of the desired nitro-product and for limiting the deactivation of sulphuric acid by continuous removal of water in a water-benzene azeotrope. The polluting disposal of sulphuric acid was also substituted by off-site regeneration of the spent acid by specialized firms. However, this regeneration requires cost- and energy-intensive recovery, purification and re-concentration steps. Moreover, the isomer distribution in the nitration final mixture is often not optimal. For instance, the mixture obtained from toluene nitration contains 55–60% of *ortho*-MNT, 3–4% of *meta*-MNT and 35–40% of *para*-MNT.

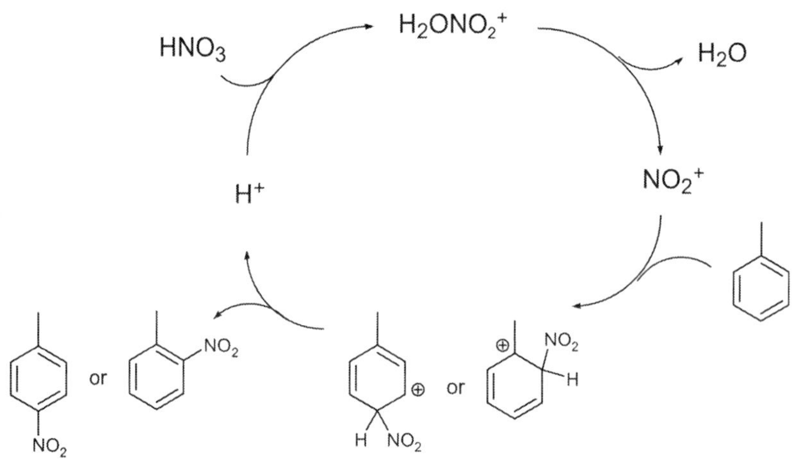

Fig. 8.7 Nitration of toluene. Reaction mechanism

Nevertheless, the greatest demand is for the last isomer and any adjustment for shifting such distribution is not easy.

Thus the substitution of solid acid catalysts for sulphuric acid appears an attractive solution, because they can be easily handled, can be regenerated in situ and can, in principle, improve the regioselectivity of the reactions (especially high with shape-selective zeolites). Because of that, in the last 20 years a significant research effort has been devoted to the development of nitration methods catalysed by zeolites [33, 34].

The mechanism proposed for aromatic nitration by nitric acid in presence of sulphuric acid and often accepted in presence of solid acid catalysts is described in Fig. 8.7. The first step is the protonation of nitric acid by a stronger acid. The second step is the dehydration of the protonated nitric acid with formation of a nitronium ion (NO_2^+). The attack of the aromatic substrate by this electrophile species (step 3) which is considered as the kinetic limiting step, forms an intermediate named Wheland or σ complex. The loss of a proton from this complex leads to the nitroaromatic product (step 4). However, in the nitration of highly reactive substrates, the formation of the σ complex involves two successive steps, with the formation first of an additional intermediate (π complex). Thanks to the deactivating effect of the nitro-group, polynitration is largely unfavoured.

With all the solid catalysts, including zeolites, the rate of nitration is lower than with sulphuric acid. This is due to two factors: (1) the concentration of protonic sites is much lower and (2) some chemisorption on the zeolitic active sites of water contained in the nitrating agent or/and produced by nitration (Fig. 8.7), leading to their deactivation. To maintain the activity of the solid catalyst in nitration with nitric acid, it is therefore essential to remove water from the reaction system by azeotropic distillation of the reaction mixture, by chemical trapping (acetic anhydride is typically added) or by operating in vapour-phase. Another way is to substitute nitric acid by nitrating agents, such as acyl or alkyl nitrates, nitrogen dioxide or dinitrogen pentoxide. However, these nitrating agents are more expensive and, in addition, hazardous and/or with limited selectivity.

Most reports focussed essentially on the selectivity of nitration: minimisation of by-product formation and control of the regioselectivity, with determination of the optimal catalyst and operating conditions. When considering the molecular size of the nitro-products resulting from nitration of monosubstituted benzenic substrates, such as toluene, a high selectivity to the less hindered *para* isomer could be expected with medium-pore size zeolites, such as MFI (and this behaviour was actually observed in the synthesis of dialkylbenzenes). However, this shape selectivity was not observed in liquid- (and gas-) phase nitration of toluene with nitric acid over H-MFI, because of a rapid deactivation of the catalyst due to the blockage of the access to the channels by bulky o-MNT molecules, trapped at the channel intersections. As a consequence, nitration of toluene over H-MFI zeolite was essentially due to the unselective homogeneous nitration [35]. On the other hand, whatever the operating conditions, the more selective catalyst resulted to be, unexpectedly, BEA zeolite i.e., a large pore zeolite. As high *para/ortho* ratio values were found in both liquid phase and within the BEA micropores, nitration mainly occurs within the pores, through a transition-state selective process [36]. However in 2-MNT nitration, the *para/ortho* ratio was found in the liquid phase twice as high as within the micropores (17 against 8.5), which suggests, at least for the formation of dinitrotoluene (DNT), an additional effect of product shape selectivity. It should be underscored that the high *para* selectivity of vapour-phase toluene nitration with dilute nitric acid over H-BEA was previously explained by a faster desorption of *p*-MNT on the basis of molecular modelling studies [37]. Furthermore, for nitration with acetyl nitrate over H-BEA, it was proposed that the high *para* selectivity did not result from classical transition state selectivity, but was induced by the reaction of the arene with a surface-bound acetyl nitrate complex [36]. This hypothesis is supported by ^{15}N-NMR measurements, which show that this complex is present only on the *para* selective BEA zeolite. However, this explanation cannot be retained for the high regioselectivity also found over the BEA zeolite with other nitrating agents.

Since nitration of activated aromatic compounds is very easy, it is preferably carried out under mild conditions in order to avoid secondary undesired reactions. Phenol nitration was performed over several zeolite catalysts in batch reactor at room temperature using dilute nitric acid (30%). The best results were obtained over a H-BEA zeolite in presence of tetrachloromethane as a solvent: nitrophenol yield of 93% after a reaction time of 2 h, no formation of dinitrophenol (as it typically occurs with concentrated nitric acid) and high selectivity to the desired *ortho* isomer (ortho/para ratio of 8.7). Tentatively, the authors attribute such high *ortho* selectivity to a preferred orientation of phenol molecules inside the BEA micropores that increases and enhances the accessibility of *ortho* position to nitronium ion [38].

Like acylation products (Sect. 8.2.2), nitro products are bulkier and more polar than the substrates and, most likely, the competition among these molecules for accommodating into the zeolite micropores has a significant effect on the reaction rate. Unfortunately, there are no kinetic data reported in the literature and, hence, no direct proofs of this possible autoinhibition. However, the analysis of the compounds retained within the micropores during 2-MNT nitration shows that the DNT/MNT ratio is 1.5–4.5 times greater in the micropores than in the liquid phase [36]. This relatively large gap, although the difference in polarity is much less pronounced than between aromatic hydrocarbon substrates and the corresponding nitro compounds (e.g., between toluene and MNT), is a strong argument in favour of the autoinhibition hypothesis. Another argument in favour of autoinhibition is the quasi asymptotic plateau in conversion observed in many nitration processes, even when they were carried out in absence of water. The desorption of 2,6-DNT (the bulkier nitro product of 2-MNT nitration) was compared over a series of H-BEA samples. A clear correlation was obtained between the rate of desorption and the activity of the samples for 2-MNT nitration (Fig. 8.8). The faster the desorption, the higher the activity, which means that nitration is limited by product desorption [36]. This conclusion is an additional clue in favour of autoinhibition.

As in acylation reactions, the long residence time of the polar product molecules within the zeolite micropores favours their further transformation. Product molecules can undergo two types of reactions: polynitration and oxidation, this latter reaction being particularly important at high temperature (especially under vapour-phase conditions). Since both types of resulting products are bulky and very polar, they are retained within the pores, leading eventually to condensation products. Thus, during vapour-phase nitration using dilute nitric acid, the following compounds were found to be trapped in the BEA zeolite micropores: 4-nitrobenzoic acid (the major product), benzaldehyde, DNT (in small amounts) and anthraquinone (in trace amounts). These products block the access of the reactant molecules to the zeolite pores, leading to the catalyst deactivation. Part of these compounds can be removed by solvent treatment, the residual part being eliminated by combustion. Another possible cause of deactivation is the slow and irreversible dealumination and degradation of the zeolite framework.

Therefore, as in aromatic acylation, the key point to develop an effective nitration process is to favour the desorption of the bulky and polar desired products, which can be done by applying the rules detailed in Sect. 8.2.2. and by removing efficiently water. All the authors agree with the choice of H-BEA zeolite as optimal catalyst, because it possesses generally a small particle size, an adequate (three-dimensional) pore structure and a proper pore aperture size, allowing an easy desorption of the desired products. The desorption can be additionally facilitated by the presence of mesopores created by dealumination. The literature shows that small-scale processes can be easily developed for the selective synthesis of many nitroaromatic products of industrial importance. However, to develop processes on a larger scale, an additional significant research effort is needed on the following topics: use of reactors adapted to commercial applications (mainly for liquid-phase nitration for which most of the experiments were carried out in simple batch

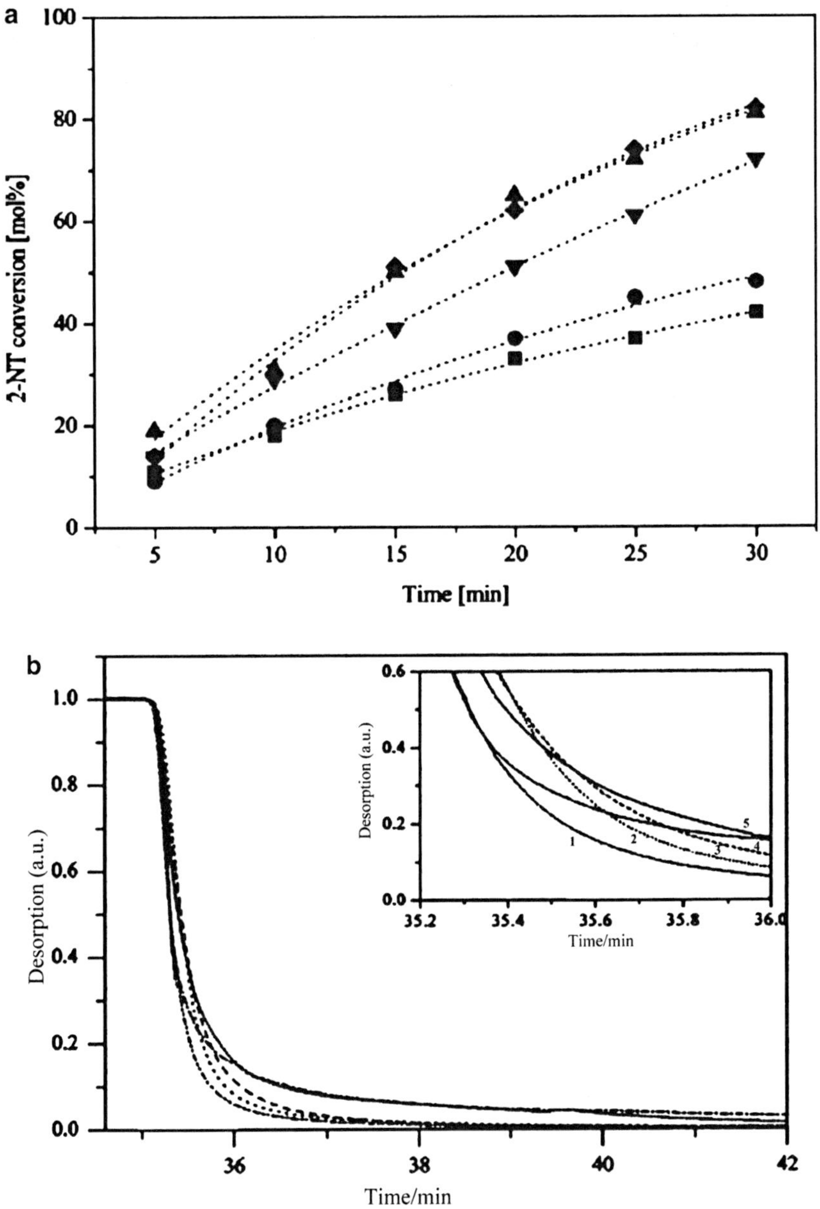

Fig. 8.8 (**a**)Conversion of 2-nitrotoluene (2NT) over five different H-BEA samples and (**b**) desorption (normalized values) of 2,6-dinitrotoluene product (26DNT) from the same samples (♦1; ▲2; ▼3; ●4; ■5). Reproduced (adapted) from [36], with permission from Elsevier

reactors), determination of kinetic models and thorough investigation of the deactivation processes and of regeneration techniques.

8.2.4 Halogenation of Aromatics

Halogenated aromatic compounds constitute an important class of fine and specialty chemicals [39, 40]. Intermediates such as chloro-benzenes or toluenes are widely used in the chemical industry. Chlorinated aromatic compounds are also excellent insecticides, herbicides and fungicides. Bromine-containing compounds are chiefly employed as fire retarding or extinguishing chemicals.

Arene halogenation can lead to addition products (e.g., $C_6H_6Cl_6$ from benzene) or to nuclear or side-chain substitution products. Most of the commercial processes are performed with molecular halogens, either in the absence of catalysts through a radical pathway (addition and side chain substitution) or in the presence of Lewis acid catalysts such as $FeCl_3$ or $AlCl_3$ (nuclear substitution) through an electrophilic mechanism.

This paragraph will be limited to the synthesis of nuclear-chlorinated aromatics. The mechanism of this reaction is shown in Fig. 8.9 on the example of toluene chlorination and by considering "Cl^+" as the electrophilic species. In halogenoarenes, the contribution of resonance and inductive effect of the halogen have opposite effects. As the latter is predominant, the halogen substituents deactivate (weakly) the aromatic ring. Therefore, polysubstitution can be observed in particular with Lewis acid catalysts. However, with the other catalysts, this polysubstitution can often be limited by operating with an over-stoichiometric value of the substrate/halogenating agent ratio. Furthermore, despite their deactivating effect,

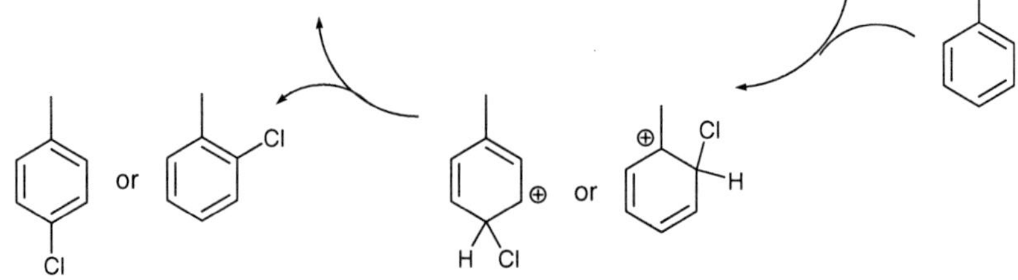

Fig. 8.9 Chlorination of toluene. Reaction mechanism

halogen substituents orient the electrophilic substitution in *ortho* and *para* positions.

Like in acylation and for similar reasons, the substitution of the conventional Lewis acid catalysts by solid acids and especially by shape-selective zeolites was highly desirable and the active research carried out in this field has led to the development of commercial processes for the synthesis of mono- and dichlorotoluenes. To develop commercially efficient zeolite catalysts for nuclear halogenation of aromatic compounds, the following conditions must be satisfied [39]:

1. The contribution of free-radical pathways with production of addition and side-chain substituted compounds must be negligible.
2. Hydrogen chloride that is generated by halogenation (Fig. 8.9) should not damage seriously the zeolite framework.
3. Halogenation should occur within the zeolite micropores to take profit of the shape-selective properties.

Conditions 1 and 2 restrict the operation to the liquid phase. Indeed, at low temperatures, free-radical pathways are limited and they are practically suppressed when the reaction is carried out in the absence of light and in the presence of free-radical scavengers such as O_2. Furthermore, structural damage of the zeolite is more easily prevented in liquid than in vapour phase, e.g., by reaction with a base like $NaHCO_3$ or with an adsorbent like the Na-A zeolite or by continuously purging the reactor with dried air. An additional benefit of the air purge is to eliminate the free-radical reactions [41].

Only the liquid-phase chlorinations will be considered here. Molecular chlorine is the traditional reagent for chlorinating aromatic compounds. In the absence of catalysts, it reacts readily through an electrophilic mechanism with activated aromatic substrates only. Liquid chlorinated agents such as sulphuryl chloride (SO_2Cl_2), *tert*-butyl hypochlorite, etc. which are more easily handled than Cl_2 can also be used. They are however less reactive and, hence, react readily with activated substrates only. At low temperatures, SO_2Cl_2 is the true electrophilic species whereas at high temperatures, it decomposes into SO_2 and Cl_2 which becomes the chlorinating species [39].

Large-pore zeolites of the FAU (X and Y), BEA, MOR, LTL types, a medium- (MFI) and a small- (LTA) pore size zeolites have been tested, particularly by Japanese and Indian researchers, in the liquid phase chlorination with Cl_2 of toluene, naphthalenes, xylenes, chlorobenzenes and chlorotoluenes. The *para*/*ortho* ratios of the reaction products were somewhat higher than with the conventional $FeCl_3$ catalyst. The nature of the exchanged cation and of the zeolite were shown to significantly influence the *para* selectivity, the monodimensional LTL zeolite (Table 8.2) and the K^+ cation (Table 8.3) giving the best results. Two important observations are: (a) the low selectivity of the MOR zeolite which, like LTL, is monodimensional and (b) the appearance of the positive effect of potassium only for high values of the exchange rate. In chlorination of biphenyl, the higher selectivity to the desired 4,4′-dichlorobiphenyl (which is a precursor of high-performance polymers) was also obtained with the K exchanged LTL zeolite.

Table 8.2 Regioselectivity of the zeolite-catalysed chlorination of ortho-xylene with Cl_2 at 253K. Influence of the zeolite

Zeolite	K-MFI	K-FAU (X)	K-FAU (Y)	K-BEA	K-MOR	K-LTL
4/3	1.45	1.75	1.5	1.65	1.55	11.7

Table 8.3 Influence of the exchanged cations

Cations[a]	H(76) K(24)	Li(80) K(20)	Na(80) K(20)	K(99)	Ca(83) K(17)	Sr(79) K(21)	Ba(78) K(22)
4/3	1.3	1.55	2.25	11.7	1.45	1.65	1.55

[a]The numbers in parentheses correspond to the percentages of exchange

Table 8.4 Influence of the solvent on the 4-chloro-o-xylene/ 3-chloro-o-xylene ratio (4/3)

Solvent	$ClCH_2CH_2Cl$	$ClCH_2CH_2Cl$	$CHCl_3$	CCl_4	CCl_4	None	None
Temp. (°C)	80	40	40	70	40	80	40
4/3	11.7	2.6	2.35	3.6	1.35	2.0	1.9

The authors concluded that the factors orienting effectively the substitution are the ionic radius, the charge and the spatial arrangement of the exchanged cations and their force fields [39].

Aliphatic carboxylic acids, such as chloroacetic acid or their salts, but not aromatic carboxylic acids, were shown to cause a significant increase in the *para* selectivity of chlorination of activated or deactivated substrates over the K-LTL catalyst [42]. To explain this positive effect, it was cautiously proposed that the interaction between the acid and the K cations limits the diffusion of the products, especially of the bulkiest *ortho* isomer. Chlorinated solvents were also found to affect the *para* selectivity, a significant positive effect being observed at the higher reaction temperatures (Table 8.4). This effect is particularly pronounced with 1,2-dichloroethane as solvent and can be cumulated with the positive effect of aliphatic acids [43].

Sulphuryl chloride was substituted for Cl_2 in the chlorination of toluene [41] and cumene in batch reactors. As with Cl_2, the K-LTL zeolite was demonstrated in cumene chlorination to be more active and more *para* selective and the best results were obtained at 80°C with 1,2-dichloroethane as solvent (*para*/*ortho* ratio=39). The recycle of the catalyst shows a relatively fast decrease in activity. This deactivation was shown to be similar to that caused by treatment of the catalyst by HCl solutions and thus attributed to framework dealumination and degradation [43]. Furthermore, in addition to the proposal of a kinetic model based on curious assumptions (the catalyst would activate the substrate and not the chlorinated agent), a method was developed to follow the catalyst deactivation during toluene chlorination [41]. The reaction rate was shown to decrease rapidly with time, becoming close to zero after 80 min. This deactivation was accompanied by a significant decrease in the *para*/*ortho* chlorotoluene ratio (from 3.6 to 2.0). The oxidation treatment of the deactivated sample for 12 h at 500°C allows only a partial regeneration and no restoration at all of the *para* selectivity. The reversible activity loss was proposed to be due to the accumulation of highly chlorinated products within the zeolite micropores and the irreversible deactivation attributed to dealumination and ultimately degradation by the HCl product of the zeolite framework. It could, however, be reminded that the causes of deactivation can be easily prevented.

In conclusion, many experiments carried out generally in batch reactors have clearly demonstrated the significant environmental benefit of the substitution of Lewis metal acids by zeolite catalysts for the *para* selective nuclear chlorination of aromatic compounds. For most of aromatic substrates and chlorinating agents, the superior performance of the K-LTL zeolite in terms of activity and selectivity is well established. However, the structural reasons of this observation remain somewhat mysterious. Similarly, more elaborated explanations of the positive effect of chlorinated solvents and aliphatic carboxylic acids, based, for instance, on the relative polarity and the size of all the molecules involved in the reaction, have to be advanced. A most likely weakness of the LTL zeolite (well demonstrated for other one-dimensional zeolites such as MOR) is the easy blockage of the channels by heavy reaction products and then, as a consequence, a very fast loss in activity. Even though, this high sensitivity to deactivation was recently confirmed [41], the nature of the responsible carbonaceous compounds remains to be established. Moreover, there are no investigations focused on the development of optimal ways to limit this deactivation and to regenerate the activity. A primary reason for that is the general use of batch reactors, which are less suitable than flow reactors for kinetic studies and for the investigation of deactivation phenomena.

8.3 Acid-Catalysed Rearrangements

Acid-catalysed rearrangements play an important role not only in the field of refining and petrochemicals, but also in the transformation of specialty and fine chemicals [44]. In refining and petrochemical processes, acid solid catalysts and, especially, shape-selective zeolites have substituted the corrosive and non-regenerable conventional catalysts, such as $AlCl_3$ and H_2SO_4, which were previously employed. In contrast, even if zeolites and other molecular sieves have largely demonstrated their high potential for catalysing in a clean and selective way the rearrangement of functionalised substrates, the conventional homogeneous catalysts are still widely used.

The following types of rearrangement will be briefly considered here on the basis of review papers recently published: (a) skeletal isomerisation of olefins, on the examples of *n*-butenes and α-pinene rearrangement [45, 46]; (b) isomerisation of functionalised saturated substrates, such as epoxides [47, 48]; (c) arene isomerisation, such as the Fries rearrangement [49]; and (d) ring enlargement reactions, such as the Beckmann rearrangement [50].

8.3.1 Skeletal Isomerisation of Olefins

Skeletal isomerisation of alkenes is catalysed by protonic acid solids. One of the most extensively investigated is the conversion of *n*-butenes into *iso*-butene, a raw material used in the production of various important chemicals: methyl *tert*-butyl ether (MTBE) (an additive component for gasoline, recently questioned owing to ground-water contamination), alkylated phenols, cresols, polyisobutene, etc. The discovery by shell researchers of the remarkable activity, selectivity and stability of the H-FER zeolite [51] which contains 10MR channels connected by narrower 8MR channels (Fig. 8.10), constituted a major breakthrough in the field. Indeed, in contrast to other proposed catalysts, this zeolite is capable of operating in absence of water at a relatively low (hence thermodynamically favourable) temperature.

Several studies have been carried out to understand the origin of the remarkable catalytic properties of H-FER. One key observation is that, at initial reaction times, on the fresh zeolite, the isomerisation of butene is accompanied by the formation of large amounts of side-products derived from disproportionation, namely, propene and pentenes. In fact, the high selectivity to *iso*-butene appears only after an initial fast formation of carbonaceous compounds within the micropores [52, 53]. Two main different explanations were advanced, based on the probable mechanisms of *iso*-butene, propene and pentene formation:

1. Over the fresh zeolite, *n*-butene transformation occurs for a major part through a bimolecular pathway, according to which the octene intermediates are transformed into two unsaturated C_4 molecules (*iso*- or/and n-$C_4^{=}$) or into propene and pentene (Fig. 8.11), and, for a minor part, through a monomolecular pathway with selective formation of *iso*-butene. Carbonaceous deposits then cause a fast deactivation of the bimolecular pathway without affecting the monomolecular one [54, 55].
2. Only bimolecular mechanisms are responsible for *iso*-butene formation [53, 56, 57]. Over the fresh zeolite, *n*-butene is only transformed through the bimolecular pathway described above with simultaneous formation of *iso*-butene, propene and pentenes. Subsequently, the increase in selectivity to *iso*-butene caused by carbonaceous deposits is related to the development of a

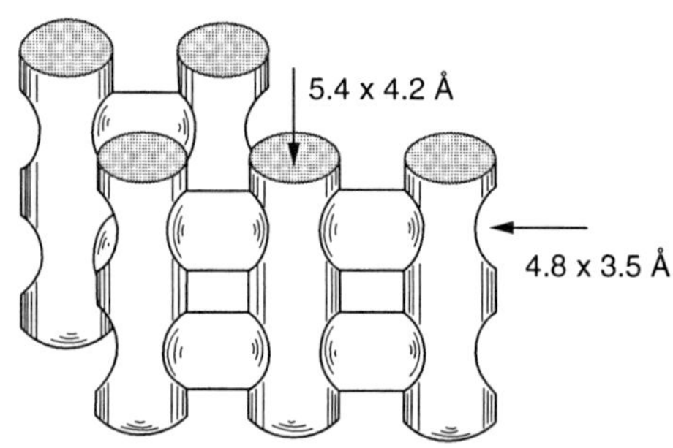

Fig. 8.10 Pore structure of the FER zeolite

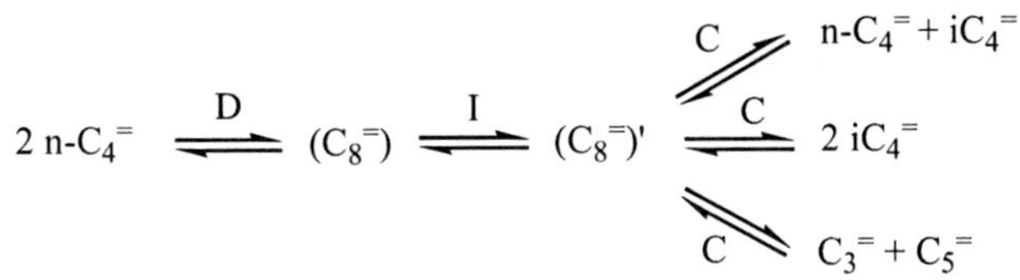

Fig. 8.11 Skeletal isomerisation of *n*-butene. Bimolecular mechanism. *D* dimerisation; *I* isomerisation; *C* cracking

Fig. 8.12 Autocatalytic mechanism of *n*-butene isomerisation into *iso*-butene

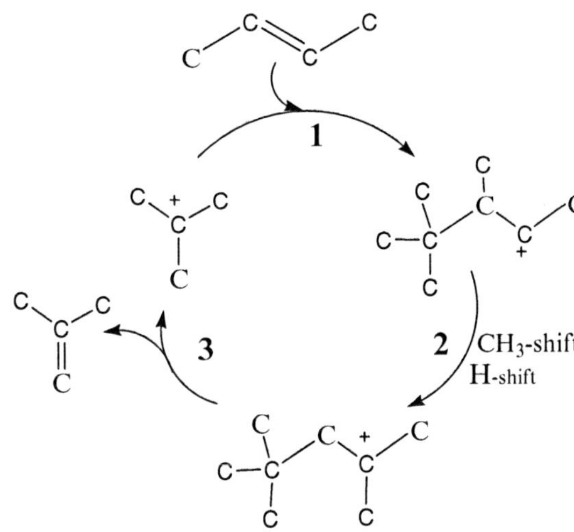

new isomerisation mechanism. This mechanism, that is very selective to *iso*-butene, has the particularity to involve, as active sites, the carbenium ions formed from *iso*-alkene products which desorb slowly from the 10MR channels, especially isobutene. It is therefore an autocatalytic process (Fig. 8.12). Coke precursors such as cyclopentadienyl species [58] and even coke molecules (methylpolyaromatic species) adsorbed on protonic sites [59] were also proposed as active species.

Both proposals can successfully explain not only the positive effect of carbonaceous compounds on the selectivity to *iso*-butene, but also various other observations, such as the drastic change with carbonaceous compounds in the isotopic composition of *iso*-butene resulting from isomerisation of ^{13}C-labelled butanes, the positive effect on the selectivity to *iso*-butene of the Si/Al ratio or of a decrease in *n*-butene pressure, etc. There is however a strong argument against the first hypothesis: the proposed monomolecular mechanism is unlikely, for it formally involves very unstable primary carbenium ions as intermediates or, rather, as transition states (Fig. 8.13). Conversely, all the steps involved in the bimolecular and pseudo-monomolecular processes are known to be highly probable. In particular, the steps of the autocatalytic process where tertiary and secondary carbocations are

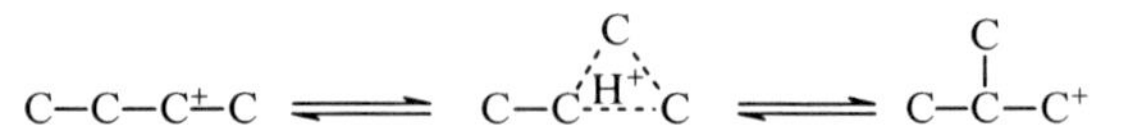

Fig. 8.13 Monomolecular mechanism of *n*-butene isomerisation into *iso*-butene

involved (Fig. 8.12) are known to occur over very weak acid sites or/and at low temperatures. Moreover, only the second proposal can account for the unexpected significant increase with the amount of carbonaceous compounds of the activity of the protonic sites for isobutene formation (expressed as their turnover frequency).

α-Pinene is a very reactive substrate and it can isomerise at low temperatures, even on relatively weak acid catalysts [45, 46]. The reaction proceeds through two parallel and irreversible routes leading to monocyclic products or to bi- and tricyclic ones, respectively. A complex reaction mixture is generally formed. The main reaction steps are shown in Fig. 8.14. The tertiary pinanyl carbocation formed by protonation of α-pinene rearranges into precursors of bi- and tricyclic terpenes (essentially camphene and tricyclene) and of monocyclic species (limonene and derivatives). Weak acids were shown to favour the formation of camphene, whereas strong acids that of monocyclic terpenes. Furthermore, limonene, which is more reactive than camphene, isomerises readily into terpinene and terpinolene, which can be subsequently disproportionated into *p*-menthenes and *p*-cymene. These secondary reactions become particularly significant at high conversions (>90%) and when pinene isomerisation is carried out under too severe conditions. Therefore, an adequate choice of operating conditions and catalyst acidity is crucial to obtain selectively camphene and limonene (the desired products).

Camphene is used as solvent for varnish in the automobile industry, food additive as well as intermediate for production of camphor, fragrance materials, acrylates, terpene-phenol resins, etc. Limonene is used as a component of flavours and fragrances, as additive in cleaning and cosmetic products, food and pharmaceuticals [46].

Industrially, the isomerisation of α-pinene is operated above 373 K over weakly acidic titanium oxide catalysts. Before use, titanium oxide is treated with sulphuric acid in order to form a layer of titanic acid on the surface [60]. The yield in the main products, i.e., camphene, tricyclene and limonene, is of ca. 75–80%, but the reaction rate is low. It is why a significant effort was made on the research of more active and selective catalysts. Zeolitic catalysts were widely investigated, because their acidity, porosity and shape-selectivity properties can be easily tuned and adjusted. In particular, over microporous molecular sieves, bimolecular side-reactions, such as disproportionation, can be minimised or, even, inhibited, leading to higher selectivities. This is experimentally confirmed and, over a FAU zeolite, lower amounts of heavy by-products are formed than over an amorphous silica-alumina at similar α-pinene conversions (ca. 98%): 12% compared to 60%, respectively [61].

The isomerisation of α-pinene was generally investigated in liquid phase at 323–393 K using a batch reactor. Under these conditions, some limitations in the diffusion of pinene and the related products within the micropores are expected,

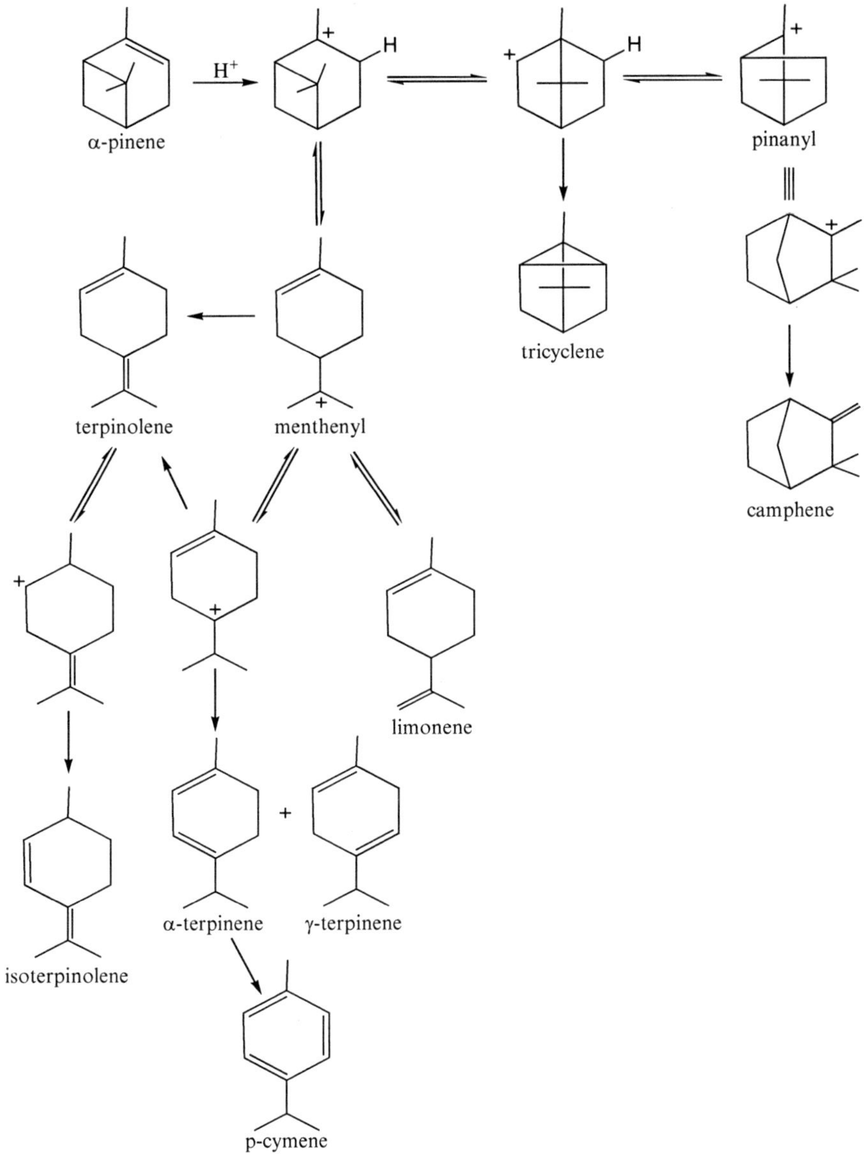

Fig. 8.14 Reaction pathway for the acid-catalysed isomerisation of α-pinene

especially with average-pore size zeolites and zeotypes. For this reason, the choice is oriented to large-pore zeolites (FAU, BEA, MOR) and AFI silicoaluminophosphates (SAPO-5). The best results are obtained at 393 K, with a HCl-dealuminated mordenite sample with a bulk Si/Al ratio of ca. 50: pinene conversion of 79% with a selectivity to camphene of 32%, to limonene of 24% and to terpinenes and terpinolene altogether of 16% [61]. The good catalytic behaviour of this sample can be related to the creation of internal mesopores by dealumination which transform the one-dimensional pore system in a quasi three-dimensional one, limiting significantly the pore blockage by heavy products and, hence, the deactivation [22]. However, the conversion reaches a plateau after one hour of reaction and it suggests that such pore blockage is not completely suppressed.

Similar values of camphene and limonene yields were obtained at 323 K over clinoptilolite, a natural average-pore size zeolite with the HEU topology. Better results are found at 363 K over a slightly HCl-dealuminated H-FER sample (Si/Al of 13 instead of 9 for the parent sample) [62]. As in the case with H-MOR, the dealumination treatment creates internal mesopores which can make quasi three-dimensional the pore system of this average-pore size zeolite. In agreement with a decrease in the deactivation rate, values of α-pinene conversion close to 100% could be attained, whereas with the parent sample, a plateau was found at 70% conversion. At α-pinene conversion close to 100% over the dealuminated sample, the yields in camphene and limonene were of 47% and 35% respectively, which are slightly higher than over the commercial TiO_2 catalyst.

8.3.2 Epoxide Isomerisation

Various aldehydes, ketones, esters, etc. that are employed directly or as intermediates in the synthesis of fine chemicals can be prepared by epoxidation of olefins followed by rearrangement of the epoxide in the presence of acidic or basic catalysts. This second step is still often carried out in liquid-phase batch reactors using metal halides or mineral acids as catalysts. Substitution of these homogeneous systems by heterogeneous ones seems very attractive both from the economical and environmental point of view. Moreover, for relatively large-volume production, economical continuous processes (in liquid but also in gas phase) can be developed [49, 63].

With protonic acid catalysts, epoxide rearrangement involves three successive steps: (1) protonation on the oxygen atom and formation of the oxonium ion; (2) rearrangement of the activated epoxide into a carbocation; and (3) deprotonation and formation of the desired product (Fig. 8.15).

The rearrangement of the carbocation and the aldol condensation of the carbonyl compounds into heavy molecules, which are the precursors of coke, are at the origin of various undesired side-products. As a consequence, the selectivity and the

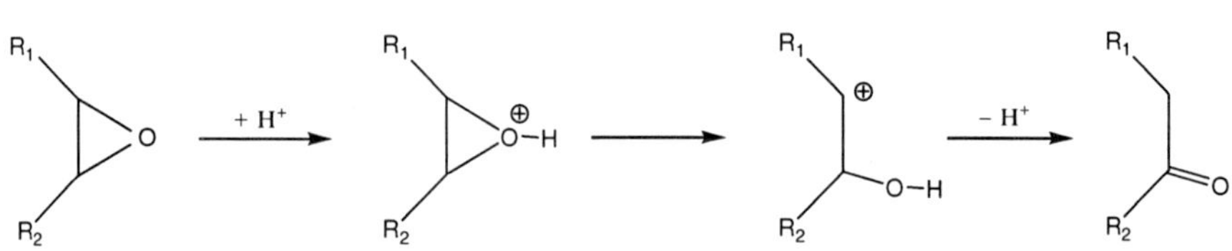

Fig. 8.15 Acid-catalysed rearrangement of epoxides

stability of conventional solid acid catalysts are generally low. It is why the research was essentially focussed on zeolitic catalysts, whose shape-selectivity features could limit, or even suppress, the undesired formation of side-products. However, there is little advantage in the use of zeolites in the rearrangement of small aliphatic epoxides. Indeed, the formation, via hydride shift, of various isomers of the desired product and their desorption from the zeolite micropores are not sterically limited and a mixture of aldehydes and ketones is generally obtained.

Only the rearrangements of styrene and α-pinene oxides will be considered here. The resulting aldehydes are widely used in the fragrance industry: alkyl phenyl acetaldehydes can be used as fragrances with hyacinth or rose odour and campholenic aldehyde is a key molecule for the synthesis of various sandalwood-like chemicals.

The isomerisation of different styrene oxides is carried out in a fixed-bed reactor, in gas phase at 573 K over various small-, average- and large-pore zeolites or zeotypes and on amorphous metal oxides. The MFI zeolites are more selective or/ and more stable than other heterogeneous catalysts. With these average-pore size zeolites, the formation of the bulky by-product triphenylbenzene as well as of the carbonaceous deposits responsible for deactivation ("coke") is sterically inhibited. The selectivity also depends on the catalyst acidity. Thus, a 100% yield in phenyl acetaldehyde can be obtained by using a Cs-doped B-MFI, i.e., a MFI catalyst with very weak acid sites and even with basic sites. This catalyst was demonstrated to be very efficient for the selective rearrangement of bifunctional compounds too, such as phenyl glycidic acid methyl ester.

Under acid conditions, α-pinene oxide (PinOx) rearranges easily to the desired campholenic aldehyde and also gives rise to a wide series of by-products (up to 200). The most common ones are presented in Fig. 8.16. A high selectivity to

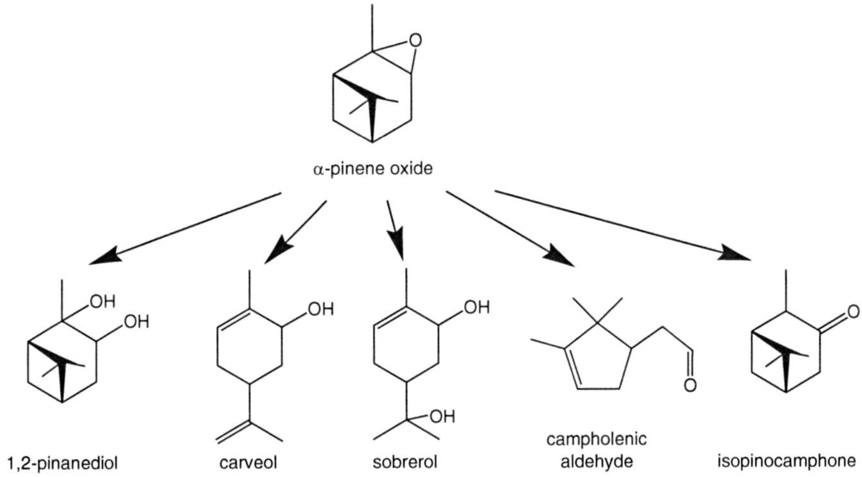

Fig. 8.16 Some products obtained in the acid-catalysed rearrangement of α-pinene oxide

campholenic aldehyde (ca. 85%) can be obtained through the conventional homogeneous process ($ZnBr_2$ as catalyst and benzene as solvent). Nevertheless, the major drawback of this process is due to the water pollution by zinc halides, causing severe problems in sludge treatment for killing the bacteria. So, a significant effort was devoted to the research of a selective heterogeneous process.

Over H-USY zeolites (ultra-stable Y zeolites with FAU morphology) with high amounts of mesopores and at complete conversion of PinOx (ca. 100%), selectivity as high as 80% is obtained over 24 h in batch reactor with epoxide to catalyst mass ratio of 7.5 [64]. A pretreatment of the catalyst with diluted HCl, followed by washing and calcination at 823 K, is crucial to have enhanced activity without loss in selectivity. The reaction must be carried out at unusually low temperature (273 K) to reduce the uncontrolled by-product formation and thus to maximize the selectivity to the desired aldehyde. The catalyst can be reused with no loss in performance, provided it is regenerated under air atmosphere at 823 K. Analogously, PinOx is successfully transformed into campholenic aldehyde over Ti-BEA too [65]. This titanosilicate zeolite was tested both in liquid and in vapour-phase. In liquid-phase batch tests, selectivities of up to 89% can be achieved in acetonitrile, thanks to its weakly basic character that minimizes unwanted side-reactions promoted by stronger acid sites. However, 1,2-dichloroethane is the most suitable solvent in terms of low catalyst deactivation and turn-over numbers of 576 (at 81% selectivity) can be obtained.

Actually, since PinOx gives rise easily to undesired bimolecular reactions, very low intraporous reactant concentrations are ideal for high yields of campholenic aldehyde. Thus, thanks to the hydrophobic character of Ti-BEA, non-polar solvents can compete with PinOx molecules for adsorption within the zeolite micropores and allow an optimal local low concentration of reactant. Conversely, under solventless vapour-phase conditions, such oligomerisation processes prevail and the selectivity to campholenic aldehyde decreases abruptly. A competitive inert co-adsorbate, such as methylcyclohexane or 1,4-dimethylcyclohexane, has to be added to the reactor feed to limit the intraporous PinOx concentration. With these additives, aldehyde selectivities as high as 85% can be attained at industrially practicable temperatures [49].

8.3.3 Beckmann Rearrangement

In 2003, a two-step commercial process based on microporous molecular sieve catalysts was developed by Sumitomo Chemical Co. for the synthesis of ε-caprolactam (CL). CL is used in very large amounts (ca. 4 Mtons/year) for the production of nylon-6 fibers and resins. This process is a combination of the Enichem process of cyclohexanone ammoximation catalysed by a MFI titanosilicate (*vide infra* Sect. 8.6.3) with the vapour-phase Beckman rearrangement of the resulting cyclohexanone oxime (COX) over a highly siliceous H-MFI catalyst at 623–673 K. This new process is more economical and environmentally friendly than the conventional route which used corrosive sulphuric acid or oleum as catalyst for the Beckman rearrangement and produced between 2 and 4.5 tons of ammonium sulphate by-product per ton of CL [47, 48].

A large variety of acidic oxides: non-zeolitic materials, such as alumina, boria, etc., and zeolitic ones with small, average and large pores, were tested as catalysts in the gas-phase rearrangement of COX into CL. An unexpected conclusion was that, in order to perform selectively this reaction, very weak protonic sites (e.g., zeolite silanol groups) were more efficient than stronger sites (e.g., bridging hydroxyl groups). Thus, as early as 1986, Sato et al. [66] found that extremely highly siliceous H-MFI samples, with therefore practically no bridging hydroxyl groups, were active in the vapour-phase Beckmann rearrangement with a high selectivity to ε-caprolactam (75–80%) at total conversion of COX. These highly active and very selective siliceous MFI samples (Si/Al>100,000) were shown to have virtually no acidity: no acidity detected by NH_3 TPD and only a slight decrease in COX conversion and in CL yield by adding NH_3 to the COX reactant as well as no activity for methanol transformation into dimethyl ether [67].

The relative contribution of the different types of silanols of high-silica MFI samples was established by Heitmann et al. [68]: the silanol nests are the most suitable sites rather than vicinal and, especially, terminal silanols. Furthermore, the selective blockage of terminal silanols by silylation or by etherification with methanol was shown to enhance significantly the selectivity to CL without any loss in activity. The FT-IR spectroscopic study of adsorption and reaction of COX on a silicalite S-1 sample (with MFI topology) confirms the catalytic role of the silanol nests. It is proposed that Beckmann rearrangement requires the cooperation of two neighbouring hydroxyls of these ≡SiOH nests, allowing the proper reactant-active site configuration and favouring the stabilization of the intermediate [69].

Despite the very large difference in strength of the protonic acid sites involved in liquid and in gas phase Beckmann rearrangement, the mechanism (Fig. 8.17) is admitted to be the same in both conditions. A strong argument in favour of this conclusion is that, like in liquid phase, the rearrangement is stereoselective in the

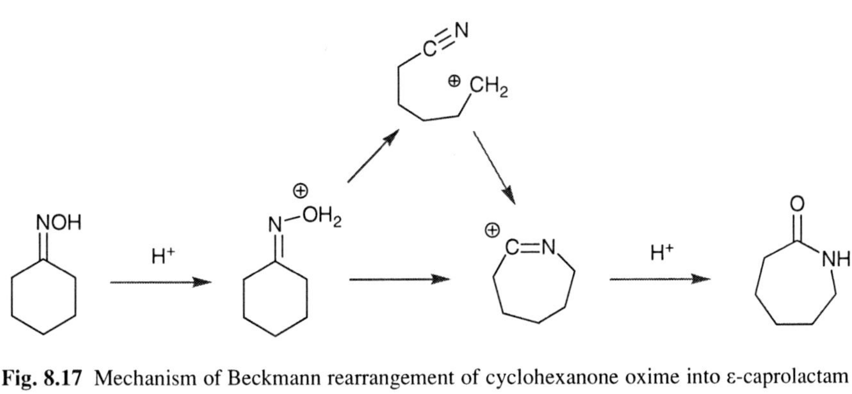

Fig. 8.17 Mechanism of Beckmann rearrangement of cyclohexanone oxime into ε-caprolactam

heterogeneously catalysed reaction: it is always the alkyl group in *anti* position to the hydroxyl group of the oxime that migrates [67].

As COX and especially CL molecules are relatively bulky in comparison to the pore apertures of MFI, the location of the active sites was a matter of debate. Some authors conclude that COX, but not the bulkier CL molecules, could enter the MFI micropores and it means that the rearrangement occurs on the outer surface of the crystals only. Conversely, other authors conclude that both molecules enter the micropores and, hence, that rearrangement can take place inside them. However, it is likely that desorption of CL from these micropores is diffusion limited and therefore, the Beckmann rearrangement is essentially located in the outer rim of the crystals. Both proposals (either rearrangement on the outer surface or within the pores, but with diffusion limitations) can explain the increase in activity and selectivity observed when the crystal size of the MFI catalysts is decreased.

A maximum in catalyst stability and selectivity [70] can be observed in the range of the process temperatures (623–673 K). The strong adsorption of CL would account for the low stability and selectivity observed at low temperatures, the long contact time favouring the formation of side products. At high temperatures, secondary reactions of the COX reactant and of the CL product (i.e., decomposition into 5-hexenenitrile and into cyclohexanone and polymerisation) become important, with, as a consequence, low selectivity and stability values.

Catalyst deactivation is essentially due to pore blockage by nitrogen-containing species resulting from condensation of caprolactam [71] or of unsaturated nitrile by-products [72]. The catalyst activity can be restored by combustion of the trapped species at elevated temperatures. In order to conduct continuously both the reaction and the regeneration at the same time, the commercial process is operated with a fluidised bed system [73]. The fluidised-bed reactor has additional advantages over fixed-bed reactors: it can be operated at short contact times, favourable to high caprolactam selectivity; nanosized catalyst particles can be used without any sensible pressure drop; the temperature of the exothermic Beckmann rearrangement can be easily controlled.

8.3.4 Fries Rearrangement

Hydroxyarylketones can result from the simple acid-catalysed isomerisation of arylesters via the Fries rearrangement (Fig. 8.18). Generally, this rearrangement is combined with the esterification of the starting hydroxyarenes, allowing the one-pot two-step synthesis of hydroxyarylketones. Most studies over acid zeolites deal with the gas- or liquid-phase synthesis of *ortho*- and *para*-hydroxyacetophenones (HAP), which are respectively precursors of salicylic acid and of paracetamol. The gas-phase reactions are carried out in fixed-bed reactors, whereas the liquid-phase ones essentially in batch reactors. The *o*-HAP isomer can be selectively produced in gas-phase and the p-HAP in liquid phase [25, 50].

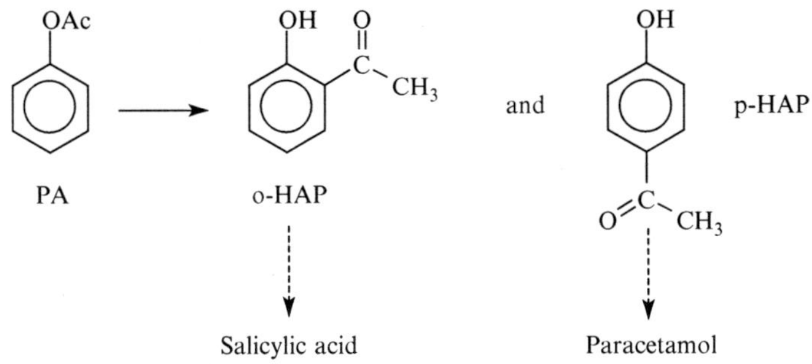

Fig. 8.18 Fries rearrangement of phenyl acetate (PA) into hydroxyacetophenones (HAP)

8.3.4.1 Gas-Phase Reactions

The rearrangement of phenylacetate (PA) was investigated at 673 K over H-FAU and H-MFI zeolites. Phenol (P) is the main product, the desired HAP products being formed in minor amounts only. With both zeolites and especially with H-FAU, *ortho*-HAP was favoured over the *para* isomer. The following reaction scheme is proposed: PA dissociates into P and ketene, *o*-HAP results partly from intramolecular rearrangement of PA, partly from C acylation of P by PA, *p*-HAP from this latter reaction only. A very fast deactivation can be observed, which was mainly ascribed to the trapping of carbonaceous deposits ("coke") within the zeolite micropores.

However, both the catalyst stability and selectivity are considerably improved by operating at lower temperatures and especially by substituting a phenol-acetic acid mixture for PA. The average-pore size MFI zeolite is more stable than the large pore FAU and BEA zeolites and this behaviour is ascribed to steric limitations in the formation of detrimental carbonaceous compounds. Moreover, by decreasing the size of the crystallites, an increase in both stability and selectivity to *o*-HAP is observed [74–76] and it suggests that some limitations in desorption of o-HAP and coke precursors are present. These limitations are then confirmed by the results of the analysis of "coke." Indeed, carbonaceous deposits are constituted by the most polar compounds among reactant and products, namely phenol, in large amounts, HAP, acetoxyacetoacetophenones (AXAP) and heavier products, such as 2-methylchromone and 4-methylcoumarine (resulting from successive transformations of *o*-AXAP) or methyl naphthols (from addition of acetic acid to phenol).

The reaction scheme was established at the optimal operation temperature (280°C) by studying the effect of the contact time (and hence conversion) on the product yields. At short contact times, there is a fast O-acetylation of phenol (PA formation):

$$\mathrm{P + AcOH \rightleftarrows PA + H_2O} \tag{8.1}$$

accompanied by a much slower (20 times) C-acetylation with the sole formation of *o*-HAP. As a consequence of the high rate of reaction (8.1), thermodynamic equilibrium among P, PA, acetic acid and water is rapidly established and, at higher contact times, the products result essentially from the transformations of PA alone or of PA with P. *o*-HAP was demonstrated to result from acetylation of P by PA:

$$\mathrm{PA} + \mathrm{P} \rightleftarrows \mathrm{P} + \mathrm{HAP}\ (\text{essentially } o-\mathrm{HAP}) \tag{8.2}$$

and from a slower intramolecular transformation:

$$\mathrm{PA} \rightleftarrows o-\mathrm{HAP} \tag{8.3}$$

The high selectivity of reaction (8.2) to the bulkier *ortho* isomer, that is unexpected over a shape selective MFI zeolite, is attributed to the high stability of the transition state [77]. Further kinetic modelling [78] confirms the proposed reaction scheme and, moreover, suggests that high o-HAP yield and selectivity can be obtained by reacting over H-MFI catalysts mixtures of phenol and acetic acid with a high AcOH/P ratio, which was then experimentally confirmed.

8.3.4.2 Liquid-Phase Reactions

Various zeolites, especially MFI and BEA have been used as catalysts for the transformation of PA in batch reactor. As in the gas phase, HAP (*ortho* and *para* isomers), *p*-AXAP and P are the main products and deactivation is relatively rapid. *o*-HAP, P and p-AXAP are primary products and are directly obtained from PA. The formation of these products is explained by a selective intramolecular rearrangement of PA into *o*-HAP (reaction (8.3)) and by PA autoacylation. Part of P results from PA dissociation. *p*-HAP is a secondary product and it appears only at 2% conversion. As was demonstrated by adding phenol to PA, *p*-HAP results essentially from the acylation of P with PA (reaction (8.2)). By exploiting this side reaction, it is possible to synthesize selectively *p*-HAP [53, 79]. Thus, with H-BEA zeolite, the yield and selectivity to *p*-HAP passes from 5 and 28% with cumene as solvent, to 24 and 60%, respectively, when the reaction is directly carried out in phenol [79].

The catalyst stability can be significantly improved by continuous extraction of the catalyst [80], by using a polar solvent such as sulfolane [75], by increasing the temperature or by operating in a flow reactor [81]. This suggests that PA transformation is limited by desorption of the products that are more polar than PA. As a consequence, the reaction probably occurs in the outer rim of the crystals, which explains why H-BEA zeolites, generally synthesized under the form of small crystallites (ca. 20 nm), are considered the best catalysts for PA transformation into HAP products.

In addition to their positive effect on the catalyst stability, polar solvents orient the selectivity of PA transformation to the formation of *para* products. Thus, the (*p*-HAP+*p*-AXAP)/*o*-HAP ratio was equal to 7.5 with sulfolane as solvent and only to 0.8 with dodecane. Polar solvents limit also the formation of heavy reaction products responsible for deactivation. All this could be related to the competition among sulfolane, PA and reaction products for the occupancy of the zeolite micropores and for chemisorption on the protonic sites.

8.4 Synthesis of Amines

8.4.1 Aliphatic Amines

Methylamines (MA) are major chemical intermediates with a yearly production in the order of 10^7ton. Monomethylamine (MMA) is used in the production of solvents (N-methylpyrrolidone), pesticides, etc. Dimethylamine (DMA) is employed in the production of solvents (dimethylformamide, dimethylacetamide), of chemicals for water treatment, etc.; trimethylamine (TMA) finds application in animal feed supplements, such as choline chloride, and in ion exchange resins [82, 83]. Historically, the production of MA is carried out by vapour-phase ammonolysis of methanol with ammonia over a silica-alumina catalyst:

$$\begin{aligned} &NH_3 + x\,CH_3OH \rightarrow MA + x\,H_2O \\ &(x = 1,2,3;\ \ MA = MMA \text{ or } DMA \text{ or } TMA) \end{aligned} \tag{8.4}$$

(x=1, 2, 3; MA=MMA or DMA or TMA)under the following operating conditions: adiabatic fixed-bed reactor, temperature of 673–693 K, pressure of ca. 20 bar, molar NH_3/MeOH ratio of 1–4. The product distribution which approaches thermodynamic equilibrium (MMA:DMA:TMA of 10:18:72 wt.% at 673 K with NH_3/MeOH ratio of 1) is completely unbalanced with respect to the market demand (20:65:15 wt.%). DMA production can be enhanced by recycling TMA with excess NH_3, which is costly and energy intensive due to the requirement of large-size separation and recycling units. Therefore, the substitution in methanol ammonolysis of a shape-selective zeolite catalyst for silica-alumina seems to be the obvious solution to satisfy the high demand for DMA at a more limited cost.

Three different reactions can be observed during methanol ammonolysis:

1. Sequential methylation of ammonia to mono-, di- and tri-methylamines:

$$NH_3 \underset{-H_2O}{\overset{+CH_3OH}{\longrightarrow}} MMA \underset{-H_2O}{\overset{+CH_3OH}{\longrightarrow}} DMA \underset{-H_2O}{\overset{+CH_3OH}{\longrightarrow}} TMA \tag{8.5}$$

2. Transmethylation or disproportionation:

$$\begin{aligned} 2MMA &\rightarrow DMA + NH_3;\, MMA + DMA \rightarrow TMA + NH_3 \text{ and } 2DMA \\ &\rightarrow TMA + MMA \end{aligned} \tag{8.6}$$

3. Side reactions leading to small amounts of undesired products: dimethylether, aldehydes, carbonaceous compounds, etc.

At the stationary state, all the protonic acid sites are occupied by ammonium and methylammonium ions. Reactions 8.5 proceed in two steps, the second one being rate determining: (a) alkylation of ammonium ions by methanol via formation of a co-adsorption complex, proton transfer from the ammonium ion to methanol, liberation of water and formation of a C–N bond and (b) ammonia-mediated release into the gas phase of the formed amines (sorbed as alkylammonium ions). Two mechanisms were proposed for this release: an adsorption-assisted desorption mechanism, in which the methylammonium ion is simply replaced by either an ammonium or another methyl ammonium ion and a "scavenging" mechanism in which the sorbed methylammonium ion remains at the active site and serves as a methyl source for disproportionation [84, 85]. As a consequence, in absence of steric constraints, the MA formed can re-equilibrate during diffusion through the pores.

The great deal of research carried out on methanol ammonolysis over zeolites led to the development of commercial processes for the selective production of DMA by Nitto Chemical (now Mitsubishi Rayon) in 1984 and by Mitsui Toatsu in 1997 [86]. Unexpectedly, both processes use as catalyst a H-MOR zeolite with channels apparently too large ($0.65\ nm \times 0.70\ nm$) for a shape-selective formation of DMA. Indeed, when one considers the sizes of the MA molecules (MMA: $0.37\ nm \times 0.39\ nm \times 0.44\ nm$; DMA: $0.39\ nm \times 0.47\ nm \times 0.60\ nm$; TMA: $0.39\ nm \times 0.54\ nm \times 0.61\ nm$), small pore-size zeolites, whose apertures are between 0.35 and 0.50 nm, seem more suitable to hinder the formation and the diffusion of TMA within the micropores [82, 87]. In agreement with that, high values of selectivity to DMA and low values to TMA, are found with some small-pore zeolites and zeotypes and the best results are obtained with RHO, KFI (ZK-5), LEV (Nu3) and CHA [82, 87].

Various treatments were developed to improve the selectivity to DMA and to limit the relatively high production of dimethyl ether. These treatments were also tested with medium- and large-pore zeolites. Among these, the most successful zeolite is the large-pore MOR zeolite. It is known for the high strength of its acid sites and it contains two types of channels: large channels, in which reactant and product molecules can diffuse, connected by small channels, in which only small molecules such as NH_3 can diffuse. The treatments have two main objectives: (1) to suppress the acidity and, hence, the activity of the non-shape-selective outer surface, e.g., by depositing or grafting an inert compound (via CVD methods) or by selectively dealuminating the outer surface and (2) to tune the channel size or, more precisely, the diffusion path of amine molecules by exchange with bulky cations (e.g., Na^+ used with the Mitsubishi catalyst), silylation (used by Mitsui Toatsu) or controlled dealumination by hydrothermal treatment [88].

Methanol ammonolysis over the MOR catalysts must be operated at lower temperature (573 K) than over silica alumina catalyst (>673 K). Indeed, the product shape selective properties of the mordenite catalysts decrease when increasing temperature. High temperatures mean easy diffusion of bulkier molecules and migration of the cations eventually introduced in order to control the channel size. High reaction temperatures have also a significant negative effect on the catalyst stability. The framework degradation, due to the aggressive reaction environment (presence of H_2O, NH_3, etc.), is significantly increased as well as the formation of carbonaceous compounds ("coke"), which is the main cause of deactivation. Aldehydes (mainly formaldehyde) were demonstrated to be the main precursors of coke. The presence of these compounds in methanol recycled from the same or from other plants (e.g., for dimethylformamide manufacture) has a dramatic effect on catalyst stability. Thus, an increase in the aldehyde content from 0.15 to 0.4 $g h^{-1} kg_{catalyst}{}^{-1}$ causes a decrease by a factor 2–4 of the catalyst life. It is therefore crucial to control and limit the aldehyde content of the feed.

8.4.2 *Heterocyclic Ring Formation: Synthesis of Pyridines*

Pyridine bases constitute the most important family of heterocyclic compounds in terms of commercial applications. These compounds and especially pyridine (A) and picolines (the three regioisomers: α-picoline, β-picoline and γ-picoline; B, C, D, respectively) are largely used not only as solvents and acid-scavengers, but also as key precursors to a wide range of agrochemicals (insecticides, herbicides and fungicides), pharmaceuticals (e.g., antituberculosis drugs), polymers and resins [89]. Originally isolated from coal tar, they have been produced for more than 50 years by the gas-phase heterogeneously catalysed Chichibabin reaction in fluidised- or fixed-bed reactors. The amorphous silica-alumina, which was firstly used as catalyst, was progressively substituted by shape-selective zeolites for the production of pyridines and alkylpyridines from a range of carbonyl-containing feeds. Thus, pyridine (A) and β-picoline (C) are produced together by condensing acetaldehyde, formaldehyde, and ammonia over a MFI zeolite at 673–773 K. Also, small amounts of α- and γ-picolines (B and D) as well as of higher alkylated pyridines, especially 3,5-dimethyl pyridine (3,5-lutidine) and nitriles can be formed. The balance equations for the formation of A and C are presented in reactions (8.7) and (8.8)

$$2CH_3CHO + CH_2O + NH_3 \rightarrow A + 3H_2O + H_2 \tag{8.7}$$

$$2CH_3CHO + 2CH_2O + NH_3 \rightarrow C + 4H_2O \tag{8.8}$$

These equations reveal no details on the complexity of these cascade reactions that involve various intra- and inter-molecular condensations. Some hydrogen-transfer or dehydrogenation steps participate in pyridine formation too (reaction (8.7)). The reaction schemes were proposed generally on the simple basis of the variation of the product distribution with the feed composition. However, less speculative information was also drawn from the use of isotopically labelled compounds. Thus, in a recent paper, Calvin and coworkers [90] conclude, from ^{13}C incorporation patterns, that, over a silica-alumina catalyst, A and C are formed (reactions (8.7) and (8.8)) through two main pathways involving a common propenimine intermediate that condenses with acetaldehyde, acrolein, and propionaldehyde (or their imines).

Since 1980, Mobil has demonstrated the superiority of the H-MFI zeolites over silica-alumina catalysts for synthesizing pyridine from acetaldehyde, formaldehyde and ammonia. Higher yields (10–20% more) are obtained in products A+C thanks to the reduction in the formation of heavy alkylpyridines, as well as higher A to C ratio. Other zeolites were similarly investigated. The large pore MOR and FAU zeolites gave lower yields to A+C than MFI and MEL but the higher yield was obtained with MWW (MCM-22) catalyst. Treatment of MFI with tetraethylorthosilicate in chemical vapour-deposition conditions was tested in order to passivate the non-selective sites of the outer surface of the crystals [91]. No increase in the pyridine yield (hence, of the shape selectivity) was observed and such behaviour was attributed to the very small concentration of active sites on the outer surface of the chosen sample.

With H-MFI, the higher yields to A+C were found on samples with high Si/Al ratios (from 75 to 200). As these samples contain practically no Lewis acidic sites, the catalytic activity can be ascribed to protonic sites, found either in the bridging OH groups present in small amounts or in the silanol groups demonstrated as active in the Beckmann rearrangement. The superior performance of the low aluminium samples can be related to the easier desorption of the reaction products, which, therefore, undergoes limited secondary reactions.

Hydrogen-transfer reactions catalysed by acidic sites might avoid the formation of H_2 indicated in reaction (8.7). However, at the relatively narrow channel intersections of the MFI zeolite, these reactions, that involve bulky bimolecular intermediates, are likely limited by steric constraints. Therefore, in order to favour the formation of pyridine (reaction (8.7)), various promoter metals such as Tl, Pb, Co, etc., were added to the catalysts. Indeed, these metals are able to cycle between different oxidation states and they can help removing hydrogen from the reaction system. Table 8.5 shows that, in the transformation of a 2:1:4 acetaldehyde – formaldehyde – ammonia mixture in a fixed bed reactor at 450°C, the exchange of MFI by such metals causes a significant increase in the rate of pyridine

Table 8.5 Comparison of the yields in the main products of transformation of a 2:1:4 acetaldehyde, formaldehyde, ammonia mixture over a series of heterogeneous catalysts in a fixed bed at 723K. Adapted from Ref. [89]

Catalyst	Pyridine+picolines[a]	Pyridine[a]	Picolines[a]	Pyridine/β-picoline ratio
Silica-alumina	52	33	19	3.0
H-MFI	61	42	19	3.8
Tl-MFI	81	63	18	7.0
Pb-MFI	79	60	19	7.5
Co-MFI	78	57	21	8.1

[a]Yields (%) of products

formation (reaction (8.7)). In contrast, no change is observed in the rate of picoline formation, which does not require dehydrogenation or hydrogen-transfer reactions (reaction (8.8)).

Catalyst deactivation is essentially due to the formation and retention of heavy basic secondary products ("coke"). Fluid bed technology allows continuous reaction and regeneration, whereas with fixed-bed reactors intermittent regeneration cycles are necessary. Extended life has been demonstrated with doubly metal-promoted MFI catalysts. The additional promoters are noble metals, such as Pt, Pd, Rh or Ru. They are widely employed to reduce the temperature of combustion of coke and, thanks to this, they limit the catalyst degradation during the regeneration step. If hydrogen is fed together with the usual reactant mixture, these promoters are able to hydrogenate the coke precursors, thus limiting the catalyst deactivation. Another way to extend significantly the catalyst life is to carry out the combustion of coke in air added with an alcohol (preferentially methanol because of its high polarity). According to Shimizu et al. [91], this positive effect is due to: (i) a reduction of the acid strength, leading to an easier desorption of heavy basic products and (ii) a cooperation of alcohol with oxygen during coke combustion.

8.5 Diels-Alder Cycloadditions

The Diels-Alder reaction is an important tool for the synthesis of cyclic compounds. Thanks to this reaction, it is possible to build up a high added-value molecule (or a precursor of it) in only one synthetic step and, sometimes, from accessible and relatively simple reactants. Various kinds of inorganic materials with acidic character, such as silicas, aluminas, clays, etc., have been used as heterogeneous catalysts for this reaction, but the use of microporous (and mesoporous) molecular sieves allows, in general, milder conditions and, mainly, improved regio- and stereoselectivity.

The acid sites present in (or grafted on) molecular sieves typically catalyse the Diels-Alder cycloaddition by coordinating to the dienophile and therefore, by activating it to the nucleophilic attack by the diene (in Fig. 8.19, the case of the reaction between cyclopentadiene and maleic anhydride). The predominant frontier molecular orbitals involved in the reaction are the HOMO of the diene and the LUMO of the dienophile. The interaction between the acid centre of the catalyst and the carbonyl group (in the case of Fig. 8.19) of the dienophile has the effect of lowering the energy level of its LUMO, enabling a more efficient overlapping with the HOMO of the diene. However, some authors noted that, especially with zeolites, the concentration effect inside the micropores plays a role on its own as well. In fact, even in absence of acidic sites, zeolites can promote gas-phase Diels-Alder transformations. Such effect is not zeolite-specific, but any adsorbent with affinity for dienes, such as carbon-based porous materials, displays similar effects [92, 93].

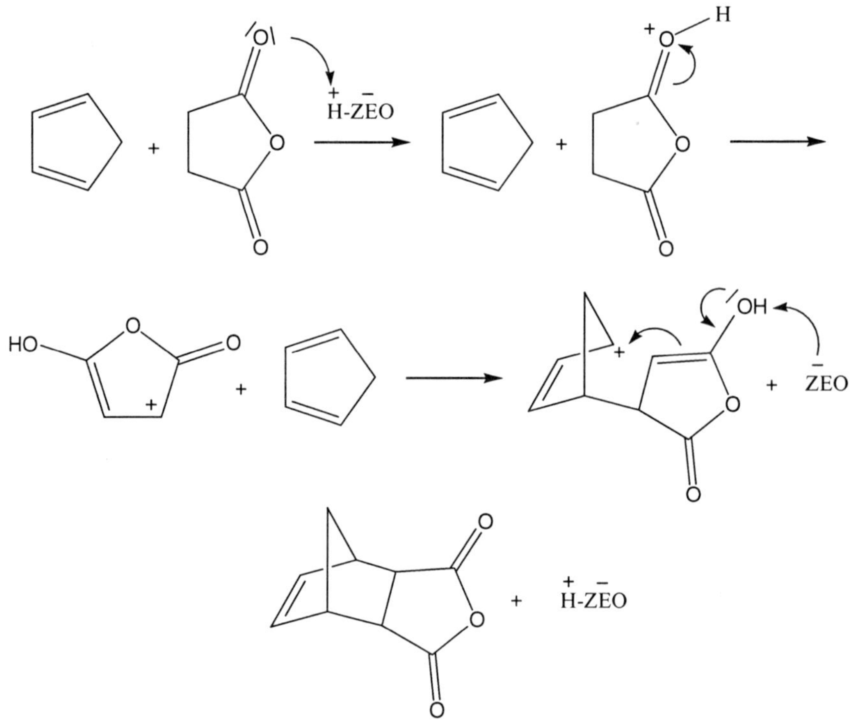

Fig. 8.19 Diels-Alder cycloaddition between cyclopentadiene and maleic anhydride over an acidic molecular sieve

The advantage of using zeolitic catalysts for this sort of reaction can be different case to case. For instance, the peculiar structure of FAU zeolite can stabilize the presence of Cu(I) species (in a Cu(I)-Y zeolite), which are optimal sites for the cycloaddition of butadiene to 4-vinylcyclohexene with high selectivity [94]. Otherwise, the strict steric constraint imposed by the microporous channel system can lead to higher chemo- and regioselectivities than those recorded with conventional homogeneous acid catalysts. So, the addition of cyclopentadiene to *cis*-cyclooctene is performed over Na-FAU (Na-Y) zeolite with low conversions, due to the poor diffusion of the substrates in the porous network, but with excellent selectivity and the almost exclusive formation of the mono-substituted norbornene (an interesting monomer for polymerisation reactions; Fig. 8.20) [95]. Analogously, over a H-BEA zeolite it is possible to add isoprene to methyl acrylate and obtain the desired 1,4-disubstituted cyclic acid (Fig. 8.21) with exceptional results (100% yield and 99% regioselectivity), [96]. Interestingly, other zeolites or homogeneous Lewis acids tested gave poorer results, $AlCl_3$ being the best homogeneous system with a maximal yield and regioselectivity of 83 and 95%, respectively.

An attractive approach to the search for the optimal catalyst for this reaction is based on the so-called "bio-inspired" molecular imprinting, according to which the

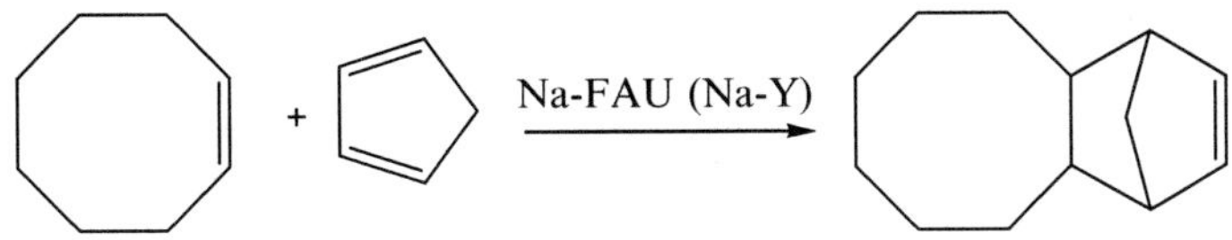

Fig. 8.20 Diels-Alder cycloaddition between cyclopentadiene and *cis*-cyclooctene

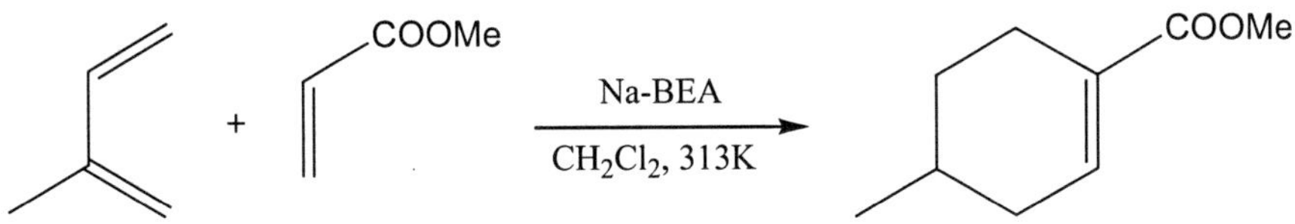

Fig. 8.21 Diels-Alder cycloaddition between isoprene and methyl acrylate

catalytic site is designed and shaped around a particular transition state ensemble, as it happens for enzymes. When a template, which resembles the shape of the Diels-Alder transition state, is used for the synthesis of a zeolite, an improved selectivity of the final catalyst should be obtained after removal of the template. Following this approach, an ITQ-7 zeolite a three-dimensional molecular sieve with 12-membered pores (ISV structure) was prepared using, as a structure directing agent, a quaternary ammonium species, synthesised by Diels-Alder addition. Once the template is eliminated, the catalyst shows a remarkable activity with respect to other zeolites with similar topology, but prepared in a conventional way, without tailored active sites [97].

The use of mesoporous molecular sieves displays some peculiar features too, sometimes due to larger pore size of these materials, sometimes due to the less marked acid character of their active sites compared to zeolitic catalysts. In particular, aluminium-containing mesoporous silica exhibits better performance with large organic molecules than H-MFI or H-FAU [98, 99]. In the reaction of cyclopentadiene with crotonaldehyde, aluminum *iso*-propoxide grafted by post-synthetic modification onto MCM-41 shows the highest activity with respect to directly prepared Al,Si-MCM-41. The results are comparable to the use of homogeneous *p*-toluenesulfonic acid and even better than with conventional $AlCl_3$. Likewise, mesoporous aluminosilicate, ion-exchanged with $ZnCl_2$, is effective in the Diels-Alder addition of cyclopentadiene to methylacrylate with good yields (ca. 90%) and excellent selectivity [100]. In fact, the amount of undesired side-products, due to polymerisation of the alkenes, is lower than with conventional homogeneous BF_3-OEt_2 Lewis acid catalyst, thanks to the milder acidity of the ordered mesoporous aluminosilicate, which is more suitable for this reaction.

Lastly, the immobilisation in ordered porous matrices of systems that are active, in homogeneous phase, for the stereoselective Diels-Alder cycloaddition is a viable strategy to obtain heterogeneous catalysts with unexpected characteristics, also in terms of enantioselectivity (Sect. 8.7). For example, when a phenyl-substituted bis (oxazoline) copper(II) complex is non-covalently immobilised into a FAU (Y)

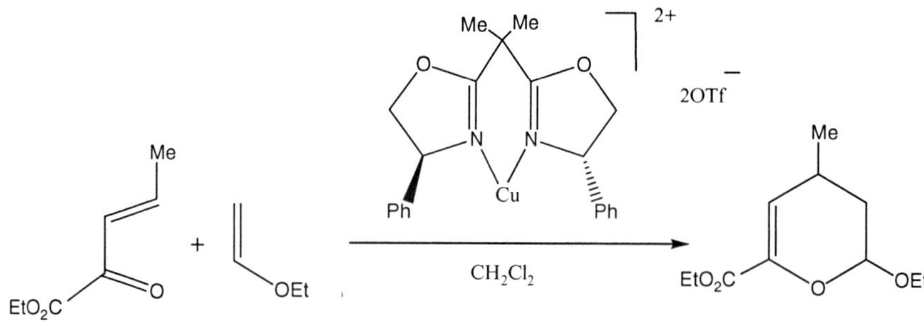

Fig. 8.22 Hetero-Diels-Alder cycloaddition over bis(oxazoline)-Cu(II)/FAU zeolite

zeolite, it is active for the hetero-Diels-Alder addition reported in Fig. 8.22, with an enantiomeric excess (e.e. 41%) which is significantly higher than that obtained with the homogeneous catalyst (e.e. 20%) under comparable conditions [101]. Curiously, the configuration of the product obtained over the heterogeneous catalyst is opposite to that obtained in homogeneous phase. Such result appears to be clearly due to the use of zeolitic materials and to the pore confinement effect linked to it.

8.6 Redox Reactions

The general name of *redox-active molecular sieves* comprises a wide series of porous solids in which the catalytically active site is able to induce a redox transformation in the substrate. According to the definition, the active site itself must not necessarily undergo a change in its oxidation state. Rather, it causes a change in the oxidation state of the reactant, by catalysing the action of the oxidising (or reducing) agent. In addition, zeolites and zeotypes, with their ordered porous network, introduce steric effects, which are important for successful syntheses, and possess dispersed and isolated catalytic centres in well-defined molecular environment, which is a noteworthy feature to have a stable, reusable and durable catalyst. Thanks to all these characteristics, redox-active molecular sieves are often considered the "inorganic equivalent" of enzymes, whose key to success is represented by having isolated and identical active sites in a proper chemical surroundings with adequate geometry, polarity and steric features.

Typically, redox-active centres are added to microporous or mesoporous solids following four main synthetic pathways: (1) by synthesis through isomorphous substitution; (2) by ionic exchange; (3) by post-synthesis modifications; and (4) by encapsulation.

1. By introducing a redox-metal precursor into the synthesis mixture of the molecular sieve (e.g., during the hydrothermal synthesis of a zeolite), it is possible to replace isomorphously some lattice atoms (Si, Al or P) with redox-active centres. Thanks to this technique, the chemical environment of the site is

strongly dependent on physical-chemical characteristics of the main matrix. However, some centres can be "buried" within the walls of the solid and they can be not accessible and available for catalysis. Moreover, the presence of redox-metal precursor in the synthesis mixture may affect critically some preparation phases (gelation, crystallization, etc.) and it is then often necessary to set up a completely new synthesis protocol of the desired material. All this can be very time-consuming as many factors affect the crystallisation of the molecular sieves.

2. The ionic exchange is possible whenever the solid possesses anionic sites, as it takes place with aluminosilicates (zeolites) and silico-aluminophosphates (SAPOs). The negative charges can be compensated by metallic cationic species with redox properties. Nevertheless, this preparation method can lead, as a drawback, to sites with a poorly defined chemical surrounding environment, mainly when the diameter of some hydrated cations is too large to have access to the porous network. Again, due to the high mobility of the exchangeable ions, labile species prone to leaching or to aggregation into inactive oxidic clusters can occur, especially when they are employed under liquid-phase conditions.
3. By means of post-synthesis modifications it is possible to obtain relatively easily redox molecular sieves from well-known solids (some of them can be also commercially available). The redox-active metal is generally added to the support as a precursor that can be deposited irreversibly onto the surface or, sometimes, into the solid matrix. With zeotypes, redox metals, such as Ti, V or Cr, can be incorporated by post-synthesis isomorphous substitution under vapour-phase conditions. Conversely, the use of mesoporous solids, with wide and highly accessible pores, has opened the way to the deposition of bulky and highly functionalised precursors, which can diffuse and then be grafted or tethered onto the inner surface of the solid (Fig. 8.23). Redox-active metals can thus be added to the solid obtaining a single-site catalyst, in which the active site has a definite chemical surroundings, as in conventional homogeneous

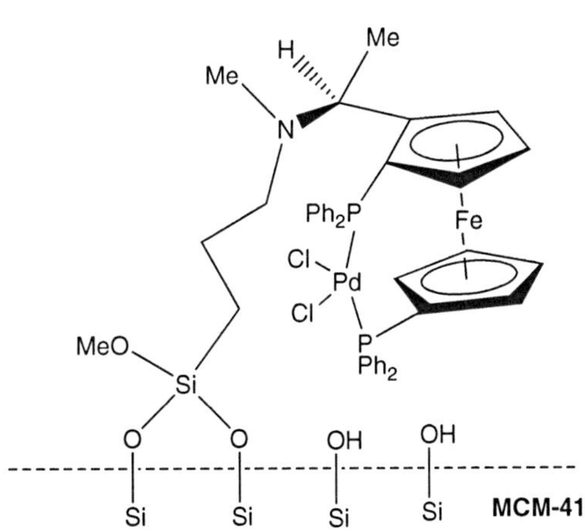

Fig. 8.23 A chiral palladium complex tethered onto mesoporous MCM-41. Reproduced (adapted) from [160], with permission from RSC

systems, with all the advantages of heterogeneous systems, namely easy separation, recovery and recyclability.

4. Encapsulation covers a large selection of methods for trapping catalytically active species within the pores of inorganic ordered supports and it allows to keep unaffected the optimal performances of the original homogeneous catalysts and, in some cases, thanks to a positive cooperative effect, to have a system with improved characteristics. Large complexes, such as porphyrines, phthalocyanines or transition-metal Schiff bases complexes are accommodated within or directly synthesised in cages of the porous solid (e.g., supercages of cubic or in the hypercages of hexagonal faujasite). Typically, such encapsulated metal complexes are assembled in situ by intrazeolite synthesis and complexation and, for this reason, they are often referred to as "ship-in-the-bottle complexes." Then, once formed, the metal complexes are spatially entrapped in the molecular sieve, since they are too bulky to diffuse out, whereas the reactants and the products can freely move through the pores. In order to maximise the catalyst lifetime, it is crucial that the complexes and the ligands are highly stable under the reaction conditions.

Since redox molecular sieves found their most relevant application in selective oxidation processes, in next sections, the attention will be paid onto this class of organic transformation.

8.6.1 Oxidation Over Redox Molecular Sieves

8.6.1.1 Oxidants

The oxidation of organic substrates leads to the production of major industrial valuable products, such as ketones, aldehydes, esters, carboxylic acids, alcohols, etc. Nevertheless, mainly in fine chemicals manufacture, oxidations are still carried out with huge stoichiometric amounts of conventional oxidants, such as chromium (VI) species, permanganate or organic peracids. Excessive quantities of toxic metals imply the unavoidable production of large amounts of toxic wastes that need proper disposal. Thanks to redox molecular sieves, it is possible to use oxidants that are economically and environmentally more sustainable. Oxidants that contain a high percentage of active oxygen and that give rise to harmless (or easily recyclable) side products are molecular oxygen, hydrogen peroxide, nitrous oxide and alkyl hydroperoxides (typically *tert*-butylhydroperoxide; TBHP).

Molecular oxygen is in principle the optimal oxidant, but aerobic oxidation is difficult to tune and the reactions must be performed often at low conversions to avoid over-oxidations. Actually O_2 is a triplet biradical and free-radical chain autoxidation pathways can dominate, leading to a final poor selectivity, especially when complex fine chemicals are involved as substrates. Hydrogen peroxide can oxidise organic compounds with an atom efficiency of 47% and with the generation of water as the only co-product. It is relatively cheaper (<0.5 EUR kg^{-1} for 100% H_2O_2) with respect to conventional lab-scale oxidants, but more expensive than O_2 or air [102]. Nonetheless, H_2O_2 is not easily miscible with non-polar solvents and cannot be used under strictly anhydrous conditions. This can be a major drawback for many fine chemistry applications. Nitrous oxide is widely considered as an "ideal" monooxygen donor, just yielding dinitrogen as a single side product. Thanks to its high stability, it can be applied in a broad range of reaction conditions. At industrial scale, N_2O can be obtained as a zero-cost by-product from the adipic acid synthesis process, but, for fine chemical applications, as it is a gas, it is not readily suitable for standard batch liquid-phase reaction equipment. Alkyl hydroperoxides (principally, TBHP) are largely used for small-scale industrial oxidations. They have good thermal stability ($t_{1/2}$=520 h at 403 K for a 0.2 M solution in benzene) and show fewer handling risks than 70% H_2O_2 or peracetic acid [103]. TBHP is reasonably versatile, as it is soluble in water as well as in non-polar solvents. Nevertheless, the atom efficiency of organic hydroperoxide is rather poor and they are scarcely used with medium-pore zeolites, as they are not easily accommodated in micropore channels.

8.6.1.2 Reaction Mechanisms

Oxidations catalysed by redox-active molecular sieves follow generally three reaction pathways: (1) one-electron free-radical oxidation, (2) oxygen transfer involving a peroxometal species, and (3) oxygen transfer involving an oxometal species.

A typical example of the first case is the cobalt-catalysed hydrocarbon oxidation, where the metal acts as a one-electron oxidant and the key steps involve the formation of alkoxy and alkylperoxy radicals (Fig. 8.24). This kind of mechanism is supposed to take place with cobalt- or manganese-containing aluminophosphates (AlPOs). In oxygen transfer pathways, the oxygen atom is provided heterolytically by an oxygen donor (e.g., H_2O_2 or ROOH), that generates either a peroxometal complex or an oxometal complex, as the active oxidising species (Fig. 8.25a, b). In the peroxometal route the metal site does not undergo any change in oxidation state during the catalytic cycle and therefore, it increases the oxygen-transfer ability of the coordinated peroxo group by acting as a Lewis acid (Fig. 8.25a). Conversely, in the oxometal route, the metal site itself acts as the oxidising species: it undergoes a two-electron reduction and is then re-oxidised by the oxygen donor (Fig. 8.25b).

i.) ROOH + Co(II) → RO•+ Co(III)OH
ii.) RO• + RH → R• + ROH
iii) ROOH + Co(III)OH → ROO• + Co(II) + H2O
iv) R• + O2 → ROO•
v) ROO• + RH → R• + ROOH

Fig. 8.24 Mechanism of cobalt-catalysed free-radical hydrocarbon oxidation

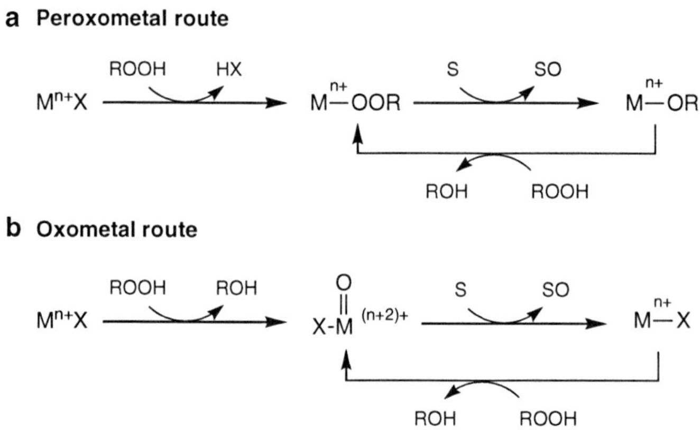

Fig. 8.25 Catalytic oxygen transfer involving (**a**) peroxometal species or (**b**) oxometal species

Early transition metals with a d^0 electronic configuration, such as Ti(IV), Zr(IV), Mo(VI), W(VI) or Re(VII), are relatively poor oxidants and react generally via the peroxometal pathway. On the contrary, many late and/or first-row transition metals, such as Cr(VI), Mn(V) or Os(VIII), that are good oxidants in their highest oxidation state, operate via the oxometal pathway. Some other metals, such as V(V), can react via both routes depending on the substrate and on the reaction conditions.

8.6.1.3 Main Catalysts

Titanium-containing molecular sieves and, above all, TS-1 have received so far the largest attention by researchers. These solids are employed for a wide series of selective oxidations with H_2O_2 under mild conditions, namely alkene epoxidation, alcohol oxidation, ketone ammoximation, oxidation of sulphur compounds, etc. and, currently, TS-1 is the only redox molecular sieve that finds wide commercial use as a catalyst at industrial scale. The remarkable efficiency of TS-1 for selective oxidation with diluted aqueous hydrogen peroxide is attributed to: (1) the isolation of titanium sites, preventing the undesired decomposition of H_2O_2, which is induced and catalysed by pairs of adjacent titanium sites, and (2) the hydrophobic character of the lattice, enabling the preferential adsorption of the hydrophobic substrates in the zeolite micropores in the presence of water [104]. Nevertheless, one of the main disadvantages of TS-1 is the relatively narrow micropore network (pore diameter 0.55 nm × 0.51 nm) and its use is thus restricted to substrates whose kinetic diameter is smaller than 0.55 nm (e.g., cyclohexene cannot be epoxidised within its pores). To overcome this drawback, many research teams incorporated titanium into (or added it onto) molecular sieves with larger pores and most detailed studies focus on BEA, MWW, MOR or ISV zeolites or on ordered mesoporous materials, mostly of the MCM-41 and SBA-15 type [105]. Among these large-pore systems, Ti-BEA zeolite displayed interesting performances not only in epoxidation, but also in alcohol oxidation as well as in hydroxylation of alkanes and aromatics [106, 107]. Mesoporous Ti-containing materials, however, often show smaller activity and poorer structural stability with respect to TS-1 [108, 109]. In addition to titanium, numerous reports have claimed the preparation of a large number of new solids, differing for the framework and for heteroatom substitution. Metals such as Zr, V, Cr, Fe, Co and Sn have been successfully added in microporous and mesoporous sieves [110]. They are active and selective for carrying out oxidations in liquid phase using H_2O_2 or organic peroxides as oxidants and they often provide a functionality which is complementary to that offered by Ti-containing materials. For instance, Cr-based silicates are suitable for allylic oxidation of alkenes and arene side-chain oxidation, whereas Ti-based ones are not. In some cases, anyway, heavy doubts exist on whether the activity has to be attributed to surface sites or to metal ions released in solution, particularly for materials containing V, Cr or Fe [111]. These elements are excellent promoters of radical oxidations with H_2O_2 even at ppm levels, and their activity in solution can be mistaken for that of heterogeneous sites. Strong experimental evidences in this regard are thus always necessary for assessing the real heterogeneous nature of the catalyst.

Among non-transition metal-based molecular sieve, the only redox-active catalyst which has shown so far a confirmed heterogeneous character and a remarkable stability and recyclability is a Sn(IV)-substituted BEA zeolite developed by Corma and co-workers [112]. In this zeolite isolated tetrahedral Sn sites are homogeneously inserted into the zeolite framework and they show noteworthy and peculiar activity for the Baeyer-Villiger oxidation of cyclic ketones, aromatic aldehydes and α,β-unsaturated aldehydes. This reaction has great commercial value with applications spanning from antibiotics, to steroids and many aspects of agrochemistry. The singular behaviour of Sn-BEA with H_2O_2 was first evidenced in comparison with Ti-BEA. When these two zeolites were studied in the oxidation of dihydrocarvone under the same conditions, major amounts of the corresponding cyclic lactone were observed over the former, whereas the latter gave rise to large amounts of the epoxidised product [110], Ti-containing zeolites being inactive in Baeyer-Villiger oxidations (Table 8.6). Combined spectroscopic, mechanistic and kinetic studies demonstrated that the active site is due to the presence of partially hydrolysed

Table 8.6 Baeyer-Villiger oxidation of dihydrocarvone with different catalysts. Adapted from [110]

Oxidant	Conversion (mole%)	Selectivity to products (mole%)		
	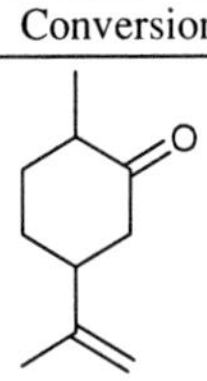	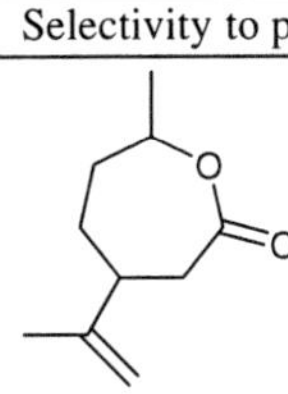	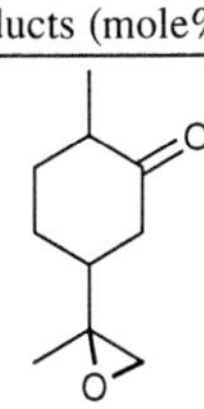	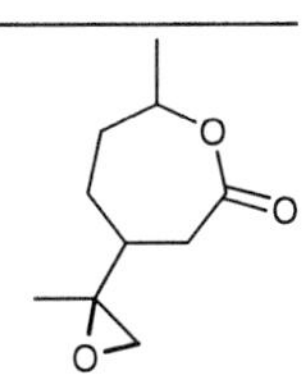
Sn-BEA/H_2O_2	68	100	0	0
Ti-BEA/H_2O_2	46	0	79	0
m-CPBA[a]	85	11	71	18
CH_3ReO_3[b]/H_2O_2	9	30	33	20

[a]*meta*-Chloroperbenzoic acid; [b]Methyltrioxorhenium

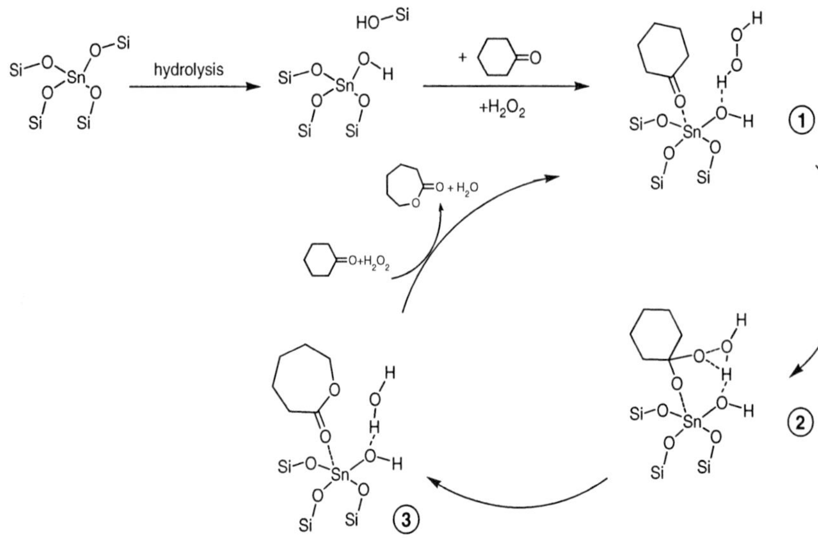

Fig. 8.26 Mechanism of the Baeyer-Villiger oxidation of cyclohexanone catalysed by Sn-BEA zeolite

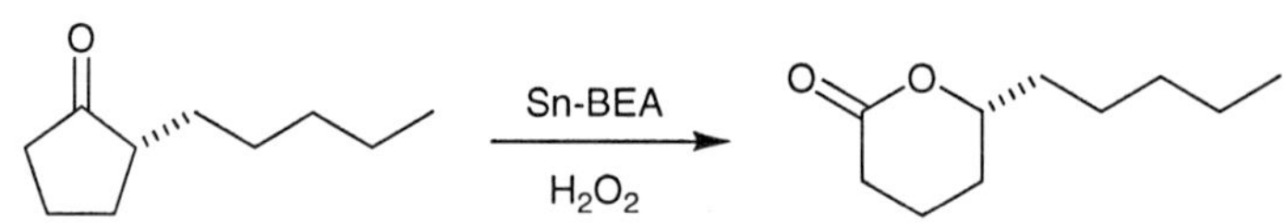

Fig. 8.27 Oxidation of *R*-delfone into δ-decalactone

framework Sn centres, where a Lewis acid site and an adjacent basic site on the oxygen atom of the Sn-OH group formed upon hydrolysis are simultaneously present [112]. Such acid-base pair is needed to accomplish the following proposed catalytic mechanism (Fig. 8.26): (1) coordination of the cyclic ketone to the Sn centre and of H_2O_2 to "stannol" (Sn-OH) site; (2) oxygen transfer from H_2O_2 to the coordinated ketone; and (3) desorption of the lactone and water. The determining step for chemoselectivity to Baeyer-Villiger oxidation is the first one, in which the oxygen atom to be inserted is placed on the proper line of approach for the attack to the activated coordinated carbonyl group. Sn-BEA was also tested in the oxidation of delfone into δ-decalactone (an aroma for the flavours and fragrances industry; Fig. 8.27). The desired product was obtained in high yields (86%), with very high turn-over numbers (ca. 10,000) and with negligible catalyst deactivation. Over this substrate, using the *R*-isomer of delfone (the most useful for flavouring purposes), it was shown that the enantiomeric configuration of the migrating carbon atom is retained and this is an important feature when enantiopure ketones have to be oxidized [112].

In alternative, zeolite-encapsulated metal catalysts are generally stable towards metal leaching, provided the organic complex is resistant to oxidative or solvolytic degradation. They give interesting results mainly in alkane oxidation, alkene epoxidation and phenol oxidation and several oxygen donors can be used, such as O_2, H_2O_2, PhIO, NaClO, or organic peroxides. In particular, they exhibit promising performances not only in terms of regio- and chemoselectivity, but also of stereoselectivity oxidation of fine chemicals. For instance, encapsulation of metal-salicylethylendiiminate (metal-salen) complexes in a ultrastabilized Y zeolite provides catalysts that show improvements in conversion, selectivity and diastereoisomeric excess (over the homogeneous complex) in the epoxidation of limonene and α-pinene, the solids being reusable and stable to leaching [113]. In this case, the high added-value of the products would account for the use of a sophisticated oxidation catalyst, but the width of the zeolite openings could be a remarkable issue when bulkier and more richly functionalised substrates were taken into account.

8.6.2 Epoxidation

Epoxides are useful intermediates for the production of several fine and specialty chemicals both at laboratory and industrial scale. Although alkene epoxidation is generally considered a well-established methodology, a truly versatile and eco-friendly heterogeneous system to be applied in fine chemicals synthesis is still vague and conventional processes based on hazardous and aggressive oxidants are thus widely used.

Titanium silicalites with MFI and MEL structures (TS-1 and TS-2, respectively) are the reference materials for this kind of reaction. Most of the theoretical, mechanistic, spectroscopic and kinetic studies on the reactivity have been carried out over these catalysts and sometimes the results obtained over TS-1 were extended to titanium-based molecular sieves with larger pore systems. However, to avoid any over-simplification, it is worth highlighting that the peculiar chemical environment around Ti sites in TS-1 (the hydrophobic character, the site isolation, the geometrical constraint, etc.) is unique and can be hardly found in other kinds of ordered materials.

Ti(IV) sites catalyse epoxidation via a oxygen transfer with peroxometal complex formation (Fig. 8.28). H_2O_2 binds to Ti(IV) forming a hydroperoxide species. Then, the distal peroxidic oxygen, activated by the proximity to Ti, is transferred to the alkene double bond, with contemporary formation of a Ti-alkoxide, that can undergo a new catalytic cycle, and a water molecule [104]. With an alkylhydroperoxide as oxidant a comparable mechanism is operative, but the related alcohol is co-produced. Consistently with a heterolytic pathway, the reaction takes place with complete retention of the C=C bond configuration.

Ti-MFI (TS-1) is suitable for the epoxidation of simple linear alkenes, whereas Ti-BEA with a wider pore network (0.76×0.64 nm) can be used with branched or cyclic olefins, such as 4-methyl-1-pentene or 1-methylcyclohexene (Table 8.7). In all these examples, the reaction is carried out in polar solvents. However, as a general rule, thanks to the hydrophobic character of Ti-MFI, a protic solvent as

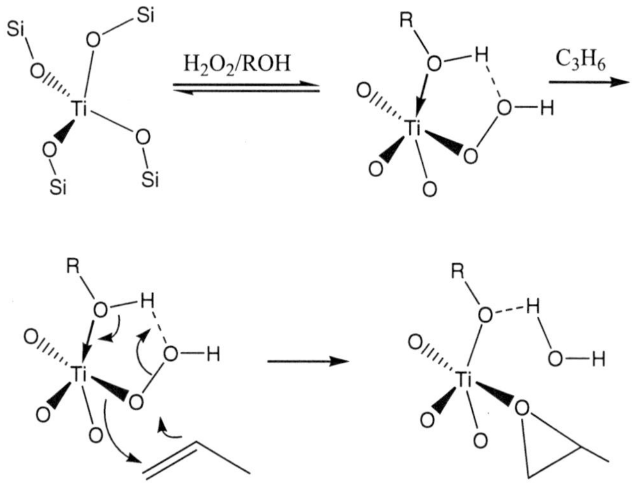

Fig. 8.28 Mechanism of epoxidation over Ti(IV)-substituted molecular sieves

Table 8.7 Epoxidation reaction with aqueous H_2O_2 catalysed by Ti-BEA and Ti-MFI. Adapted from [176]

Substrate	Catalyst	Substrate conversion (mole%)	Selectivity to epoxide (mole%)
1-Hexene	Ti-MFI	14	95
	Ti-BEA	12	4
2-Hexene	Ti-MFI	25	100
	Ti-BEA	13	4
1-Octene	Ti-MFI	11	80
	Ti-BEA	9	6
4-Methylpentene	Ti-MFI	n.d.[a]	n.d.
	Ti-BEA	14	14
Cyclohexene	Ti-MFI	n.d.	n.d.
	Ti-BEA	18	0
1-Methylcyclohexene	Ti-MFI	0	n.d.
	Ti-BEA	13	1

[a]Not determined

methanol is a good medium for epoxidation, whereas with Ti-BEA (or other more hydrophilic molecular sieves) aprotic solvents, such as acetonitrile or acetone, are preferable. In addition, since many of these large-pore zeolites contain Al sites, they also possess a Brønsted acidic character that give them a potential for bifunctional cascade catalysis (*vide infra* Sect. 8.8). In fact, the formed epoxide ring can be promptly activated by such acidity and ring-opening products (diols or hydroxyethers) or rearrangement derivatives (aldehydes, ketones) are formed, causing a sharp decrease in selectivity to epoxides. A way to minimize acid-catalysed undesired reactions is to employ highly concentrated 85% H_2O_2 or with the urea-H_2O_2 adduct. With this stratagem, a 95% yield in epoxide was obtained from the acid-sensitive 3-methyl-2-buten-1-ol [114].

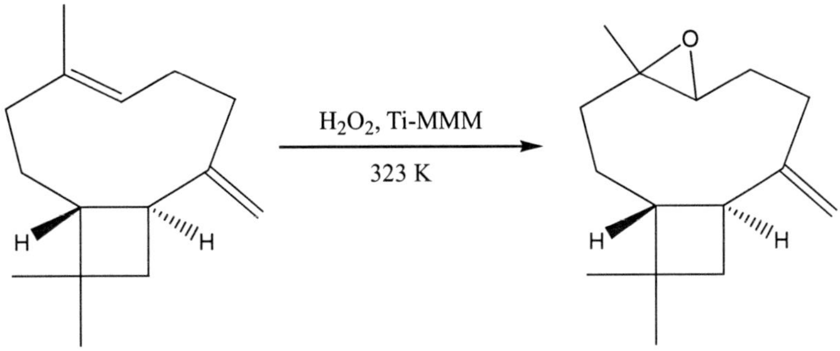

Fig. 8.29 The epoxidation of (-)-caryophyllene over Ti-MMM catalyst

Bulkier unsaturated substrates, such as cyclic terpenes, fatty acid derivatives or polycyclic aliphatic compounds can be epoxidised preferentially (or only) over mesoporous molecular sieves. Therefore, Ti(IV) has been added to silica structures with different geometries and features, such as MCM-41, TUD-1, SBA-15, ITQ-2 and so on. For instance, the naturally occurring terpene, (-)-caryophyllene (Fig. 8.29) is oxidised with H_2O_2 over Ti-MMM (an ordered mesoporous silica) to the 4,5-monoepoxide derivative with a selectivity of 80% at 75% of alkene conversion in 4 h, whereas in presence of TS-1 no reaction is observed [115]. However, in many cases, when the pore diameters are too large and no proper "sieving" effect occurs, the siliceous materials act as simple supports for the active sites and the need for an ordered array of pores does not seem essential. The performances obtained over Ti-containing mesoporous solids are often fully comparable to those obtained over Ti-containing amorphous non-ordered silicas [116–118]. The highly hydrophilic nature of MCM-41-based solids can be detrimental if relatively non-polar olefins have to be epoxidised and, in the attempt to create a more hydrophobic material, the surface of a Ti-MCM-41 was silylated [119]. The silylated catalyst showed a greater activity in the epoxidation of cyclohexene compared with the unsilylated one. Nevertheless, if substrates with intermediate polarity or with various functionalities are considered, a hydrophilic environment around Ti(IV) sites can give improved results [120].

Together with simple alkenes, unsaturated alcohols have been extensively studied to study whether the hydroxy group has any rate- or selectivity-enhancing effect on the epoxidation of the C=C bond. Some evidences were collected over Ti-MFI, but sometimes results were controversial due to major steric requirements around the Ti centres [121]. Later, some tests over mesoporous Ti-catalysts, where steric constraints are less crucial, confirmed that there is a marked directing effect of the alcoholic group in terms of reactivity (allylic alcohols are more readily epoxidised than pure alkenes) and of selectivity (*syn* epoxidation of cyclic allylic alcohols) (Fig. 8.30).

Metals other than titanium have been much less studied. For example, V-AlPOs with AFI structure (VAPO-5) catalyses selectively the epoxidation of allylic

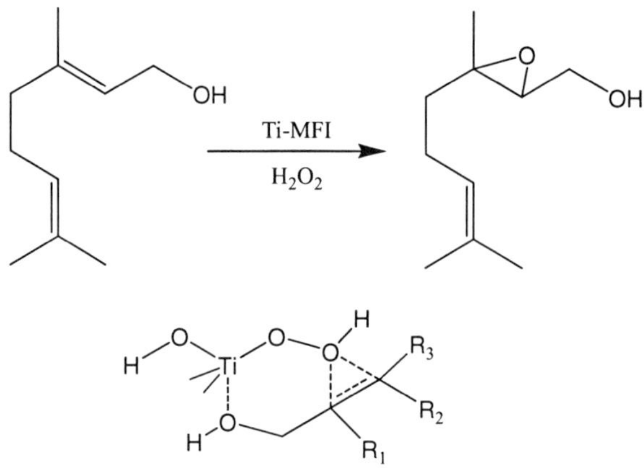

Fig. 8.30 Hydroxy-assisted regioselective epoxidation of geraniol

alcohols with TBHP [122]. However, especially in presence of protic species, attention must be paid to the leaching of active homogeneous V species. Again, Zr-MCM-41 or Mn-MCM-41 were active in the oxidation of *trans*-stilbene to *trans*-stilbene oxide in liquid phase [123]. Other examples include Ru-based porphyrin complexes immobilised via tethering over a mesoporous MCM-41 silica structure [124]. The preparation of these complexes is not easy and they use an unusual oxidant, such as 2,6-dichloropyridine N-oxide. Nevertheless, these systems show a remarkable stability (Ru-Por-MCM-41 is still very active after 11,700 catalytic cycles in the epoxidation of styrene) and superior stereospecificity in the epoxidation of *cis*-alkenes. In these examples the mesoporous molecular sieve acts sometimes as a pure support onto which the well-studied homogeneous metal complex is heterogenised and thanks to which the active metal can be recovered and reused. However, in many other cases, metal complexes confined in molecular sieves display better performances than their homogeneous counterparts since "confinement effects" may affect favourably the catalytic transformation by (1) protecting the active site from deactivation processes, (2) stabilising eventual labile reaction intermediates inside the porous network, (3) increasing the local concentration of the concurring reactants and, in particular, and (4) controlling and directing the product selectivity (in terms of regioselectivity and stereoselectivity, as well) [125, 126].

8.6.3 Ammoximation

Titanium-containing molecular sieves are optimal catalysts for the conversion of carbonyl compounds into the corresponding oximes in presence of NH_3 and H_2O_2. In particular, the most relevant application is not in the field of fine chemistry: the ammoximation of cyclohexanone to COX, a key step in the manufacture of ε-caprolactam and hence of nylon-6. In the EniChem-Sumitomo ammoximation process cyclohexanone reacts with NH_3 and H_2O_2 over TS-1 to give the related oxime with excellent conversions (99.9%) and very good yields (98.2%) [127]. This is an alternative to the conventional ammoximation route based on hydroxylammonium sulphate that involves the co-production of large amounts of ammonium sulphate and the use of hazardous chemicals, such as oleum, halides and nitrogen oxides.

The oxidation of NH_3 to hydroxylamine is the key step of the whole process. Tetrahedral Ti(IV) sites with two labile coordinative positions in TS-1 are supposed to be the active sites over which H_2O_2 oxidises NH_3. Then, NH_2OH desorbs and reacts non-catalytically with the ketone in the bulk solution to form the oxime [128]. For this reason, the reaction is free from steric limitations typically linked to the use of a microporous zeolitic catalyst and it is applicable to any (bulkier) ketone or aldehyde. The ammoximation of aldehydes is faster than that of ketones and the selectivity to oximes is even higher (Table 8.8).

TS-1 was, therefore, applied to the synthesis of *para*-hydroxyacetophenone oxime too, which is then converted, via Beckmann rearrangement, to paracetamol, a widely-used analgesic (Fig. 8.31) [129]. Good results were obtained over TS-1, i.e., 100% selectivity to oxime at ca. 50% conversion of ketone, whereas lower yields were observed over Ti-Al-BEA and Ti-ZSM-48 zeolites. With regard to the stability of the TS-1 catalyst, the main causes for deactivation are: (1) the dissolution of Si species in the basic ammonia solution, leading to the migration of Ti and its accumulation on the external surface of the solid; (2) the change of coordination geometry (from tetrahedral to octahedral) of Ti centres upon removal out of the framework; and (3) the irreversible adsorption of reaction by-products leading to the blockage of active sites.

Table 8.8 Ammoximation of aldehydes and ketones over Ti-MFI. Adapted from [177]

Substrate	Time (h)	Conversion (mole%)	Selectivity (mole%)
Acetone	6.0	79.7	98.1
3-Hexanone	4.0	70.4	98.1
Methyl isobutyl ketone	3.0	98.0	99.5
Cyclohexanone	4.0	98.2	96.4
p-Tolualdehyde	2.0	97.0	97.7
Benzaldehyde	2.5	97.0	99.4

Catalyst 1.5 g Ti-MFI (Si:Ti=29), substrate:H_2O_2:NH_3 ratio 1:1.2:2.0, solvent *tert*-butanol, 343 K

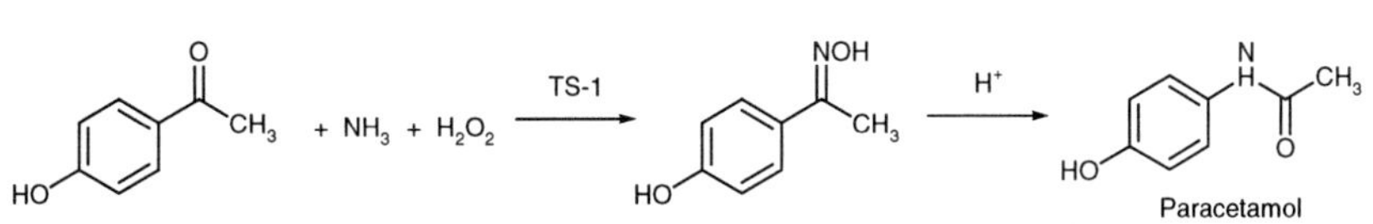

Fig. 8.31 Paracetamol synthesis via ammoximation

8.6.4 Oxidation of C–H Bond

The insertion of an oxygen atom into a C–H bond is not easy and, for this purpose, strongly electrophilic oxidants or highly active radical species are employed. In very general terms, the reaction can follow a heterolytic (Fig. 8.32a) or a homolytic pathway (Fig. 8.32b). In the former, a singlet oxygen atom (^{1}O) from an electrophilic oxidant is inserted into the C–H bond, whereas in the latter the reaction is driven by a triplet oxygen atom (^{3}O) and it involves the formation of radical intermediates via hydrogen abstraction [177]. According to the nature of the oxidant, of the reactants and, often, of the reaction medium, the reaction can follow preferentially one of these two pathways. Therefore, by finely tuning the reaction conditions it is often possible to keep the transformation in the desired direction and hence, to enhance the selectivity to the class of the required products.

8.6.4.1 Oxidation of Aromatic C–H

So far, redox-active molecular sieves find main application in aromatic oxidation of relatively simple substrates, such as monocyclic or bicyclic arenes or phenols. The information obtained over these examples can give an idea of the potential behaviour over more functionalised fine chemicals molecules.

Titanium-MFI zeolites (TS-1) catalyse the oxidation of the aromatic nucleus, to give phenols, and the conversion is typically kept low, to minimise over-oxidation of the substrate (phenols are more readily epoxidisable than the parent aromatic

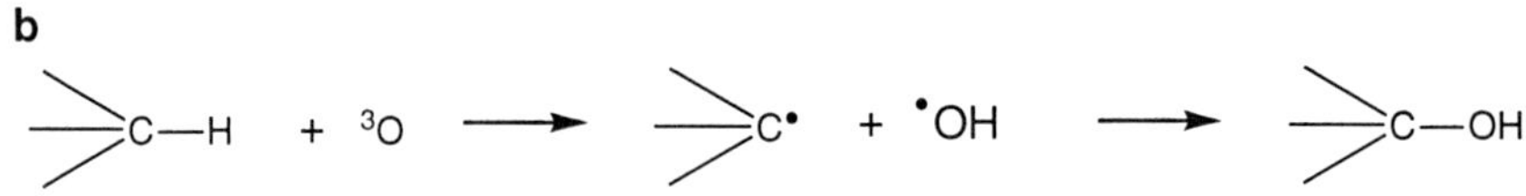

Fig. 8.32 Pathways for the insertion of an oxygen atom into a C–H bond

Table 8.9 Oxidation of toluene over Ti-MEL and V-MEL. Adapted from [177]

Catalyst	Conversion (mole%)	Selectivity to products (mole%)				
		o-Cresol	*p*-Cresol	Benzylalcohol	Benzaldehyde	Others
Ti-MEL	39.6	36.0	59.0	–	–	5.0
V-MEL	35.1	20.0	17.0	8.0	52.0	3.0

Catalyst 0.1 g Ti-MEL (Si:Ti=77), V-MEL (Si:V=79), substrate:H_2O_2 ratio 3:1, solvent acetonitrile, 333 K, 8 h

compounds). Conversely, vanadium- and tin-containing MEL zeolites promote mainly the oxidation of side-chain substituents, leading to the formation of aromatic ketones and alcohols (Table 8.9).

According to an electrophilic mechanism, the oxidation of aromatic C-H bonds is hindered when electron-withdrawing groups, such as Cl, NO_2, etc. are present on the aromatic ring. However, over microporous zeolitic catalysts the performance is due to a combination of electronic and steric factors. So, for instance, over TS-1, the hydroxylation rate follows the order: benzene>toluene>*para*-xylene>*para*-methylethylbenzene. Whenever mesoporous catalysts are used, steric effects become less important than electronic ones. However, only a relatively narrow series of substrates has been studied over mesoporous sieves and it is difficult to make direct comparisons.

The oxidation of phenol to hydroquinone with hydrogen peroxide is the first example of industrial exploitation of TS-1 on a large scale. With this process, introduced by EniChem company in 1986, hydroquinone and pyrocatechol are currently produced. The zeolite process has numerous advantages over the previous conventional ones using peroxoacids and Co^{2+}, Fe^{2+}, etc. as catalysts. When the reaction conditions are optimised, 94% selectivity based on phenol and 84% selectivity based on H_2O_2 at 30% phenol conversion are obtained, with almost equal yields of catechol and hydroquinone [130]. This process found later an effective and elegant integration in the organic synthesis of fine chemicals, as in the Rhodia process for the manufacture of vanillin, a major flavour ingredient [131]. Globally, one equivalent of phenol, H_2O_2, methanol, formaldehyde and O_2 are converted into one equivalent of vanillin and three equivalents of water (Fig. 8.33).

When phenol is the desired product rather than dihydroxyarenes, benzene can be selectively oxidised by H_2O_2 over TS-1 using sulfolane as a solvent.

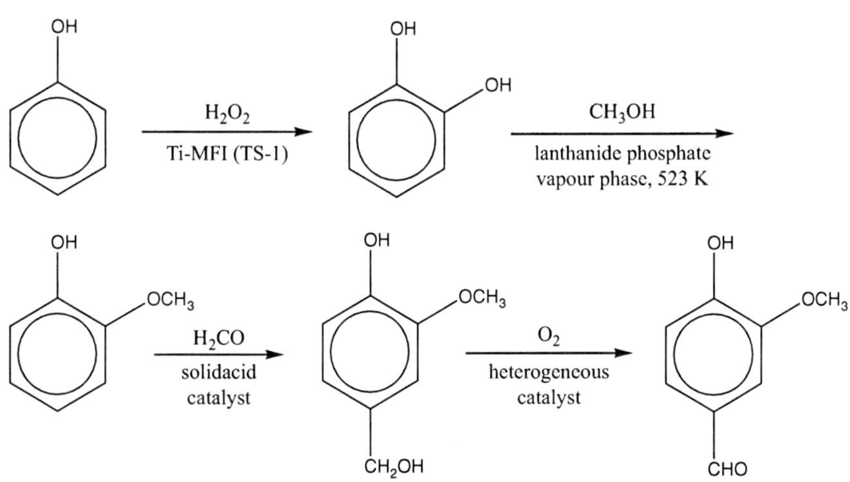

Fig. 8.33 Rhodia vanillin process

The remarkable selectivity to phenol is due to the formation of a bulky sulfolane-phenol adduct that cannot enter the micropores of TS-1. Further oxidation of phenol to give dihydroxyarenes, quinones, tars, etc. is therefore hindered. Moreover, a post-synthesis treatment with NH_4HF_2 and H_2O_2 of TS-1, removes the more exposed, and therefore, less selective titanium sites and the resulting catalyst, called TS-1B, attains selectivity values on benzene up to 85% and on H_2O_2 up to 61% [132].

Bulkier aromatics are converted mainly over mesoporous redox materials, typically Ti- or V-containing ones. Trimethylphenol (TMP) is oxidised over Cu (II)-substituted molecular sieves to 2,3,5-trimethyl-1-4-benzoquinone (TMBQ), a key intermediate in the production of Vitamin E. Over MCM-41 substituted with Cu(II) and small amounts of Al(III), 98% of TMP conversion and more than 85% TMBQ selectivity were obtained in 20 min when acetonitrile was the solvent and *tert*-butylhydroperoxide the oxidant [133]. Ti-containing ordered mesoporous silicas, such as Ti-MMM (a mesoporous material with an ordered array of hexagonal pores) and Ti-MMM-2 (a stable form of hexagonal mesoporous material), showed promising results in the same reaction as well. TMP conversion and TMBQ selectivity values as high as 98 and 76%, respectively, were obtained with H_2O_2 over Ti-MMM-2 [134]. However, in these cases too, structural ordering of the mesoporous Ti-silicates is not essential and optimal performances can be attained also over non-ordered mesostructured Ti-containing silicas [135]. Likewise, the oxidation of 1-naphthol afforded 1,4-naphthoquinone (84%), 1,4-dihydroxy-naphthalene (11%) and 1,2-naphthoquinone (5%) over V-MCM-41 with high efficiency in H_2O_2 [136]. Hydrothermally stable mesoporous titanium molecular sieves, as Ti-MMM-2, are promising systems for the liquid-phase selective oxidation of 2-methyl-1-naphthol with aqueous hydrogen peroxide in acetonitrile to give 2-methyl-1,4-naphthoquinone (menadione, a key intermediate in the synthesis of vitamins of the group K). Quantitative conversion of naphthol (>98%) and high selectivity to menadione (75%) are obtained at 353 K after 30 min of reaction. Ti-MMM-2 does not show titanium leaching and it can be recycled several times, without any remarkable loss in activity and selectivity [137]. Noteworthy, the Ti-MMM-2/H_2O_2 system has an excellent E-factor (environmental factor) of 0.3 with respect to the conventional CrO_3-based process, for which the E-factor is 18.

The effect of the incorporation of various first transition metals (Ti, V, Cr, Mn, Fe, Co, Ni, Cu and Zn) in MCM-41 was studied in the liquid-phase oxidation of anthracene into 9,10-anthraquinone [138]. Cr-MCM-41 showed the best performance with TBHP at 353 K, giving a conversion of 79% and selectivity of 98%. In this case, high yields were obtained by using aprotic solvents having low dielectric constants and, hence, minimal solute-solvent interactions.

8.6.4.2 Oxidation of Aliphatic C–H

Most of the model and applicative studies about aliphatic C–H oxidation have been carried out with *n*-alkanes. These investigations are often not applicable to the synthesis of fine chemicals, as these molecules bring always several different functional groups which are more easily oxidisable than C-H bonds. However, in the field of alkane oxidation, a particular and illustrative example worth citing is the use of Fe(III)-containing AlPOs with ATO structure (Fe-AlPO-31) for the oxidation of cyclohexane into adipic acid and cyclohexanone. The reaction is carried out in one step under solvent-free conditions and yields of 65, 15 and 5% in adipic acid, cyclohexanone and cyclohexanol, respectively, are attained in 24 h at 100–130°C under moderate pressure (15 bar), glutaric and succinic acids being some minor side products [139]. The main interesting feature, together with the suppression of solvent and of corrosive promoters, is that dry air is used as oxidant. Nonetheless, such optimised process is not straightforwardly applied to the oxidation of other bulkier and/or more functionalised molecules.

Conversely, the C-H bond activation at the benzylic position by transition metal complexes under mild conditions is an important transformation in the synthetic fine organic chemistry. Chromium(VI) catalysts are extensively used for the transformation of alkylaromatics to aromatic ketones and acids in homogeneous phase, but such reaction conditions are totally unsustainable with regard to the human health and the environment. As seen above, vanadium- and tin-silicalites promote primarily the side-chain oxidation of alkylarenes. Then, Co-containing AlPOs with ATS structure (Co-AlPO-36) can be employed for the solvent-free liquid-phase oxidation of *para*-xylene with air to give toluic acid, 4-formylbenzoic acid and the fully oxidised terephthalic acid. The yields of the reaction are not excellent, but the absence of other relevant side-products and the sustainability of the process are its main advantages [140].

Analogously, Mn(III)-framework substituted AlPOs with AFI structure (Mn-AlPO-5) can be applied in conjunction with acetylperoxyborate (APB), as a source of active oxygen, to the selective oxidation of picolines, pyrazines and pyridazines to the related carboxylic acids, with a limited production of N-oxide by-products. In particular, 4-picoline can be converted into isonicotinic acid in good yields (ca. 70%) and remarkable selectivities (>90%) in absence of organic solvents [141].

Alternatively, for bulkier substrates, chromium-containing MCM-41 proved to be an efficient catalytic method for the benzylic oxidation of a wide series of alkylarenes to the corresponding carbonyl derivatives in good yields, with concentrated TBHP (70%) as oxidant. In particular, benzylic $-CH_3$ groups are oxidised only in moderate yields, while the benzylic $-CH_2$-groups are smoothly converted into ketones in good yields [142, 143]. In these examples, chromium sites undergo redox cycles involving Cr(III), Cr(V)-oxo and Cr(VI)-peroxo species, thus playing a direct role in the oxidation mechanism. However, when Cr-containing materials are employed, careful tests for checking the real heterogeneity of the catalyst are needed, in order to exclude any role of minimal amounts of metal leached out from the solid. In fact, some catalysts can be recycled several times without any remarkable loss of activity and without any detectable change in active metal content, as often confirmed by elemental analysis. Yet, tiny amounts of metal in homogeneous phase can be frequently responsible for the observed catalytic activity and can be found as contaminants in the final product (especially when the active metal is toxic, as in the case of Cr or V) [111].

8.6.5 Oxidation of Alcohols

Alcohols are the main products of the oxidation of C–H bonds in hydrocarbons, but, as mentioned above, they can readily undergo further oxidation into the related ketones, aldehydes or carboxylic acids. In particular, titanium-containing zeolites (Ti-MFI and Ti-BEA) promote the selective oxidation of alcohols with H_2O_2 under mild conditions. From primary alcohols the corresponding aldehydes are obtained, but, if the conditions are not thoroughly tuned, carboxylic acids and esters can be also formed at relatively high conversions. Secondary alcohols give in high yields the corresponding ketones, which in general are stable under the mild conditions required and do not undergo further oxidative degradation. Tertiary alcohols, as expected, are not oxidised, but they can be slowly converted to the corresponding alkyl hydroperoxides. In comparison, secondary alcohols are oxidised faster than primary ones [144] and, thanks to the remarkably low reactivity of methanol, it can also be used as a solvent in TS-1 oxidations. So, over TS-1, the oxidation of propylene glycol produces hydroxyacetone in high selectivity [145]. Analogously, 2-phenylethane-1,2-diol undergoes chemoselective oxidation of the secondary alcohol group, affording β-hydroxyacetophenone as main product (73%) and mainly benzaldehyde as by-product (20%), from oxidative cleavage of C –C bond.

By analogy to homogeneous metal complexes, a cyclic intermediate (Fig. 8.34), where both the alcohol and the hydroperoxide are bound to the Ti centre, is supposed to be responsible for the oxidation reaction [107, 146]. Furthermore, the nature of the active site can play a critical role in the chemoselectivity of the transformation. A typical example is the oxidation of allylic alcohols: Ti-based catalysts give mostly the epoxidation of the C=C bond, whereas V-based catalysts promote the oxidation of the alcohol group (Fig. 8.35).

Typically, heterogeneous systems for the oxidation of alcohols with H_2O_2 include early transition metals with d^0 configuration. However, in contrast with homogeneous systems where Mo(VI), W(VI) or V(V) species are the most active, the most reliable heterogeneous catalysts are usually based on Ti(IV). Actually, the heterogeneous systems containing Mo, W, V or Cr are often prone to leaching under the reaction conditions, whereas Ti-based ones are not.

In the field of mesoporous molecular sieves, one of the most attractive areas is the heterogenisation of organometallic catalysts, which are particularly active and selective in homogeneous phase. For instance, perruthenate species can be immobilized within an amine-modified MCM-41 for the oxidation of different allylic alcohols with molecular oxygen [147]. Over such a system, a large series of

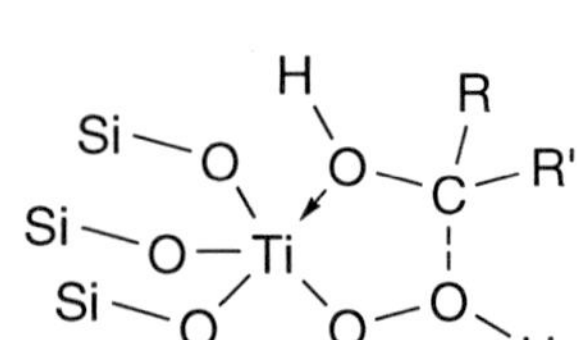

Fig. 8.34 Intermediate species in alcohol oxidation over Ti-zeolites

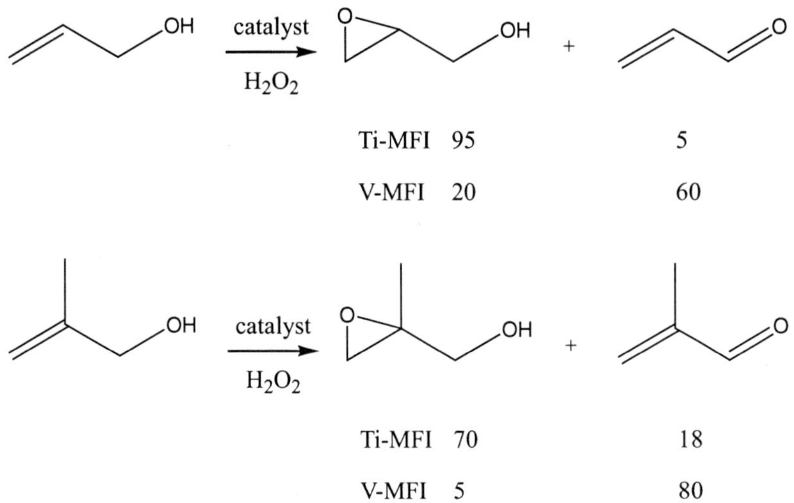

Fig. 8.35 Different chemoselectivity in oxidation of allylic and methallylic alcohols

Fig. 8.36 Oxidation of (6-chloropyridin-3-yl)methanol over perruthenate-containing MCM-41

phenylmethanols are quantitatively converted into the corresponding aldehydes in few hours without overoxidation. It was also used for the quantitative oxidation of (6-chloropyridin-3-yl) methanol into the related aldehyde, which is an important intermediate in the synthesis of the natural analgesic epibatidine (Fig. 8.36). Noteworthy, the reaction was also run over a silica aerogel-supported perruthenate, but much longer reaction times were needed, showing a peculiar effect of the ordered structure of the mesoporous molecular sieve.

Analogously, immobilization of TEMPO (2,2,6,6-tetramethyl-1-piperidinyloxy radical), which is widely used as a stable radical or a radical trapping agent, on the surface of MCM-41 is an interesting strategy for designing a versatile heterogeneous catalyst for fine chemicals oxidation (Fig. 8.37). Primary alcohols are selectively oxidised to either carboxylates or aldehydes [148]. In the oxidation of α-methyl glucoside, 1-O-methyl glucuronate is obtained with high selectivity in 20 min when NaOCl is used as a primary oxidant at 0°C. Conversion and selectivity of 50 and >95%, respectively, were reported in 1,3-pentanediol oxidation to γ-valerolactone. If an amide-tether is used instead of an ether-tether, the conversion can reach 99% with the same selectivity.

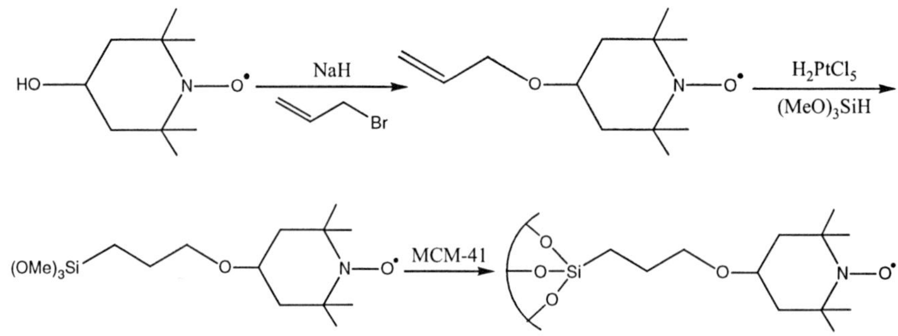

Fig. 8.37 Immobilization of TEMPO over MCM-41 molecular sieve

8.7 Stereoselective Reactions

Stereoselective synthesis of fine chemicals over heterogeneous catalysts has received much attention in last decade not only for the potential applications in pharmaceutical industry, where the stereo- and enantiopurity of each active product must be always carefully controlled and verified, but also in parallel fields, in which the use of single stereoisomers is definitely increasing, such as the synthesis of agrochemicals, flavours and fragrances. Homogeneous catalysis has reached remarkable results in this field as acknowledged by the Nobel Prize for Chemistry attributed in 2001 to Sharpless, Knowles and Noyori for their advances in asymmetric catalysis. A wide series of homogeneous catalysts is thus now available for almost every sort of diastereo- and enantioselective transformation. Nonetheless, heterogeneous catalysts possess the intrinsic advantage of being easily separated and, often, having better handling properties, especially for industrial purposes.

Microporous and mesoporous molecular sieves, thanks to their periodic ordered structure and their uniform pore size distribution, are, in principle, ideal host materials for catalytic centres with stereoselective properties. The high surface area inside their porous networks does not act simply as an "inert" support, but it affects the catalytic behaviour of the active sites through electronic and steric factors. So, when the catalysts are properly designed and prepared, they can show improved performances not only in terms of activity and chemoselectivity, but also of stereoselectivity. In particular, the confinement effect plays a relevant role due to the fact that the pores of the catalyst and the substrates have similar dimensions and that the weak interactions (steric or non-covalent ones) between the catalytic sites and the reactant molecules are, therefore, exalted. In addition, in a confined and restricted environment inside the pores the actual concentrations of the reactants are higher than in bulk. All these cooperative effects can be effectively exploited to address the transformation towards the desired stereoisomer [149]. Such effects, however, may lead to disadvantages too. A tight confinement into a narrow space can cause very poor catalytic activities due to diffusion limitations and steric constraints within the small pores, thus limiting the range of effective substrates that can be used.

Generally, the ideal heterogeneous stereoselective catalyst not only should be mechanically, chemically and thermally stable under the working conditions and easily recovered without loss of activity, but also should perform as efficiently as (or better than) its homogeneous counterpart, with high stereoselectivity (and especially enantioselectivity) to minimize the costly final separation steps and the excessive use of valuable starting reactants. However, a good stereoselectivity of the catalyst can be sometimes even more crucial than a good activity or stability, when fine chemicals of very high added value are considered.

Some few representative examples of diastereoselective and enantioselective heterogeneous systems relying on ordered molecular sieves are reported hereafter.

8.7.1 Catalysts for Diastereoselective Synthesis

Since two diastereoisomers possess different thermodynamic properties, the formation of one defined isomer can be enhanced (or inhibited) by substituting a homogeneous by an ordered porous heterogeneous catalyst. The peculiar morphology of a zeolite, a zeotype or a mesoporous material is often sufficient to obtain a highly diastereoselective system, in which the solid surface where the active sites are inserted in (or grafted on), acts as a rigid "macroligand" that influences electronically and sterically the synthesis. In some cases, the molecular sieve serves as a fine shape-selective catalyst, in which only some diastereoisomers with well-defined steric requirements can be obtained. In other cases, the presence of the solid matrix suppresses the occurrence of undesired non-selective side reactions (such as radical pathways) leading to complex mixtures of stereoisomers.

For instance, the stereochemistry of the acid-catalysed cyclization of [8,8,8-D_3] geranyl acetate is modified by passing from a homogeneous acid catalyst ($ClSO_3H$) to a FAU (Y) zeolite [150]. The confined environment within the FAU zeolite significantly alters the reaction energy profile for the cyclization of the model terpenoid and, during the intrazeolitic transformation, the *gem*-allylic methyl group adopts a preferential diastereoselective disposition in the cyclization product (64% d.e.). In contrast, in homogeneous phase, the reaction proceeds with negligible diastereoselectivity (d.e. $<5\%$) (Fig. 8.38). Another remarkable example is the acid-catalysed cyclisation of citronellal into isopulegol (Fig. 8.39). The reaction

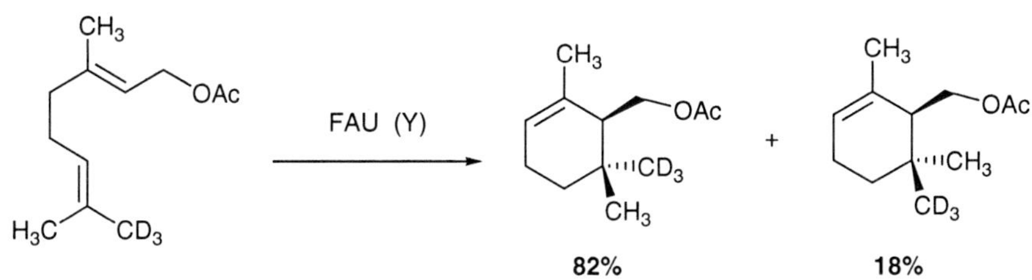

Fig. 8.38 Cyclisation of [8,8,8-D_3]geranyl acetate within FAU zeolite. Note the preferential configuration of the $-CD_3$ group

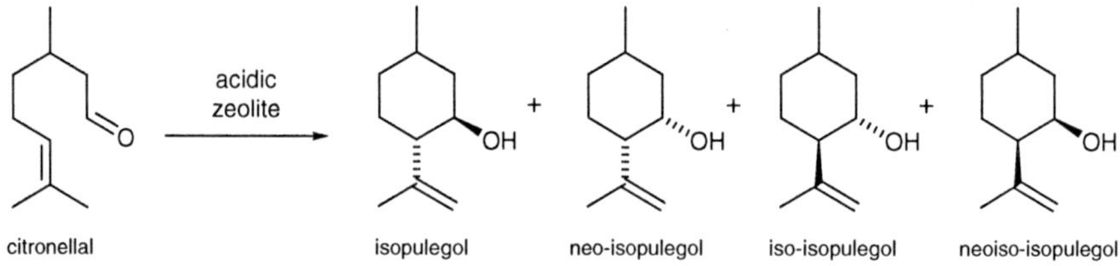

Fig. 8.39 Cyclisation of citronellal into the four diastereoisomers of isopulegol

Fig. 8.40 OH-assisted epoxidation of allylic alcohols over Ti-MFI (TS-1)

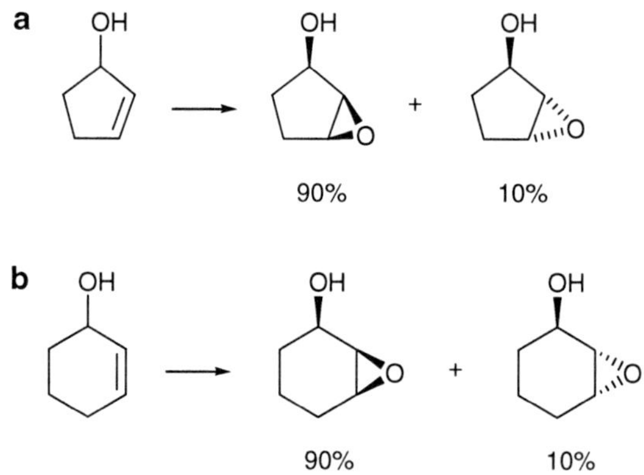

takes place over many acidic micro- or mesoporous molecular sieves, such as Al-BEA, Ti-BEA, Zr-BEA, H-FAU and Al-MSU (a mesostructured material) with high chemoselectivities (>90%), but the highest (93%) diastereoselectivity to the desired (±)-isopulegol (it can be easily hydrogenated to the valuable (±)-menthol) was recorded only over Zr-BEA [151]. Actually, the moderate strength of Zr sites limits the formation of side-products by dehydration and the steric confinement, exerted by the narrow BEA pores containing the Zr(IV), dictates the stereoselectivity, enhancing the production of (±)-isopulegol, where all the substituents are in equatorial position.

Likewise, in the epoxidation of allylic alcohols over titanium-based zeolites (Ti-MFI and Ti-BEA), the oxygen transfer from H_2O_2 to the alkenol occurs preferentially in *syn* with respect to the OH-moiety and *cis*-epoxyalcohols are selectively obtained (ca. 90% d.e.) over the *trans*-epoxyalcohols from cyclopent-2-en-1-ol and cyclohex-2-en-1-ol (Figs. 8.40a, b, respectively) [152]. The high diastereoselectivity is undoubtedly due to an OH-assisted mechanism, where the alcoholic group interacts with the Ti sites and directs the stereoselective transfer of the peroxy oxygen [153] and it is comparable to that observed over homogeneous Ti catalysts. However, with a zeolitic catalyst the undesired oxidation of the alcoholic function to aldehyde is very limited. Furthermore, the role of other potential docking sites for the OH-group, such as the ≡Si–OH groups proximal to Ti sites (especially in silica-based mesoporous catalysts) was not excluded by several authors [154].

8.7.2 Catalysts for Enantioselective Synthesis

In contrast to the previous case, enantioselective synthesis needs that a chiral inducer is present somehow in the catalyst to discriminate between the enantiomers. For that, two strategies are usually employed: (1) the addition of chiral auxiliaries to well-established heterogeneous catalysts and (2) the heterogenisation of homogenous catalysts with demonstrated activity and enantioselectivity [155].

The first approach is generally followed with amorphous inorganic oxides (silica, silica-alumina, etc.) or layered materials (clays, hydrotalcites, etc.), but some important examples are also present with micro- and mesoporous molecular sieves. The chiral inducer is typically a molecule or a ligand with low molecular mass and it is added mainly to the already formed porous catalyst by post-synthetic modifications (see Sect. 8.6). For instance, a chiral acidic zeolite can be prepared by loading molecules of *R*-1,3-dithiane-1-oxide into the supercages of FAU (Y) zeolite. The obtained catalyst is suitable for the kinetic resolution of butan-2-ol by dehydration and the reactivity of S-butan-2-ol is 39 times higher than that of the *R*-isomer [156]. The same approach can be applied to metals in low-oxidation state. So, by binding the chiral phosphorous ligand (R)-BINAP (2,2′-bis(diphenylphosphanyl)-1,1′-binaphthyl) to rhodium centres over MCM-41, a catalyst for the asymmetric hydroformylation of styrene is prepared and comparable enantiomeric excess (up to 50%) and activity to the homogeneous counterpart can be achieved [157]. For the first class of catalysts, the stability and the robustness of the chiral auxiliaries are the main issues. In some cases, furthermore, the preparation methodologies do not appear fully reproducible.

The second approach is by far the most exploited and a very large series of catalysts were prepared following such preparation pathway [157–159]. In this way, the very fine and well-established control of the stereoselectivity, peculiar of homogeneous catalytic systems, can be transferred and applied to a solid catalyst. The immobilisation of the homogeneous catalyst can occur by: (1) covalent binding, (2) ion-pair formation (when the complex to be immobilised is ionic species), (3) adsorption (due to weak hydrophilic/hydrophobic interactions, etc.), and (4) entrapment (see also Sect. 8.6). Zeolitic materials with large pores or cavities (such as the supercages of FAU) as well as ordered mesoporous oxides are the preferential choice as host solids for anchoring bulky asymmetric complexes. The field of application of immobilised chiral catalysts is actually huge and spans from asymmetric oxidation to asymmetric C–C bond formation, from asymmetric ring opening of epoxides to asymmetric hydroformylation. However, the asymmetric oxidation, over both heme- (such as metallophthalocyanines or metalloporhyrins) and non-heme complexes (such as transition metal Schiff-base compounds) and the asymmetric hydrogenation of prochiral alkenes, ketones and imines are the domains where the most promising examples are found.

Above all, a series of elegant examples based on covalently-immobilised chiral hydrogenation catalysts clearly illustrate the improvement in performances due to the confinement effect within the pores. When a chiral Pd-containing catalyst derived from 1,1′-bis(diphenylphosphino)ferrocene is anchored onto the inner walls of MCM-41, the resulting catalyst is able to hydrogenate ethyl nicotinate to ethyl nipecotinate with a 17% e.e. Interestingly, the use of a homogeneous silsesquioxane analogue leads to racemic products only [160]. The enhanced chiral induction is explained here in terms of confinement effects: the ordered structure of MCM-41 creates a "chiral space" where the interactions between the incoming substrate and both the pore wall and the asymmetric ligand are enhanced. Then, the anchoring of Rh- and Pd-containing (*S*)–(−)–2-aminomethyl-1-ethylpyrrolidine or (*1R, 2R*)–(+)–1,2-diphenylethylenediamine complexes to the concave surface of MCM-41 (with pores 3.0 nm wide) or onto the convex surface of a non-porous silica (Cabosil pyrogenic silica) leads to an effective heterogeneous catalyst for the asymmetric

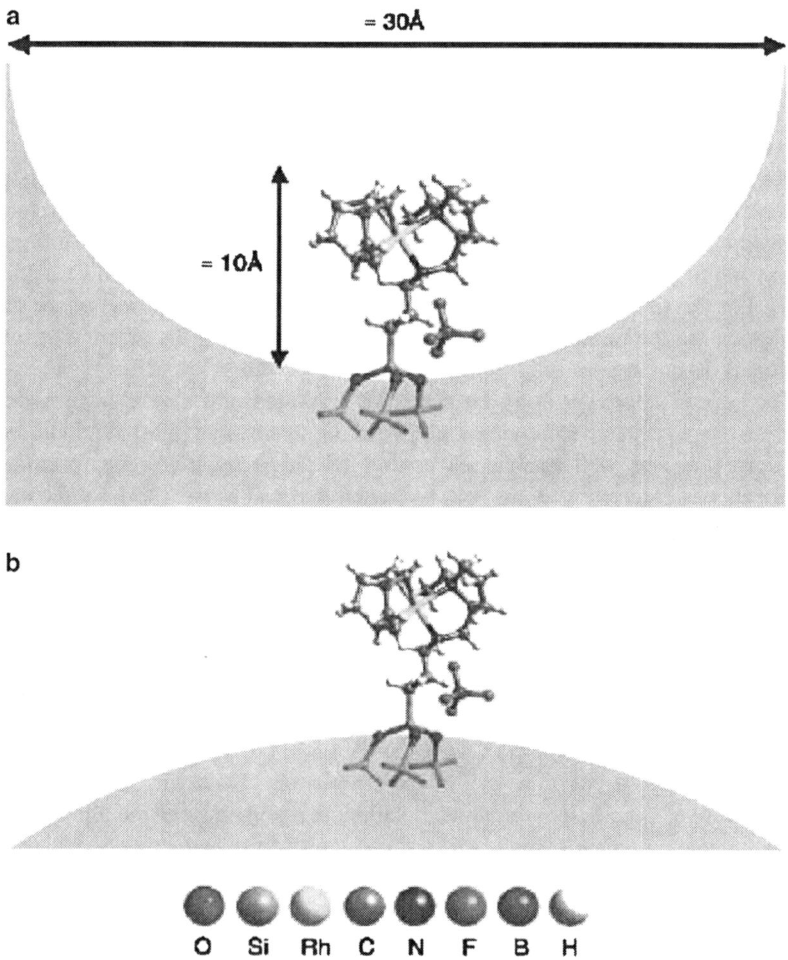

Fig. 8.41 Graphical model (to scale) showing the constraints of the catalyst once anchored on (**a**) MCM-41 and (**b**) pyrogenic Cabosil silica. Reproduced from [161], with permission from Wiley

hydrogenation of α-phenylcinnamic acid and methylbenzoyl formate [161]. A significant enhancement of the e.e. value was observed for the MCM-41-supported catalyst (with the concave surface), whereas the pyrogenic silica-supported catalyst (with the convex surface) shows the same results as those of the homogenous analogue (Fig. 8.41). Analogously, in the enantioselective epoxidation of simple alkenes (styrene, indene, etc.) over chiral Mn(III) complexes immobilised on MCM-48, the highest e.e. values were obtained when the pore size of the siliceous support is suitable and comparable with the size of the substrate. In such case, the size of the mesoporous channels of MCM-48 can be tuned by controlling the alkyl chain length of the templating agent [162]. These results confirm that not only the immobilisation on its own, but also a "closed" and well-defined confinement is essential to have a proper chiral induction on heterogeneous asymmetric reactions.

8.8 Cascade One-Pot Reactions on Bifunctional Catalysts

The synthesis of complex and multifunctionalised fine chemicals implies typically several successive steps, in which every intermediate product between the starting precursor and the target molecule is transformed, isolated and purified. Any reduction in the number of synthetic and separation stages has therefore numerous advantages: higher atom economy, lower operating costs, lower E-factors and, often, an improvement in the safety conditions [163]. The use of cascade one-pot reactions, in which various successive chemical steps are combined in the same reaction vessel without the recovery steps of intermediates, is a powerful strategy for these purposes and it mimics the behaviour of Nature, that is able to operate multi-step synthesis of large molecules in the cells of living organisms without separation of intermediate compounds.

In the field of heterogeneous catalysis, one-pot reactions usually occur over bifunctional (or multifunctional) materials, associating redox-active sites to acid/base-active centres. In some noteworthy cases, however, a unique site is able to catalyse both redox and acid/base consecutive reaction steps. Zeolites and zeotypes are versatile materials for such applications and sometimes the use of ordered and well-structured molecular sieves leads to unique performances, in terms of chemo-, regio- and stereoselectivity.

The ideal way to perform a cascade multi-step synthesis is to carry out the successive steps, not only in the same pot, but also under the same conditions. However, this favourable situation is not always possible. Actually, high yields in the target product can be obtained only when the last step is very fast and quasi-irreversibly shifted towards the formation of the final compound. In this case, the intermediate species are not detected in the bulk reaction mixture, as they are rapidly converted in further derivatives over the active sites and/or in the porous network. Conversely, if the optimal operating conditions for both the synthetic steps are different, the formation of large amounts of undesired products is likely. The one-pot reaction is thus carried out in two stages under different optimised

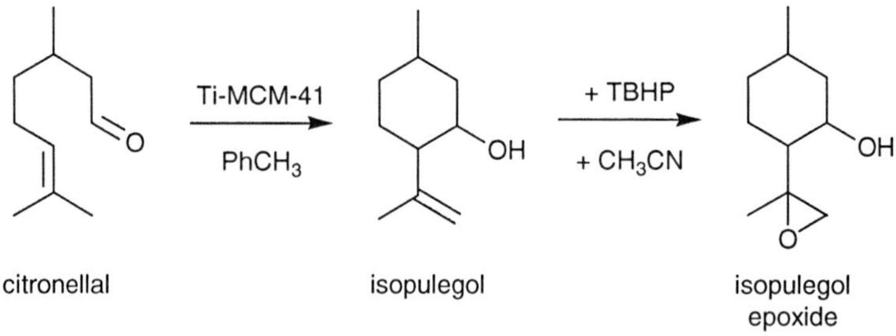

Fig. 8.42 One-pot two-step conversion of citronellal into isopulegol epoxide

conditions. An example of that is reported in Fig. 8.42. The synthesis of isopulegol epoxide (a compound with fungicidal and insect-repellent activity) is carried out over a titanium-grafted MCM-41 in two successive steps under different conditions. Citronellal is first converted into isopulegol, by acid-catalysed cyclization, which is then epoxidised to isopulegol epoxide. If the oxidant were added at the beginning of the reaction together with the other reactants, the starting unsaturated aldehyde would be directly oxidised to citronellal epoxide. To avoid it, *tert*-butylhydroperoxide is added only after the complete conversion of citronellal into isopulegol. Likewise, the cyclisation step is performed in a poorly polar solvent, such as toluene (it is an optimal medium for acid-catalysed ene-reaction), whereas the epoxidation step is performed in an aprotic polar mixture of toluene and acetonitrile (it is ideal for epoxidation over Ti-containing mesoporous catalysts). With this approach, an overall yield of 68% in isopulegol epoxide can be obtained [164].

Typically, the most relevant examples of bifunctional molecular sieves can be roughly summarized in two classes:

- molecular sieves with hydrogenation and acid (or basic) sites and
- molecular sieves with oxidation and acid (or basic) sites

In addition, some purely acidic (or basic) catalysts can catalyse two (or more) consecutive different transformations (e.g., alkylation and cyclization; esterification and rearrangement; etc.).

Some recent review papers have thoroughly described the variety of multifunctional molecular sieves [165, 166]. Hereafter, a selection of examples is reported to show the role of such catalysts in multi-step transformations.

8.8.1 Molecular Sieves with Hydrogenation and Acid (or Basic) Sites

By introducing a metal active in hydrogenation (e.g., Pd, Ni or Cu) over an acidic zeolite (such as alumino- or boro-silicates zeolites), it is possible to perform in one apparent step the acid-catalysed rearrangement of an epoxide to aldehyde and, consecutively, its hydrogenation to primary alcohol. For instance, over a

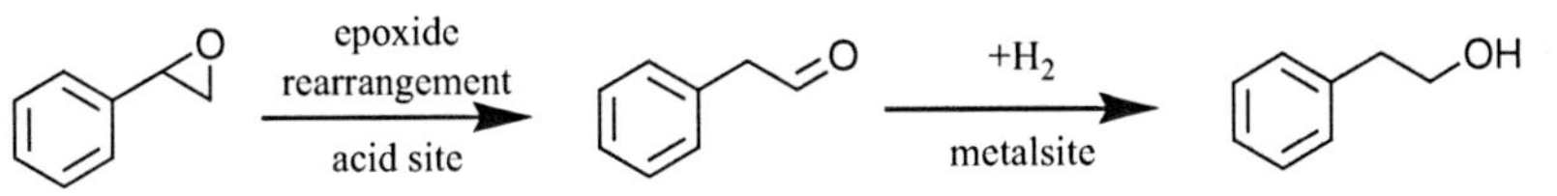

Fig. 8.43 One-pot conversion of an aromatic terminal epoxide into the related primary alcohol

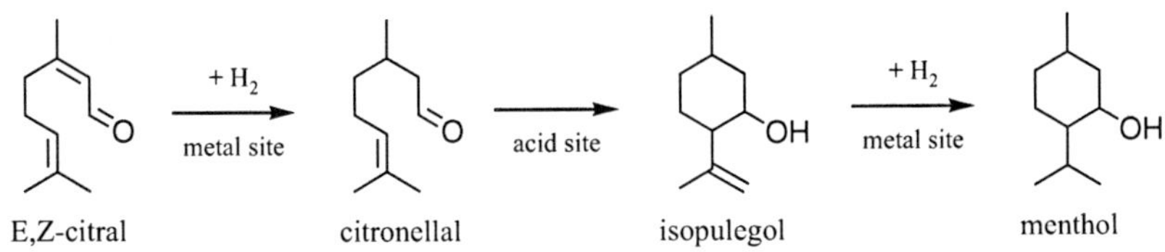

Fig. 8.44 Synthesis of menthol from citral or citronellal

Cu-containing borosilicate pentasil (MFI) zeolite, styrene oxide can be converted into 2-phenylethanol (a fragrance with sweet and floral odour) in high yields (up to 85%) (Fig. 8.43) [167].

Another example deals with the selective synthesis of menthol (a terpenic compound widely used in cosmetics and pharmaceutical applications) directly from citronellal or from citral (Fig. 8.44). Over a 3% Ir-impregnated H-BEA zeolite it is possible to carry out the cyclization of citronellal into isopulegol, catalysed by the acidic sites, as well as the hydrogenation of the latter, catalysed by the Ir sites [168]. To improve the conversion of citronellal, the reaction is first conducted under nitrogen and, then, eight bars of hydrogen (in autoclave) are added. In this way, 95% selectivity for the various menthol stereoisomers (of which 75% of the desired (-)-menthol) and complete citronellal conversion is achieved after 30 h. It is worth noting that, sometimes, under real conditions, the separation between acid/basic sites and redox sites is not sharply defined. Actually, the activity recorded during the isomerisation step (in absence of H_2) clearly increases when Ir is present onto the zeolite. Thus, not only the Brønsted acidity of the zeolite, but also the Lewis acidity of the small amounts of non-reduced Ir species can play a role in the isomerisation step.

If citral is taken as the starting compound, the transformation follows a three-step one-pot pathway: i) hydrogenation of the α,β C=C bond of citral to give citronellal; ii) isomerisation/cyclization of citronellal to isopulegol; iii) hydrogenation of isopulegol to menthol [169]. For this purpose, a single catalyst (3% Ni/Al-MCM-41) was designed to join the good selectivity of Ni in hydrogenating the α,β unsaturation in citral as well as the good activity shown by strong Lewis/weak Brønsted sites of Al-MCM-41 required for an efficient citronellal cyclisation. This system yields 90% of all the diastereoisomers of menthol at 343 K and five bars of H_2 and, in particular, produces 70–75% of racemic (±)-menthol in the final mixture after 300 min.

Otherwise, hydrogenating sites can be coupled to basic sites as well. A selective bifunctional catalyst for the low-pressure one-step synthesis of 2-ethylhexanal (a fragrance used in the composition of perfumes) from *n*-butyraldehyde is prepared by impregnating a Pd precursor onto a basic K-exchanged FAU zeolite [170]. In fact, *n*-butyraldehyde undergoes first aldol condensation over the basic sites, giving rise to the intermediate 2-ethylhex-2-enal, that is then readily reduced by H_2 to 2-ethylhexanal over the Pd sites. Under optimised reaction conditions, the final product can be obtained over a 0.5% Pd/KX zeolite with 91% selectivity at 70% conversion. In this case, the zeolite-based catalysts show better performances than non-zeolitic ones (under the same conditions 0.5% Pd/MgO displays a maximum *n*-butyraldehyde conversion of 8%).

8.8.2 Molecular Sieves with Oxidation and Acid (or Basic) Sites

As noted above, titanium-containing molecular sieves are the redox-active systems that are used more diffusively for oxidation in organic syntheses. In these catalysts, Ti centres can act at the same time as oxidising and acidic sites. In fact, Ti(δ^+) centres, inserted in (or grafted onto) the matrix of the silicate, are on their own Lewis acid sites or even mild Brønsted acid sites, due to the coordination of a molecule of water. Moreover, the acidic character can be also more marked if aluminium atoms are present in the composition of the molecular sieve. It is, therefore, difficult to obtain, in some cases, high yields in the desired oxidised product because of the acid-catalysed production of large amounts of co-products. Nevertheless, whenever these co-products are valuable compounds, the bifunctional pathway (oxidation plus acid-catalysed reaction) could be interesting for the synthetic chemist.

The bifunctional features of Ti and Al in a confined environment (such as a molecular sieve) can be exploited to have an inorganic catalyst that mimics the behaviour of the enzyme epoxidase. Linalool is oxidised and readily converted to alicyclic furans and pyrans (of interest in the flavours and fragrances industry) over Ti-containing large-pore zeolites (Ti,Al-BEA) or mesoporous materials (Ti, Al-MCM-41) because of the bulkiness of the terpenic substrates (Fig. 8.45). With these catalysts and using TBHP as oxidant high conversions of linalool with practically 100% selectivity to furans and pyrans are obtained [171]. The oxidation site (Ti) epoxidises selectively the epoxidation of the trisubstituted C=C double bond (richer in electron density than the allylic one and hence more easily epoxidised), whereas the acid site (Al) activates the epoxide ring and promotes intramolecular attack of the hydroxyl group, leading to the alicyclic products.

The intermediate epoxy derivative can alternatively rearrange into other products. This is the case in the oxidation of 2,5-dimethylfuran with H_2O_2 over Ti-MFI (TS-1). One of the unsaturations of the starting substrate is easily epoxidised and the intermediate furan epoxide immediately transforms into a *cis* dicarbonyl system (Fig. 8.46). A further isomerisation leads to the corresponding *trans* isomer. So, 2,5-dimethylfuran gives a mixture of *cis*- and *trans*-hex-3-en-2,5-dione with a selectivity of 85% at 94% conversion and a *cis* to *trans* ratio of 68:32 [172]. The formation

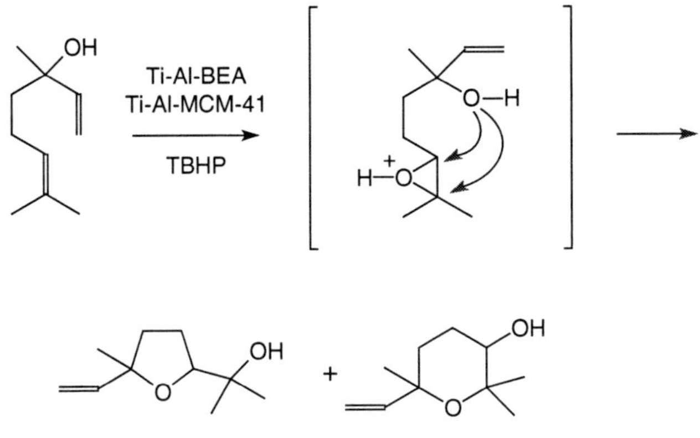

Fig. 8.45 One-pot conversion of linalool in alicyclic furans and pyrans

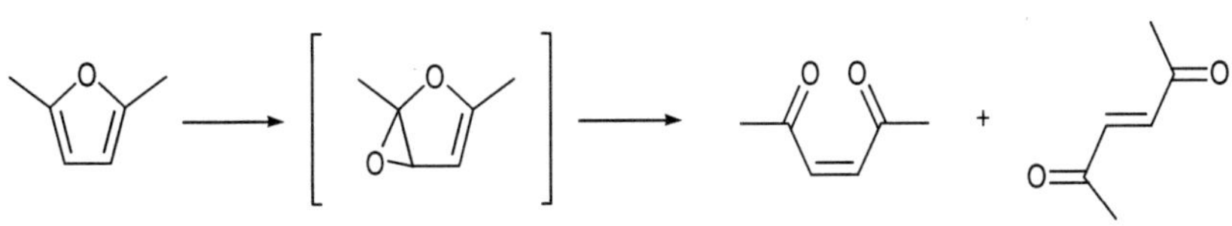

Fig. 8.46 One-pot conversion of 2,5-dimethylfuran into hex-3-en-2,5-dione

of the conjugated system in the unsaturated diketone enhances the quick rearrangement of the intermediate epoxide and, in fact, this species is not observed in the reaction medium.

8.8.3 *Multifunctional Molecular Sieves with Acid (or Basic) Sites Only*

From a formal point of view, even those catalysts which possess only catalytic centres of one kind (e.g., acidic or basic ones) can be considered as multifunctional catalysts. By means of these solids, in some elegant (and lucky) examples, complex and high added-value molecules were straightforwardly obtained from rather simple precursors.

The bulky *p*-methoxyphenyl *iso*-butyl ether (a fine chemical of interest for the fragrance industry) is quantitatively obtained from *p*-methoxybenzaldehyde and 2-butanol over a Sn- or Zr-containing BEA zeolite in one apparent step (Fig. 8.47). The aldehyde is first converted into the related *p*-methoxybenzylic alcohol by a Meerwein-Ponndorf-Verley reduction catalysed the Lewis acid centre, with the concurrent oxidation of 2-butanol into 2-butanone. Then, the intermediate alcohol reacts rapidly with the 2-butanone in excess (it is the reaction medium) to form the desired ether. Over both catalysts (Zr-BEA and Sn-BEA) *p*-methoxyphenyl *iso*-butyl

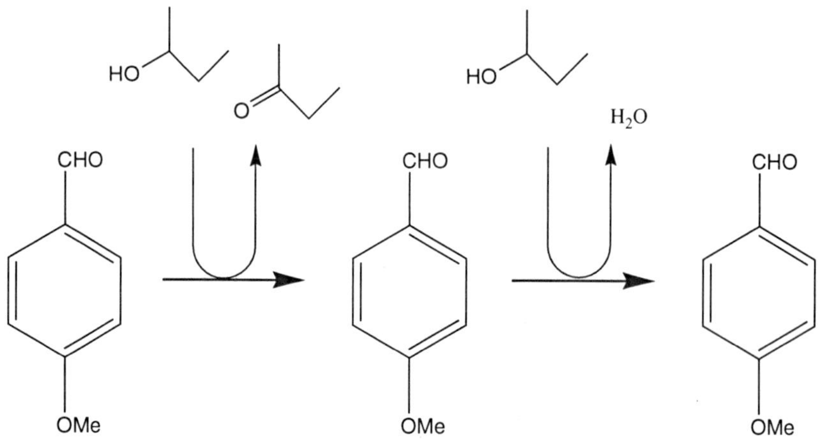

Fig. 8.47 One-pot conversion of *para*-methoxybenzaldehyde into *p*-methoxyphenyl *iso*-butyl ether

Fig. 8.48 Synthesis of triacetonamine

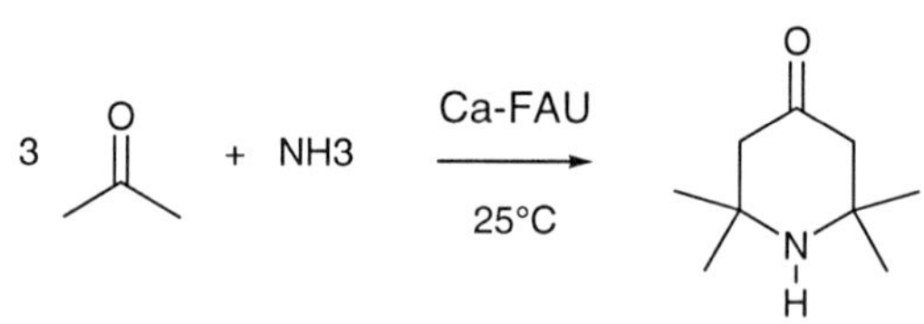

ether is obtained in 98% yield, but the Zr-containing catalyst is more active in the global process [173].

Likewise, a basic zeolite, such as Ca-FAU (Ca-Y), can act as multi-step catalyst and it is applied to the synthesis of 2,2,6,6-tetramethyl-4-oxopiperidine (triacetonamine), a key precursor in the synthesis of TEMPO (2,2,6,6-tetramethyl-1-piperidinyloxy radical; a versatile reagent for the oxidation of fine chemicals). Acetone and ammonia are the simple reactants for this synthesis (Fig. 8.48). When they are mixed in a 3:1 molar ratio with 3 wt.% of Ca-Y zeolite, moderate yields (22%) of triacetonamine are recorded, but with excellent selectivity (99%) [174]. It is, however, worth noting that in the conventional preparation the yield is hardly higher than 20% and large amounts of inorganic halides are typically used. Supposedly, the basic zeolite is able to promote both the condensation of the three acetone moieties and the ring closure with ammonia.

8.9 Conclusions

The application of micro- and mesoporous materials in the transformation of organic molecules is no more a "secret art" based on trial-and-error knowledge. Significant advances were made in the understanding of the complex interactions between the polar organic substrates and the molecular sieve catalysts. It is thus possible, on a scientific basis, optimising the catalyst features and the operating conditions, sometimes with the additional help of computational and theoretical investigations. In fact, modelling tools based on molecular mechanics and molecular dynamics, that are less time-consuming than reactivity tests, should be (and are often) used for a preliminary screening.

A key issue of most of the reactivity involved in the synthesis of fine chemicals is the kinetic limitation in product desorption. As a consequence of their long residence time within the zeolite pores, polar products are converted into non-desorbed heavy molecules (carbonaceous deposits or "coke"), which often cause deactivation. Limitations in product desorption are mainly due to the confinement of the bulky molecules in the narrow zeolite pores, rather than to their strong adsorption on the active sites.

In order to obtain a rapid and stable formation of the desired products, any sort of stratagem improving their desorption from the pores of the molecular sieve should be chosen. Measures such as the choice of the proper zeolite, the tailoring of its characteristics (use of nanosized crystals, creation of inner mesopores, optimisation of the hydrophobic- hydrophilic properties, etc.), the selection of ideal reactors as well as the optimisation of the operating conditions are all suitable for this aim. Likewise, it is necessary to finetune some critical parameters, such as temperature (high enough to facilitate product desorption without favouring the formation of by-products and carbonaceous deposits), flow rate (high enough to help desorption, but low enough to obtain the desired conversion), kind of solvent, substrate to reactant ratio and so on. In particular, the reactor type must be carefully chosen: fixed bed or batch reactor; set-up of particular reactor configurations, e.g., re-circulation, that often allows a maximisation of the yield, reactive distillation, to remove the presence of undesired or poisoning species, etc. All the possible side products should be thoroughly considered, even if they are formed in traces, such as formaldehyde in methylamine synthesis, because they can often cause a fast deactivation of the catalyst. Finally, the study of new protocols and methodologies for the mild and straightforward removal of carbonaceous deposits and therefore, for an efficient regeneration of the spent molecular sieves is a major challenge for the research teams.

With regard to the type of the molecular sieves used, only a limited number of zeolite structures with large or medium pores (essentially FAU, BEA, MOR, MWW and MFI) have proved to be suitable in the manufacture of organic chemicals so far. These zeolites are largely used in refining and petrochemical processes and because they are reasonably cheap, the methods to synthesize and to modify them with the optimal characteristics are well known and mastered.

The search for the ideal and versatile zeolitic catalyst may be long and tedious and, in this field, the example of the redox-active zeolites is paradigmatic. Among the thousands of metal-containing microporous sieves tested, only three zeolites, namely Ti-MFI (TS-1), for the oxidations with H_2O_2; Fe-MFI (Fe-ZSM-5), for the oxidation of arenes with N_2O; and Sn-BEA, for Baeyer-Villiger oxidations, have reached, or are reaching, an industrial level of development. Furthermore, whenever it is possible, essentially for industrial applications, amorphous non-ordered porous inorganic oxides (γ-Al_2O_3, SiO_2 gels or mixed oxides) are preferred to zeolitic materials because of their better commercial availability, their lower cost (mainly with respect to ordered mesoporous materials) and their better accessibility to bulky reactant fine chemicals (mostly with respect to microporous materials). Nevertheless, in some cases, the use of ordered and well-structured molecular sieves leads to unique performances, especially in the case of regioselective reactions or stereoselective syntheses over immobilised asymmetric catalyst, where the confinement effect and the steric constraints lead to the preferential formation of the desired isomer. So, even though the narrow porous system may cause some drawbacks linked to product desorption limitations, the confinement within microporous channels and cavities causes also the shape-selectivity effect that is often the main reason leading to the choice of ordered molecular sieves as catalysts in chemical synthesis. Once the efficiency and the robustness of these catalysts is confirmed, they can be employed and exploited in several different reactions over decades (as for the 25-year-old example of Ti-MFI), so that the large initial investment costs can be fully regained.

Finally, micro- and mesoporous materials can successfully act as multifunctional catalysts for one-pot multi-step reactions. In fact, in a solid matrix, two (or more) active sites with different characteristics can coexist in close proximity, but with a well-defined atomic segregation. Thus, multi-step cascade reactions can be conducted over redox-active, acidic or basic sites in the same reaction vessel with obvious economical, technological and environmental benefits.

The remarkable improvements, during last decades, in the design, preparation and development of molecular sieves with tuned and controlled porosity may open the way to the ever-increasing use of heterogeneous catalysts for the transformation of valuable fine chemicals of synthetic or natural origin, such as macromolecules, pharmaceuticals, terpenes, steroids, fatty acids and so on. As in other branches of catalysis, a strong cooperation among experts in various disciplines is needed to ensure that most of the investigations on this subject do not stop at the level of simple academic interest but, rather, carried on to set up innovative processes from laboratory up to industrial scale, following the guidelines for a sustainable and orderly development.

References

1. Guisnet M, Gilson JP (2002) Zeolites for cleaner technologies. Series: Catalytic science series, vol 3. Imperial College Press, London
2. Chen NY, Garwood WE, Dwyer FG (1989) Shape-selective catalysis in industrial applications; chemical industries. Marcel Dekker, New York, p 36
3. Song C, Garces JM, Sugi Y (2000) Shape-selective catalysis, Chemical synthesis and hydrocarbon processing, ACS Symposium Series, vol 738
4. Stoszak R (2001) In: van Bekkum H, Flanigen EM, Jacobs PA, Jansen JC (eds) Introduction to zeolite science and practice, vol 137, Elsevier, p 261 (Stud Surf Sci Catal)

5. Sheldon RA, van Bekkum H (2001) In: Sheldon RA, van Bekkum H (eds) Fine chemicals through heterogeneous catalysis. Wiley, Weinheim, p 1
6. Sheldon RA (1992) Chem Ind (London) 903
7. Sheldon RA (1994) Chem Tech (March) 38
8. Vollhardt KPC (1987) Organic chemistry. W. H. Freeman and Co., New York
9. Beck J, Haag W (1997) In: Ertl G, Knözinger H, Weitkamp J (eds) Handbook of heterogeneous catalysis. Wiley, Weinheim, p 2123
10. Meima GR, Lee GS, Garces JM (2001) In: Sheldon RA, van Bekkum H (eds) Fine chemicals through heterogeneous catalysis. Wiley, Weinheim, p 151
11. Bejblova M, Zilkova N, Cejka J (2008) Res Chem Intermed 34:439
12. Lillwitz LD (2001) Appl Catal A 221:337
13. Sumitani K, Shimada (2001) Patent JP413637, to Teijin
14. Motoyuki M, Yamamoto K, McWilliams J, Bundens (1998) Patent US5744670, to Kobe-Mobil
15. Angevine PJ, Degnan TF, Marler DO (1991) Patent US 5001295
16. Vahteristo K, Halme E, Koskimies S, Cscisery S, Laatikainen M, Niemi V (1997) Patent WO02225, to Optatech
17. Kraikul N, Rangsunvigit P, Kulprathipanja S (2006) Adsorption 12:317
18. Kraikul N, Rangsunvigit P, Kulprathipanja S (2007) Chem Eng J 131:145
19. Chobsa-Ard A, Kraikul N, Rangsunvigit P, Kulprathipanja S (2008) Chem Eng J 139:78
20. Horsley JA, Fellmann JD, Derouane EG, Freeman CM (1994) J Catal 147:231
21. Fellmann JD, Saxton RJ, Wentrcek PR, Derouane EG, Massiani P (1991) Patent US5026942, to Catalytica Inc.
22. Gnep NS, Roger P, Cartraud P, Guisnet M, Juguin B, Hamon C (1989) C R Acad Sci, Series II 309:1743
23. Marion P, Jacquot R, Ratton S, Guisnet M (2002) In: Guisnet M, Gilson JP (eds) Zeolites for cleaner technologies. Imperial College Press, London, p 281
24. Guisnet M, Guidotti M (2006) In: Derouane EG (ed) Catalysts for fine chemical synthesis. Wiley, Weinheim, p 39
25. Guisnet M, Guidotti M (2006) In: Derouane EG (ed) Catalysts for fine chemical synthesis. Wiley, Weinheim, p 69
26. Bonati MLM, Joyner RW, Stockenhuber M (2003) Catal Today 81:653
27. Bonati MLM, Joyner RW, Stockenhuber M (2007) Micropor Mesopor Mater 104:217
28. Rohan D, Canaff C, Fromentin E, Guisnet M (1998) J Catal 177:296
29. Derouane EG, Dillon CJ, Bethell D, Derouane-Abd Hamid SB (1999) J Catal 187:209
30. Moreau V, Fromentin E, Magnoux P, Guisnet M (2001) Stud Surf Sci Catal 135:4113
31. Guidotti M, Canaff C, Coustard JM, Magnoux P, Guisnet M (2005) J Catal 230:375
32. Guidotti M, Coustard JM, Magnoux P, Guisnet M (2007) Pure Appl Chem 79:1833
33. Kogelbauer A, Kouwenhoven HW (2001) In: Sheldon RA, van Bekkum H (eds) Fine chemicals through heterogeneous catalysis. Wiley, Weinheim, p 123
34. Corma A, Iborra S (2006) In: Derouane EG (ed) Catalysts for fine chemical synthesis. Wiley, Weinheim, p 105
35. Bernasconi S, Pirngruber GD, Kogelbauer A, Prins R (2003) J Catal 219:231
36. Bernasconi S, Pirngruber GD, Prins R (2004) J Catal 224:297
37. Dagade SP, Waghmode SB, Kadam VS, Dongare MK (2002) Appl Catal A 226:49
38. Dagade SP, Kadam VS, Dongare MK (2002) Catal Commun 3:67
39. Ratnasamy P, Singh AP, Sharma S (1996) Appl Catal A 135:25
40. Ratnasamy P, Singh AP (2001) In: Sheldon RA, van Bekkum H (eds) Fine chemicals through heterogeneous catalysis. Wiley, Weinheim, p 133
41. Hausladen MC, Cyganovich RC, Huang HY, Lund CRF (2001) Appl Catal A 219:1
42. Suzuki T, Komatsu C (1989) Patent US 4 831 199, to Ibahara Chemical Industry
43. Kale SM, Singh AP (2002) Appl Catal A 228:1
44. Hoelderich WF, van Bekkum H (2001) Stud Surf Sci Catal 137:821

45. Swift KAD (2001) In: Sheldon RA, van Bekkum H (eds) Fine chemicals through heterogeneous catalysis. Wiley, Weinheim, p 242
46. Corma A, Iborra S, Velty A (2007) Chem Rev 107:2411
47. Tatsumi T (2001) In: Sheldon RA, van Bekkum H (eds) Fine chemicals through heterogeneous catalysis. Wiley, Weinheim, p 185
48. Dahlhoff G, Niederer JPM, Hoelderich WF (2001) Catal Rev 43:381
49. Hoelderich WF, Laufer MC (2002) In: Guisnet M, Gilson JP (eds) Zeolites for cleaner technologies. Imperial College Press, London, p 301
50. Guisnet M, Perot G (2001) In: Sheldon RA, van Bekkum H (eds) Fine chemicals through heterogeneous catalysis. Wiley, Weinheim, p 211
51. Grandvallet P, de Jong KP, Mooiweer HH, Kortbeek AGTG, Kraushaar-Czarnetski (1992) Patent EU 501,577, to Shell
52. Xu WQ, Yin YGS, Suib SL, O'Young CL (1995) J Phys Chem 99:758
53. Guisnet M, Andy P, Gnep NS, Travers C, Benazzi E (1995) Chemical Communications
54. Houzvicka J, Ponec W (1997) Catal Rev Sci Eng 39:319
55. Meriaudeau P, Naccache C (2000) Adv Catal 44:505
56. De Menorval B, Ayrault P, Gnep NS, Guisnet M (2005) J Catal 230:38
57. De Menorval B, Ayrault P, Gnep NS, Guisnet M (2006) Appl Catal A 304:1
58. Stepanov AG, Luzgin MV, Arzumanov SS, Ernst H, Freude D (2002) J Catal 211:165
59. Andy P, Gnep NS, Guisnet M, Benazzi E, Travers C (1998) J Catal 173:322
60. Severino A, Vital J, Lobo LS (1993) Stud Surf Sci Catal 78:685
61. Lopez CM, Machado FJ, Rodriguez K, Mendez B, Hasegawa M, Pekerar S (1998) Appl Catal A 173:75
62. Rachwalik R, Olejniczak Z, Jiao J, Huang J, Hunger M, Sulikowski B (2007) J Catal 252:161
63. Hoelderich WF, Barsnick U (2001) In: Sheldon RA, van Bekkum H (eds) Fine chemicals through heterogeneous catalysis. Wiley, Weinheim, p 217
64. Hoelderich WF, Roeseler J, Heitmann G, Liebens AT (1997) Catal Today 37:353
65. Kunkeler PJ, van der Waal JC, Bremmer J, Zuurdeeg BJ, Downing RS, van Bekkum H (1998) Catal Lett 53:135
66. Sato H, Ishii N, Hirose K, Nakamura S (1986) In: Proceedings of seventh international zeolite conference, p 755
67. Ichihashi H, Ishida M, Shiga A, Kitamura M, Suzuki T, Suenobu K, Sugita K (2003) Catal Surveys Asia 7:261
68. Heitmann GP, Dahlhoff G, Hoelderich WF (1999) J Catal 186:12
69. Flego C, Dalloro L (2003) Micropor Mesopor Mater 60:263
70. Takahashi T, Nasution MNA, Kai T (2001) Appl Catal A 210:339
71. Takahashi T, Kai E, Nakao E (2004) Appl Catal A 262:137
72. Albers P, Sibold K, Haas T, Prescher G, Hoelderich WF (1998) J Catal 176:561
73. Kitamura M, Schimazu M, Yako M (1999) Patent JP 229939, to Sumimoto Chemical Co.
74. Jayat F (1996) PhD Thesis, University of Poitiers
75. Jayat F, Sabater Picot MJ, Guisnet M (1996) Catal Lett 41:181
76. Borzatta V, Busca G, Poluzzi E, Rossetti V, Trombetta M, Vaccari A (2004) Appl Catal A 257:85
77. Neves I, Jayat F, Magnoux P, Perot G, Ribeiro FR, Gubelmann M, Guisnet M (1994) J Mol Catal A 93:169
78. Guisnet M, Lukyanov DB, Jayat F, Magnoux P, Neves I (1995) Ind Eng Chem Res 34:1624
79. Heidekum A, Harmer MA, Hoelderich WF (1998) J Catal 176:260
80. Freese U, Heinrich F, Roessner F (1999) Catal Today 49:237
81. Vogt AHG, Kouwenhoven HW, Prins R (1995) Appl Catal A 123:37
82. Corbin DR, Schwarz S, Sonnichsen GC (1997) Catal Today 37:71
83. Hayes KS (2001) Appl Catal A 221:187
84. Veefkind VA, Lercher JA (1998) J Catal 180:258
85. Veefkind VA, Lercher JA (1999) Appl Catal A 181:245

86. Fujita T, Niwa K, Ogura K, Fukatsu M (1997) Patent US 5,688,854, to Nitto Chemicals
87. Jeon H-Y, Shin C-H, Jung HJ, Hong SB (2006) Appl Catal A 305:70
88. Sasaki Y, Fukatsu M (1998) Catal Surveys Japan 2:199
89. McAteer CH, Scriven EFV (2001) In: Sheldon RA, van Bekkum H (eds) Fine chemicals through heterogeneous catalysis. Wiley, Weinheim, p 275
90. Calvin JR, Davis RD, McAteer CH (2005) Appl Catal A 285:1
91. Shimazu S, Abe N, Iguchi A, Dohba M, Sato H, Hirose K (1998) Micropor Mesopor Mater 21:447
92. Dessau R M (1986) J Chem Soc, Chem Commun, 1167
93. Durand R, Geneste P, Joffre J, Moreau C (1993) In: Guisnet M, Barbier J, Barrault J, Bouchoule C, Duprez D, Pérot G, Montassier C (eds), vol 78. Elsevier, Amsterdam, p 647 (Stud Surf Sci Catal)
94. Maxwell IE, Downing RS, van Langen SAJ (1980) J Catal 61:485
95. Bergström C, Mölsä A, Yli-Kauhaluoma J (1999) World Patent 9940049
96. Eklund L, Axelsson A, Nordahl A, Carlson R (1993) Acta Chem Scand 47:581
97. Corma A (2003) Catal Rev Sci Eng 46:369
98. Onaka M, Hashimoto N, Kitabata Y, Yamasaki R (2003) Appl Catal A: Gen 241:307
99. Kugita T, Jana SK, Owada T, Hashimoto N, Onaka M, Namba S (2003) Appl Catal A: Gen 245:353
100. Onaka M, Yamazaki R (1998) Chem Lett 259
101. Wan Y, McMorn P, Hancock FE, Hutchings GJ (2003) Catal Lett 91:145
102. Noyori R, Aoki M, Sato K (2003) Chem Commun 1977
103. Brégeault JM (2003) Dalton Trans 3289
104. Clerici MG (2005) DGMK/SCI conference; "oxidation and functionalization: classical and alternative routes and sources", October 12–14, Milan, Italy
105. Kholdeeva OA, Trukhan NN (2006) Russian Chem Rev 75(5):411
106. Corma A, Camblor MA, Esteve P, Martinez A, Pérez Pariente JP (1994) J Catal 145:151
107. Corma A, Esteve P, Martínez A (1996) Appl Catal A 143:87
108. Chen LY, Chuah GK, Jaenicke S (1998) Catal Lett 50:107
109. Koyano KA, Tatsumi T (1997) Microp Mater 10:259
110. Corma A (2003) J Catal 216:298
111. Sheldon RA, Wallau M, Arends IWCE, Schuchardt V (1998) Acc Chem Res 31:485
112. Boronat M, Concepciòn P, Corma A, Renz M (2007) Catal Today 121:39
113. Schuster C, Hoelderich WF (2000) Catal Today 60:193
114. Laha SC, Kumar R (2002) J Catal 208:339
115. Kholdeeva OA, Melgunov MS, Shmakov AN, Trukhan NN, Kriventsov VV, Zaikovskii VI, Romannikov VN (2004) Catal Today 91–92:205
116. Fraile JM, Garcìa JI, Mayoral JA, Vispe E, Brown DR, Naderi M (2001) J Chem Soc, Chem Commun 1510
117. Guidotti M, Ravasio N, Psaro R, Ferraris G, Moretti G (2003) J Catal 214(2):247
118. Corma A, Moliner M, Diaz-Cabanas MJ, Serna P, Femenia B, Primo J, Garcia H (2008) New J Chem 8:1338
119. Corma A, Domine M, Gaona JA, Jordà JL, Navarro MT, Rey F, Pérez-Pariente J, Tsuji J, McCulloch B, Nemeth LT (1998) Chem Commun 2211
120. Guidotti M, Batonneau-Gener I, Gianotti E, Marchese L, Mignard S, Psaro R, Sgobba M, Ravasio N (2008) Micropor Mesopor Mater 111:39
121. Dusi M, Mallat T, Baiker A (2000) Catal Rev Sci Eng 42:213
122. Rigutto MS, van Bekkum H (1993) J Mol Catal 81:77
123. Selvaraj M, Seshadri KS, Pandurangan A, Lee TG (2005) Micropor Mesopor Mater 79:261
124. Liu C-J, Yu W-Y, Li S-G, Che C-M (1998) J Org Chem 63:7364
125. Anwander R (2001) Chem Mater 13:4419
126. Maschmeyer T, van de Water LM (2006) In: Derouane EG (ed) Catalysts for fine chemical synthesis. Wiley, Weinheim

127. Petrini G, Leofanti G, Mantegazza MA, Pignataro F (1996) ACS Symp Ser (Green Chemistry) 626:33
128. Zecchina A, Bordiga S, Lamberti C, Ricchiardi G, Lamberti C, Ricchiardi G, Scarano G, Petrini G, Leofanti G, Mantegazza M (1996) Catal Today 32:97
129. Le Bars J, Dakka J, Sheldon RA (1996) Appl Catal A: Gen 136:69
130. Clerici MG, Bellussi G (1983) Patent US 4,410,501, to Snamprogetti
131. Ratton S (1997) Chem Today 3–4:33
132. Bianchi D, Balducci L, Bortolo R, D'Aloisio R, Ricci M, Spanò G, Tassinari R, Tonini C, Ungarelli R (2007) Adv Synth Catal 349:979
133. Chou B, Tsai J-L, Cheng S (2001) Micropor Mesopor Mater 48:309
134. Trukhan NN, Romannikov VN, Paukshtis EA, Shmakov AN, Kholdeeva OA (2001) J Catal 110:202
135. Kholdeeva OA, Ivanchikova ID, Guidotti M, Ravasio N (2007) Green Chem 9-731
136. Reddy KM, Moudrakovsky I, Sayari A (1994) Chem Commun 1059
137. Kholdeeva OA, Zalomaeva OV, Sorokin AB, Ivanchikova ID, Della Pina C, Rossi M (2007) Catal Today 121:58
138. Srinivas N, Rani VR, Kishan MR, Kulkarni SJ, Raghavan KV (2001) J Mol Catal A 172:187
139. Dugal M, Sankar G, Raja R, Thomas JM (2000) Angew Chem Int Ed 39:2310
140. Thomas JM, Raja R, Sankar G, Johnson BFG, Lewis DW (2001) Chem Eur J 7:2972
141. Raja R, Thomas JM, Greenhill-Hooper M, Ley SV, Almeida-Paz FA (2008) Chem Eur J 14:2340
142. Das TK, Choudhary K, Nandanan E, Chandwadkar AJ, Sudalai A, Ravindranathan T, Sivasankar S (1997) Tetrahedron Lett 38:3631
143. Sakthivel A, Badamali SK, Selvam P (2002) Catal Lett 80:73
144. Maspero F, Romano U (1994) J Catal 146:476
145. Hayashi H, Kikawa K, Murai Y, Shigemoto N, Sugiyama S, Kawashioro K (1996) Catal Lett 36:99
146. Bellussi G, Carati A, Clerici MG, Maddinelli G, Millini R (1992) J Catal 133:220
147. Bleloch A, Johnson BFG, Ley SV, Price AJ, Shephard DS, Thomas AW (1999) Chem Commun 1907
148. Brunel D, Fajula F, Nagy JB, Deroide B, Verhoef MJ, Veum L, Peters JA, Van Bekkum H (2001) Appl Catal A 73:213
149. Thomas JM, Maschmeyer T, Johnson BFG, Shepard DS (1999) J Mol Catal A: Chem 141:139
150. Tsangarakis C, Stratakis M (2006) Eur J Org Chem 4435
151. Yongzhong Z, Yuntong N, Jaenicke S, Chuah GK (2005) J Catal 229:404
152. Kumar R, Pais GCG, Pandey B, Kumar P (1995) J Chem Soc, Chem Commun 1315
153. Adam T, Wirth W (1999) Acc Chem Res 32:703
154. Clerici MG (2000) Topics Catal 13:373
155. Blaser HU, Pugin B, Studer M (2000) In: De Vos DE, Vankelecom IFJ, Jacobs PA (eds) Chiral catalyst immobilization and recycling. Wiley, Weinheim, p 1
156. Wells RPK, Tynjälä P, Bailie JE, Willock DJ, Watson GW, King F, Rochest CH, Bethell D, Bulman Page PC, Hutchings GJ (1999) Appl Catal A: Gen 182:75
157. Li C (2004) Catal Rev Sci Eng 46:419
158. Song CE, Lee S (2002) Chem Rev 102:3495
159. McMorn P, Hutchings G (2004) J Chem Soc Rev 33:108
160. Raynor SA, Thomas JM, Raja R, Johnson BFG, Bell RG, Mantle MD (2000) Chem Commun 1925
161. Jones MD, Raja R, Thomas JM, Johnson BFG, Lewis DW, Rouzaud J, Harris KDM (2003) Angew Chem Int Ed 42:4326
162. Yu K, Gu Z, Ji R, Lou LL, Ding F, Zhang C, Liu S (2007) J Catal 252:312
163. Anastas PT, Warner J (1998) Green chemistry: Theory and practice. Oxford University Press, Oxford

164. Guidotti M, Moretti G, Psaro R, Ravasio N (2000) Chem Commun 1789
165. van Bekkum H, Kouwenhoven HW (2005) In: Cejka J, van Bekkum H (eds), vol 157. Elsevier, Amsterdam, p 311 (Stud Surf Sci Catal)
166. Guisnet M, Guidotti M (2006) In: Derouane EG (ed) Catalysts for fine chemical synthesis. Wiley, Weinheim, p 157
167. Hoelderich W, Goetz N, Hupfer L (1988) Patent DE 3801106, to BASF
168. Iosif F, Coman S, Parvulescu V, Grange P, Delsarte S, De Vos D, Jacobs P (2004) Chem Commun 1292
169. Trasarti AF, Marchi AJ, Apesteguia CR (2004) J Catal 224:484
170. Ko AN, Hu CH, Chen JY (1999) Appl Catal A: Gen 184:211
171. Corma A, Iglesias M, Sánchez F (1995) J Chem Soc, Chem Commun 1635
172. Whalen J, de Vos DE, Jacobs PA (2004) Adv Synth Catal 346:333
173. Corma A, Renz M (2007) Angew Chem Int Ed 46:298
174. Malz RE, Son Y-C, Suib SL (2004) Patent WO0228833, to Uniroyal
175. Hoelderich WF, Laufer MC (2002) In: Guisnet M, Gilson JP (eds) Zeolites for cleaner technologies. Imperial College Press, London, p 301
176. Arends IWCE, Sheldon RA, Wallau M, Schuchardt U (1997) Angew Chem Int Ed 36:1144
177. Ratnasamy P, Srinivas D, Knözinger H (2004) Adv Catal 48:1
178. Pazzucconi G, Terzoni G, Perego C, Bellussi G (2001) Stud Surf Sci Catal 135: 25–0–03

Index

C

T

CPSIA information can be obtained at www.ICGtesting.com
Printed in the USA
LVOW082059030213

318393LV00010B/201/P

9 781402 096778